The State Space Method
Generalizations and Applications

Daniel Alpay
Israel Gohberg
Editors

Linear
Operators &
Linear
Systems

Birkhäuser Verlag
Basel · Boston · Berlin

Editors:

Daniel Alpay
Department of Mathematics
Ben-Gurion University of the Negev
P.O. Box 653
Beer Sheva 84105
Israel
e-mail: dany@math.bgu.ac.il

Israel Gohberg
School of Mathematical Sciences
Raymond and Beverly Sackler
Faculty of Exact Sciences
Tel Aviv University
Ramat Aviv 69978
Israel
e-mail: gohberg@post.tau.ac.il

2000 Mathematics Subject Classification 47Axx, 93Bxx

A CIP catalogue record for this book is available from the
Library of Congress, Washington D.C., USA

Bibliographic information published by Die Deutsche Bibliothek
Die Deutsche Bibliothek lists this publication in the Deutsche Nationalbibliografie; detailed
bibliographic data is available in the Internet at <http://dnb.ddb.de>.

ISBN 3-7643-7370-9 Birkhäuser Verlag, Basel – Boston – Berlin

© 2006 Birkhäuser Verlag, P.O. Box 133, CH-4010 Basel, Switzerland
Part of Springer Science+Business Media
Printed on acid-free paper produced from chlorine-free pulp. TCF∞
Cover design: Heinz Hiltbrunner, Basel
Printed in Germany
ISBN-10: 3-7643-7370-9   e-ISBN: 3-7643-7431-4
ISBN-13: 978-3-7643-7370-2

9 8 7 6 5 4 3 2 1   www.birkhauser.ch

Contents

D.Z. Arov and O.J. Staffans

**State/Signal Linear Time-Invariant Systems Theory, Part I:
Discrete Time Systems** 115

Contents

vii

Editorial Introduction

This volume of the *Operator Theory: Advances and Applications* series (OTAA) is the first volume of a new subseries. This subseries is dedicated to connections between the theory of linear operators and the mathematical theory of linear systems and is named *Linear Operators and Linear Systems* (LOLS). As the existing subseries *Advances in Partial Differential Equations* (ADPE), the new subseries will continue the traditions of the OTAA series and keep the high quality of the volumes. The editors of the new subseries are: Daniel Alpay (Beer–Sheva, Israel), Joseph Ball (Blacksburg, Virginia, USA) and André Ran (Amsterdam, The Netherlands).

In the last 25–30 years, Mathematical System Theory developed in an essential way. A large part of this development was connected with the use of the state space method. Let us mention for instance the "theory of H_∞ control". The state space method allowed to introduce in system theory the modern tools of matrix and operator theory. On the other hand the state space approach had an important impact on Algebra, Analysis and Operator Theory. In particular it allowed to solve explicitly some problems from interpolation theory, theory of convolution equations, inverse problems for canonical differential equations and their discrete analogs. All these directions are planned to be present in the subseries LOLS. The editors and the publisher are inviting authors to submit their manuscripts for publication in this subseries.

This volume contains five essays. The essay of D. Arov and O. Staffans, *State/signal linear time-invariant systems theory, part I: discrete time systems*, contains new results in classical system theory. The essays of D. Alpay and D.S. Kalyuzhnyĭ-Verbovetzkiĭ, *Matrix-J-unitary non-commutative rational formal power series*, and of J.A. Ball, G. Groenewald and T. Malakorn, *Conservative structured noncommutative multidimensional linear systems* are dedicated to a new branch in Mathematical system theory where discrete time is replaced by the free semigroup with N generators. The essay of I. Gohberg, I. Haimovici, M.A. Kaashoek and L. Lerer, *The Bezout integral operator: main property and underlying abstract scheme* contains new applications of the state space method to the theory of Bezoutiants and convolution equations. The essay of D. Alpay and I. Gohberg *Discrete analogs of canonical systems with pseudo-exponential potential. Definitions and formulas for the spectral matrix functions* is concerned with new results and formulas for the discrete analogs of canonical systems.

Daniel Alpay, Israel Gohberg

Operator Theory:
Advances and Applications, Vol. 161, 1–47

Discrete Analogs of Canonical Systems with Pseudo-exponential Potential. Definitions and Formulas for the Spectral Matrix Functions

Daniel Alpay and Israel Gohberg

Abstract. We first review the theory of canonical differential expressions in the rational case. Then, we define and study the discrete analogue of canonical differential expressions. We focus on the rational case. Two kinds of discrete systems are to be distinguished: one-sided and two-sided. In both cases the analogue of the potential is a sequence of numbers in the open unit disk (Schur coefficients). We define the characteristic spectral functions of the discrete systems and provide exact realization formulas for them when the Schur coefficients are of a special form called strictly pseudo-exponential.

Mathematics Subject Classification (2000). 34L25, 81U40, 47A56.

Contents

1. Introduction

Canonical differential expressions are differential equations of the form

$$-iJ\frac{\partial\Theta}{\partial x}(x,\lambda) = \lambda\Theta(x,\lambda) + v(x)\Theta(x,\lambda), \quad x \geq 0, \ \lambda \in \mathbb{C}, \tag{1.1}$$

where

$$v(x) = \begin{pmatrix} 0 & k(x) \\ k(x)^* & 0 \end{pmatrix}, \quad J = \begin{pmatrix} I_n & 0 \\ 0 & -I_n \end{pmatrix},$$

and where $k \in \mathbf{L}_1^{n\times n}(\mathbb{R}_+)$ is called the potential. Such systems were introduced by M.G. Kreĭn (see, e.g., [37], [38]).

Associated to (1.1) are a number of functions of λ, which we called in [10] the *characteristic spectral functions* of the canonical system. These are:

1. The asymptotic equivalence matrix function $V(\lambda)$.
2. The scattering function $S(\lambda)$.
3. The spectral function $W(\lambda)$.
4. The Weyl function $N(\lambda)$.
5. The reflection coefficient function $R(\lambda)$.

Direct problems consist in computing these functions from the potential $k(x)$ while inverse problems consist in recovering the potential from one of these functions.

In the present paper we study discrete counterparts of canonical differential expressions. To present our approach, we first review various facts on the telegraphers' equations. By the term *telegraphers' equations*, one means a system of differential equations connecting the voltage and the current in a transmission line. The case of lossy lines can be found for instance in [45] and [18]. We here consider the case of lossless lines and follow the arguments and notations in [16, Section 2], [19, p. 110–111] and [46]. The telegraphers' equations which describe the evolution of the voltage $v(x,t)$ and current $i(x,t)$ in a lossless transmission line can be given as:

$$\frac{\partial v}{\partial x}(x,t) + Z(x)\frac{\partial i}{\partial t}(x,t) = 0$$
$$\frac{\partial i}{\partial x}(x,t) + Z(x)^{-1}\frac{\partial v}{\partial t}(x,t) = 0. \tag{1.2}$$

In these equations, $Z(x)$ represents the local impedance at the point x. *A priori* there may be points where $Z(x)$ is not continuous, but it is important to bear in mind that voltage and current will be continuous at these points.

Let us assume that $Z(x) > 0$ and is continuously differentiable on an interval (a,b), and introduce the new variables

$$V(x,t) = Z(x)^{-1/2}v(x,t),$$

$$I(x,t) = Z(x)^{1/2}i(x,t),$$

and

$$W_R(x,t) = \frac{V(x,t) + I(x,t)}{2},$$

$$W_L(x,t) = \frac{V(x,t) - I(x,t)}{2}.$$

Then the function

$$W(x,t) = \begin{pmatrix} W_R(x,t) \\ W_L(x,t) \end{pmatrix} = \frac{1}{2} \begin{pmatrix} Z(x)^{-1/2} & Z(x)^{1/2} \\ Z(x)^{-1/2} & -Z(x)^{1/2} \end{pmatrix} \begin{pmatrix} v(x,t) \\ i(x,t) \end{pmatrix} \qquad (1.3)$$

satisfies the differential equation, also called *symmetric two components wave equation* (see [16, equation (2.6) p. 362], [46, p. 256], [19, equation (3.3) p. 111])

$$\frac{\partial W(x,t)}{\partial x} = -J\frac{\partial W(x,t)}{\partial t} + \begin{pmatrix} 0 & -\kappa(x) \\ -\kappa(x) & 0 \end{pmatrix} W(x,t),$$

where

$$J = \begin{pmatrix} 1 & 0 \\ 0 & -1 \end{pmatrix} \quad \text{and} \quad \kappa(x) = \frac{Z'(x)}{2Z(x)}. \qquad (1.4)$$

We distinguish two cases:

(a) The case where $Z(x) > 0$ and is continuously differentiable on $\mathbb{R}_+$. Taking the (inverse) Fourier transform $f \mapsto \widehat{f}(\lambda) = \int_{\mathbb{R}} e^{i\lambda t} f(t)dt$ on both sides we get to a *canonical differential expressions* (also called *Dirac type system*), with $k(x) = i\kappa(x)$ and $\Theta(x,\lambda) = \widehat{W}(x,\lambda)$. The theory of *canonical differential expressions* is reviewed in the next section.

(b) The case where $Z(x)$ is constant on intervals $[nh,(n+1)h)$ for some pre-assigned $h > 0$. We are then lead to discrete systems.

The paper consists of three sections besides the introduction. In Section 2 we review the main features of the continuous case. The third section presents the discrete systems to be studied. These are of two kinds, one-sided and two-sided. Section 3 also contains a study of one-sided systems and of their associated characteristic spectral functions. In Section 4 we focus on two-sided systems and we also present an illustrative example.

In the parallel between the continuous and discrete cases a number of problems remains to be considered to obtain a complete picture. In the sequel to the present paper we study *inverse problems* associated to these first-order systems.

To conclude this introduction we set some definitions and notation. The open unit disk will be denoted by $\mathbb{D}$, the unit circle by $\mathbb{T}$, and the open upper half-plane by $\mathbb{C}_+$. The open lower half-plane is denoted by $\mathbb{C}_-$ and its closure by $\overline{\mathbb{C}_-}$. We will make use of the Wiener algebras of the real line and of the unit circle. These are defined as follows. The Wiener algebra of the real line $\mathcal{W}^{n\times n}(\mathbb{R}) = \mathcal{W}^{n\times n}$ consists of the functions of the form

$$f(\lambda) = D + \int_{-\infty}^{\infty} e^{i\lambda t} u(t)dt \qquad (1.5)$$

where $D \in \mathbb{C}^{n \times n}$ and where $u \in \mathbf{L}_1^{n \times n}(\mathbb{R})$. Usually we will not stress the dependence on $\mathbb{R}$. The sub-algebra $\mathcal{W}_+^{n \times n}$ (resp. $\mathcal{W}_-^{n \times n}$) consists of the functions of the form (1.5) for which the support of u is in $\mathbb{R}_+$ (resp. in $\mathbb{R}_-$).

The Wiener algebra $\mathcal{W}(\mathbb{T})$ (we will usually write $\mathcal{W}$ rather than $\mathcal{W}(\mathbb{T})$) of the unit circle consists of complex-valued functions $f(z)$ of the form

$$f(z) = \sum_{\mathbb{Z}} f_\ell z^\ell$$

for which

$$\|f\|_{\mathcal{W}} \stackrel{\text{def.}}{=} \sum_{\mathbb{Z}} |f_\ell| < \infty.$$

2. Review of the continuous case

2.1. The asymptotic equivalence matrix function

We first review the continuous case, and in particular the definitions and main properties of the characteristic spectral functions. See, e.g., [7], [11], [10] for more information. We restrict ourselves to the case where the potential is of the form

$$k(x) = -2ce^{ixa}\left(I_p + \Omega\left(Y - e^{-2ixa^*}Ye^{2ixa}\right)\right)^{-1}(b + i\Omega c^*), \qquad (2.1)$$

where $(a, b, c) \in \mathbb{C}^{p \times p} \times \mathbb{C}^{p \times n} \times \mathbb{C}^{n \times p}$ is a triple of matrices with the properties that

$$\cap_{\ell=0}^m \ker ca^\ell = \{0\} \quad \text{and} \quad \cup_{\ell=0}^m \operatorname{Im} a^\ell b = \mathbb{C}^p$$

for m large enough. In system theory, see for instance [30], the first condition means that the pair (c, a) is observable while the second means that the pair (a, b) is controllable. When both conditions are in force, the triple is called minimal. See also [14] for more information on these notions. We assume moreover that the spectra of a and of $a^\times = a - bc$ are in the open upper half-plane. Furthermore Ω and Y in (2.1) belong to $\mathbb{C}^{p \times p}$ and are the unique solutions of the Lyapunov equations

$$i(\Omega a^{\times *} - a^\times \Omega) = bb^*, \qquad (2.2)$$

$$-i(Ya - a^*Y) = c^*c. \qquad (2.3)$$

This class of potentials was introduced in [7] and called in [26] *strictly pseudo-exponential potentials*. Note that both Ω and Y are strictly positive since the triple (a, b, c) is minimal, and that $I_p + \Omega Y$ and $I_p + Y\Omega$ are invertible since

$$\det(I_p + \Omega Y) = \det(I_p + Y\Omega) = \det(I_p + \sqrt{Y}\Omega\sqrt{Y}).$$

Note also that asymptotically,

$$k(x) \sim -2ce^{ixa}(I_p + \Omega Y)^{-1}(b + i\Omega c^*) \qquad (2.4)$$

as $x \to +\infty$. Potentials of the form (2.1) can also be represented in a different form; see (2.22).

We first define the asymptotic equivalence matrix function. To that purpose (and here we follow closely our paper [12]) let F, G and T be the matrices given by

$$F = i \begin{pmatrix} -c & 0 \\ 0 & f_1 \end{pmatrix}, \quad T = \begin{pmatrix} ia & 0 \\ 0 & -ia^* \end{pmatrix}, \quad G = \begin{pmatrix} 0 & f_1^* \\ c^* & 0 \end{pmatrix}, \tag{2.5}$$

where $f_1 = (b^* - ic\Omega)(I_p + Y\Omega)^{-1}$.

Theorem 2.1. *Let $Q(x, y)$ be defined by*

$$Q(x, y) = Fe^{xT}(I_{2p} - e^{xT}Ze^{xT})^{-1}e^{yT}G$$

where (F, G, T) are defined by (2.5) *and where Z is the unique solution of the matrix equation*

$$TZ + ZT = -GF.$$

Then the matrix function

$$U(x, \lambda) = e^{i\lambda Jx} + \int_x^\infty Q(x, u)e^{i\lambda Ju}\,du$$

is the unique solution of (1.1) *with the potential as in* (2.1), *subject to the condition*

$$\lim_{x \to \infty} \begin{pmatrix} e^{-ix\lambda}I_n & 0 \\ 0 & e^{ix\lambda}I_n \end{pmatrix} U(x, \lambda) = I_{2n}, \quad \lambda \in \mathbb{R}. \tag{2.6}$$

Furthermore, the $\mathbb{C}^{n \times n}$-valued blocks in the decomposition of the matrix function $U(0, \lambda) = (U_{ij}(0, \lambda))$ are given by

$$\begin{aligned}
U_{11}(0, \lambda) &= I_n + ic\Omega(\lambda I_p - a^*)^{-1}c^*, \\
U_{21}(0, \lambda) &= (-b^* + ic\Omega)(\lambda I_p - a^*)^{-1}c^*, \\
U_{12}(0, \lambda) &= -c(I_p + \Omega Y)(\lambda I_p - a)^{-1}(I_p + \Omega Y)^{-1}(b + i\Omega c^*), \\
U_{22}(0, \lambda) &= I_n - (ib^*Y + c\Omega Y)(\lambda I_p - a)^{-1}(I_p + \Omega Y)^{-1}(b + i\Omega c^*).
\end{aligned}$$

See [9, Theorem 2.1].

Definition 2.2. *The function $V(\lambda) = U(0, \lambda)$ is called the asymptotic equivalence matrix function.*

The terminology asymptotic equivalence matrix function is explained in the following theorem:

Theorem 2.3. *The asymptotic equivalence matrix function has the following property: let $x \in \mathbb{R}$ and ξ_0 and ξ_1 in $\mathbb{C}^{2n}$. Let $f_0(x, \lambda) = e^{i\lambda x J}\xi_0$ be the $\mathbb{C}^{2n}$-valued solution to* (1.1) *corresponding to $k(x) = 0$ and $f_0(0, \lambda) = \xi_0$ and let $f_1(x, \lambda)$ corresponding to an arbitrary potential k of the form* (2.1), *with $f_1(0, \lambda) = \xi_1$. The two solutions are asymptotic in the sense that*

$$\lim_{x \to \infty} \|f_1(x, \lambda) - f_0(x, \lambda)\| = 0$$

if and only if $\xi_1 = U(0, \lambda)\xi_0$.

For a proof, see [10, Section 2.2].

The asymptotic equivalence matrix function takes J-unitary values on the real line:

$$V(\lambda)JV(\lambda)^* = J, \quad \lambda \in \mathbb{R}.$$

We recall the following: if R be a $\mathbb{C}^{2n \times 2n}$-valued rational functions analytic at infinity, it can be written as $R(\lambda) = D + C(\lambda I_m - A)^{-1}B$, where A, B, C and D are matrices of appropriate sizes. Such a representation of R is called a *realization*. The realization is said to be minimal if the size of A is minimal (equivalently, the triple (A, B, C) is minimal, in the sense recalled above). The McMillan degree of R is the size of the matrix A in any minimal realization. Minimal realizations of rational matrix-valued functions taking J-unitary values on the real line were characterized in [5, Theorem 2.8 p. 192]: R takes J-unitary values on the real line if and only if there exists an Hermitian invertible matrix $H \in \mathbb{C}^{m \times m}$ solution of the system of equations

$$i(A^*H - HA) = C^*JC \tag{2.7}$$
$$C = iJB^*H. \tag{2.8}$$

The matrix H is uniquely defined by the minimal realization of R and is called the *associated Hermitian matrix* to the minimal realization matrix function. The matrix function R is moreover J-inner, that is J-contractive in the open upper half-plane:

$$R(\lambda)JR(\lambda) \leq J \quad \text{for all points of analyticity in the open upper half-plane,}$$

if and only if $H > 0$. The asymptotic equivalence matrix function $V(\lambda)$ has no pole on the real line, but an arbitrary rational function which takes J-unitary values on the real line may have poles on the real line. See [5] and [4] for examples.

The next theorem presents a minimal realization of the asymptotic equivalence matrix function and its associated Hermitian matrix.

Theorem 2.4. *Let $k(x)$ be given in the form (2.1). Then, a minimal realization of the asymptotic equivalence matrix function associated to the corresponding canonical differential system is given by $V(\lambda) = I_{2n} + C(\lambda I_{2p} - A)^{-1}B$, where*

$$A = \begin{pmatrix} a^* & 0 \\ 0 & a \end{pmatrix}, \quad B = \begin{pmatrix} c^* & 0 \\ 0 & (I_p + \Omega Y)^{-1}(b + i\Omega c^*) \end{pmatrix}$$

and

$$C = \begin{pmatrix} ic\Omega & -c(I_p + \Omega Y) \\ -b^* + ic\Omega & -ib^*Y - c\Omega Y \end{pmatrix},$$

and the associated Hermitian matrix is given by

$$H = \begin{pmatrix} \Omega & i(I_p + \Omega Y) \\ -i(I_p + Y\Omega) & (I_p + Y\Omega)Y \end{pmatrix}.$$

We now prove a factorization result for $V(\lambda)$. We first recall the following: let as above R be a rational matrix-valued function analytic at infinity. The factorization

$R = R_1 R_2$ of R into two other rational matrix-valued functions analytic at infinity (all the functions are assumed to have the same size) is said to be minimal if

$$\deg R = \deg R_1 + \deg R_2.$$

Minimal factorizations of rational matrix-valued functions have been characterized in [14, Theorem 1.1 p. 7]. Assume now that R takes J-unitary values on the real line. Minimal factorizations of R into two factors which are J-unitary on the real line were characterized in [5]. Such factorizations are called *J-unitary factorizations*. To recall the result (see [5, Theorem 2.6 p. 187]), we introduce first some more notations and definitions: let $H \in \mathbb{C}^{p \times p}$ be an invertible Hermitian matrix. The formula

$$[x, y]_H = y^* H x, \qquad x, y \in \mathbb{C}^p$$

defines a non-degenerate and in general indefinite inner product. Two vectors are orthogonal with respect to this inner product if $[x, y]_H = 0$. The orthogonal complement of a subspace $\mathcal{M} \subset \mathbb{C}^p$ is:

$$\mathcal{M}^{[\perp]} = \{x \in \mathbb{C}^p \; ; [x, m]_H = 0 \; \forall m \in \mathcal{M}\}.$$

We refer to [29] for more information on finite-dimensional indefinite inner product spaces.

Theorem 2.5. *Let R be a rational matrix-valued function analytic at infinity and J-unitary on the real line, and let $R(\lambda) = D + C(zI_p - A)^{-1}B$ be a minimal realization of R, with associated matrix H. Let $\mathcal{M}$ be a A-invariant subspace of $\mathbb{C}^p$ non-degenerate with respect to the inner product $[\cdot, \cdot]_H$. Let π denote the orthogonal (with respect to $[\cdot, \cdot]_H$) projection such that*

$$\ker \pi = \mathcal{M}, \quad \operatorname{Im} \pi = \mathcal{M}^{[\perp]}$$

and let $D = D_1 D_2$ be a factorization of D into two J-unitary constants. Then $R = R_1 R_2$ with

$$R_1(z) = D_1 + C(zI_p - A)^{-1}(I_p - \pi)BD_2^{-1}$$
$$R_2(z) = D_2 + D_1^{-1}C\pi(zI_p - A)^{-1}BD_2$$

is a minimal J-unitary factorization of R. Conversely, every J-unitary factorization of R is obtained in such a way.

As a consequence we have:

Theorem 2.6. *Let $V(\lambda)$ be the asymptotic equivalence matrix function of a canonical differential expression (1.1) with potential of the form (2.1). Then it admits a minimal factorization*

$$V(\lambda) = V_1(\lambda)V_2(\lambda)^{-1}$$

where V_1 and V_2 are J-inner and of same degree.

To prove this result we consider the realization of $V(\lambda)$ given in Theorem 2.4 and note that the space $\begin{pmatrix} \mathbb{C}^p \\ 0 \end{pmatrix}$ is A-invariant and H-non-degenerate (in fact, H-positive). The factorization follows from Theorem 2.5. The fact that V_2 is inner follows from

$$H = \begin{pmatrix} I_p & 0 \\ -i(I_p + Y\Omega)\Omega^{-1} & I_p \end{pmatrix} \begin{pmatrix} \Omega & 0 \\ 0 & -\Omega^{-1} - Y \end{pmatrix} \begin{pmatrix} I_p & 0 \\ -i(I_p + Y\Omega)\Omega^{-1} & I_p \end{pmatrix}^*.$$

To prove this last formula we have used the formula for Schur complements:

$$\begin{pmatrix} A_{11} & A_{12} \\ A_{21} & A_{22} \end{pmatrix} = \begin{pmatrix} I & 0 \\ A_{21}A_{11}^{-1} & I \end{pmatrix} \begin{pmatrix} A_{11} & 0 \\ 0 & A_{22} - A_{21}A_{11}^{-1}A_{12} \end{pmatrix} \begin{pmatrix} I & A_{11}^{-1}A_{12} \\ 0 & I \end{pmatrix}$$

for matrices of appropriate sizes and A_{11} being invertible. See [20, formula (0.3) p. 3].

One could have started with the space $\begin{pmatrix} 0 \\ \mathbb{C}^p \end{pmatrix}$, which is also A-invariant and H-positive. In particular, the above factorization is not unique.

2.2. The other characteristic spectral functions

In this section we review the definitions and main properties of the characteristic spectral functions associated to a canonical differential expression.

It follows from Theorem 2.4 that $U(0, \lambda)$ has no pole on the real line and that, furthermore:

$$U_{11}(0, \lambda)U_{11}(0, \lambda)^* - U_{12}(0, \lambda)U_{12}(0, \lambda)^* = I_n$$
$$U_{22}(0, \lambda)U_{22}(0, \lambda)^* - U_{21}(0, \lambda)U_{21}(0, \lambda)^* = I_n$$

and

$$U_{11}(0, \lambda)^*U_{12}(0, \lambda) = U_{21}(0, \lambda)^*U_{22}(0, \lambda)$$

for real λ.

In particular, $U_{11}(0, \lambda)$ is invertible on the real line and $U_{21}(0, \lambda)U_{11}(0, \lambda)^{-1}$ is well defined and takes contractive values on the real line.

Definition 2.7. *The function*

$$R(\lambda) = (U_{21}(0, \lambda)U_{11}(0, \lambda)^{-1})^* = U_{12}(0, \lambda)U_{22}(0, \lambda)^{-1}, \qquad \lambda \in \mathbb{R},$$

is called the reflection coefficient function.

To present an equivalent definition of the reflection coefficient function, we need some notation: if

$$\Theta = \begin{pmatrix} A & B \\ C & D \end{pmatrix} \in \mathbb{C}^{(p+q)\times(p+q)}, \qquad A \in \mathbb{C}^{p\times p}, \quad \text{and} \quad X \in \mathbb{C}^{p\times q}$$

we set

$$T_\Theta(X) = (AX + B)(CX + D)^{-1}.$$

Note that

$$T_{\Theta_1\Theta_2}(X) = T_{\Theta_1}(T_{\Theta_2}(X)) \tag{2.9}$$

when all expressions are well defined.

Theorem 2.8. *Let* $\Theta(x, \lambda) = U(x, \lambda)U(0, \lambda)^{-1}$. *Then,* $\Theta(x, \lambda)$ *is also a solution of* (1.1). *It is an entire function of* λ. *It is* J-*expansive in* $\mathbb{C}_+$,

$$J - \Theta(x, \lambda)J\Theta(x, \lambda)^* \begin{cases} = 0, & \lambda \in \mathbb{R} \\ \leq 0, & \lambda \in \mathbb{C}_+, \end{cases}$$

and satisfies the initial condition $\Theta(0, \lambda) = I_{2n}$. *Moreover*

$$R(\lambda) = \lim_{x \to \infty} T_{\Theta(x,\lambda)^{-1}}(0), \quad \lambda \in \mathbb{R}. \tag{2.10}$$

The matrix function $\Theta(x, \lambda)$ is called the matrizant, or fundamental solution of the canonical differential expression. Its properties may be found in [22, p. 150]. For real λ the matrix function $U(0, \lambda)$ is J-unitary. Hence we have:

$$\Theta(x, \lambda)^{-1} = U(0, \lambda)U(x, \lambda)^{-1}.$$

The result follows using (2.9) and the asymptotic property (2.6).

In fact, the function R is analytic and takes contractive values in the closed lower half-plane. For a proof and references, see [10] and [13, Theorem 3.1 p 6].

Theorem 2.9. *A minimal realization of* $R(\lambda)$ *is given by*

$$R(\lambda) = -c(\lambda I_p - (a + i\Omega c^* c))^{-1}(b + i\Omega c^*). \tag{2.11}$$

See [10]. It follows in particular that the spectrum of the matrix $a + i\Omega c^* c$ is in the open upper half-plane. Note that Ω is not arbitrary but is related to a, b and c via the Lyapunov equation (2.2).

A direct proof that R is analytic and contractive in $\mathbb{C}_-$ can be given using the results in [33], as we now explain.

Definition 2.10. *A* $\mathbb{C}^{n \times n}$-*valued rational function* R *is called a proper contraction if it takes contractive values on the real line and if moreover it is analytic at infinity and such that*

$$R(\infty)R(\infty)^* < I_n.$$

The following results are respectively [33, Theorem 3.2 p. 231, Theorem 3.4 p. 235].

Theorem 2.11. *Let* R *be a* $\mathbb{C}^{n \times n}$-*valued rational function analytic at infinity and let* $R(z) = D + C(zI - A)^{-1}B$ *be a minimal realization of* W. *Let*

$$\mathcal{A} = \begin{pmatrix} \alpha & \beta \\ \gamma & \alpha^* \end{pmatrix} = \begin{pmatrix} A + BD^*(I_n - DD^*)^{-1}C & B(I_n - D^*D)^{-1}B^* \\ C^*(I_n - DD^*)^{-1}C & A^* + C^*(I_n - DD^*)^{-1}DB^* \end{pmatrix}.$$

Then the following are equivalent:

1) *The matrix function* R *is a proper contraction.*
2) *The real eigenvalues of* $\mathcal{A}$ *have even partial multiplicities.*
3) *The Riccati equation*

$$X\gamma X - iX\alpha^* + i\alpha X + \beta = 0. \tag{2.12}$$

has an Hermitian solution.

The matrix $\mathcal{A}$ is called the state characteristic matrix of W and the Riccati equation (2.12) is called its state characteristic equation.

Theorem 2.12. *Let R be a $\mathbb{C}^{n\times n}$-valued proper contraction, with minimal realization $R(z) = D + C(zI - A)^{-1}B$ and let (2.12) be its state characteristic equation. Then, any Hermitian solution of (2.12) is invertible and the number of negative eigenvalues of X is equal to the number of poles of R in $\mathbb{C}_-$.*

Consider now the minimal realization (2.11). The corresponding state characteristic equation is

$$Xc^*cX - iX(a^* - icc^*\Omega) + i(a + i\Omega cc^*)X + (b + i\Omega c^*)(b^* - ic\Omega) = 0.$$

To show that $X = \Omega$ is a solution of this equation is equivalent to prove that Ω solves the Lyapunov equation (2.3). Indeed,

$$0 = \Omega c^*c\Omega - i\Omega(a^* - icc^*\Omega) + i(a + i\Omega cc^*)\Omega + (b + i\Omega c^*)(b^* - ic\Omega)$$

$$\Longleftrightarrow$$

$$0 = -i\Omega a^* + ia\Omega + bb^* - i\Omega(a - c^*b^*) + i(a - bc)\Omega + bb^*$$

$$\Longleftrightarrow$$

$$0 = i(a^\times\Omega - \Omega a^{\times *}) + bb^*,$$

which is (2.3).

The *scattering matrix function* is defined as follows:

Theorem 2.13. *The differential equation (1.1) has a uniquely defined $\mathbb{C}^{2n\times n}$-valued solution such that for $\lambda \in \mathbb{R}$,*

$$\begin{pmatrix} I_n & -I_n \end{pmatrix} X(0, \lambda) = 0,$$

$$\lim_{x\to\infty} \begin{pmatrix} 0 & e^{ix\lambda}I_n \end{pmatrix} X(x, \lambda) = I_n.$$

The limit

$$\lim_{x\to\infty} \begin{pmatrix} e^{-ix\lambda}I_n & 0 \end{pmatrix} X(x, \lambda) = S(\lambda)$$

exists for all real λ and is called the scattering matrix function of the canonical system. The scattering matrix function takes unitary values on the real line, belongs to the Wiener algebra $\mathcal{W}$ and admits a factorization $S = S_+S_-$ where S_+ and its inverse are analytic in the closed upper half-plane while S_- and its inverse are analytic in the closed lower half-plane.

We note that the general factorization of a function in the Wiener algebra and unitary on the real line involves in general a diagonal term taking into account quantities called partial indices; see [31], [32], [34], [17]. We also note that conversely, functions with the properties as in the theorem are scattering matrix functions of a more general class of differential equations; see [41] and the discussion in [7, Appendix].

Theorem 2.14. *The scattering matrix function of a canonical system* (1.1) *with potential* (2.1) *is given by:*

$$S(\lambda) \;=\; (I_n + b^*(\lambda I_p - a^*)^{-1}c^*)^{-1}$$
$$\times(I_n - (ib^*Y - c)(\lambda I_p - a)^{-1}(I_p + \Omega Y)^{-1}(b + i\Omega c^*)).$$

A minimal realization of the scattering matrix function is given by $S(\lambda) = I_n + C(\lambda I_{2p} - A)^{-1}B$, *where*

$$A = \begin{pmatrix} a & b(ic\Omega - b^*) \\ 0 & a^{\times *} \end{pmatrix},$$

$$B = \begin{pmatrix} b \\ (I_p + Y\Omega)^{-1}(c^* + iYb) \end{pmatrix},$$
$$C = (c \quad ic\Omega - b^*).$$

Set

$$G = \begin{pmatrix} -\Omega & iI_p \\ -iI_p & -Y(I_p + \Omega Y)^{-1} \end{pmatrix}.$$

Then it holds that

$$i(AG - GA^*) \;=\; -BB^*,$$
$$CG \;=\; iB^*,$$

and thus S takes unitary values on the real line.

For a proof, see [8, p. 7]. The last statement follows from [5, Theorem 2.1 p. 179], that is from equations (2.7) and (2.8) with $H = X^{-1}$ and $J = I_p$. Since

$$X = \begin{pmatrix} I_p & 0 \\ i\Omega^{-1} & I_p \end{pmatrix} \begin{pmatrix} -\Omega & 0 \\ 0 & (\Omega + \Omega Y\Omega)^{-1} \end{pmatrix} \begin{pmatrix} I_p & 0 \\ i\Omega^{-1} & I_p \end{pmatrix}^*$$

the space $\begin{pmatrix} \mathbb{C}^p \\ 0 \end{pmatrix}$ is A invariant and H-negative. Thus Theorem 2.5 on factorizations leads to:

Theorem 2.15. *The scattering matrix function of a canonical system* (1.1) *with potential* (2.1) *admits a minimal factorization of the form*

$$S(z) = U_1(z)^{-1}U_2(z)$$

where both U_1 and U_2 are inner (that is, are contractive in $\mathbb{C}_+$ and take unitary values on the real line).

The fact that U_2 is inner (and not merely unitary) stems from the fact that the Schur complement of $-\Omega$ in H is equal to

$$-Y(I_p + \Omega Y)^{-1} - iI_p(-\Omega)^{-1}(-iI_p) = (\Omega + \Omega Y\Omega)^{-1}$$

and in particular is strictly positive.

Such a factorization result was also proved in [12, Theorem 7.1] using different methods. It is a particular case of a factorization result of M.G. Kreĭn and H. Langer for functions having a finite number of negative squares; see [39].

We now turn to the spectral function. We first recall that the operator

$$Hf(x) = -iJ\frac{\mathrm{d}\,f}{\mathrm{d}\,x}(x) - v(x)f(x)$$

restricted to the space of $\mathbb{C}^{2n}$-valued absolutely continuous functions with entries in $\mathbf{L}_2$ and such that

$$(I_n \quad -I_n)f(0) = 0$$

is self-adjoint.

Definition 2.16. *A positive function $W : \mathbb{R} \to \mathbb{C}^{n \times n}$ is called a spectral function if there is a unitary map U from $\mathbf{L}_2^n$ onto $\mathbf{L}_2^n(W)$ mapping H onto the operator of multiplication by the variable in $\mathbf{L}_2^n(W)$.*

Theorem 2.17. *The function*

$$W(\lambda) = (V_{22}(\lambda) - V_{12}(\lambda))^{-*}(V_{22}(\lambda) - V_{12}(\lambda))^{-1}$$

is a spectral function, the map U being given by

$$F(\lambda) = \frac{1}{\sqrt{2\pi}} \int_0^\infty \begin{pmatrix} I_n & I_n \end{pmatrix} \Theta(x,\lambda)^* f(x)dx. \tag{2.13}$$

A direct proof in the rational case can be found in [26]. When $k(x) \equiv 0$, we have that $W(\lambda) = I_n d\lambda$, and the unitary map (2.13) is readily identified with the Fourier transform.

Definition 2.18. *The Weyl coefficient function $N(\lambda)$ is defined in the open upper half plane; it is the unique $\mathbb{C}^{n \times n}$-valued function such that*

$$\int_0^\infty \begin{pmatrix} iN(\lambda)^* & I_n \end{pmatrix} \begin{pmatrix} I_n & I_n \\ I_n & -I_n \end{pmatrix} \Theta(x,\lambda)^* \Theta(x,\lambda) \begin{pmatrix} I_n & I_n \\ I_n & -I_n \end{pmatrix} \begin{pmatrix} -iN(\lambda) \\ I_n \end{pmatrix} dx$$

is finite for $-i(\lambda - \lambda^*) > 0$.

In the setting of differential expressions (1.1), the function N was introduced in [27]. The motivation comes from the theory of the Sturm-Liouville equation. The Weyl coefficient function is analytic in the open upper half-plane and has a nonnegative imaginary part there. Such functions are called Nevanlinna functions.

Theorem 2.19. *The Weyl coefficient function is given by the formula*

$$\begin{aligned} N(\lambda) &= i(U_{12}(0,\lambda) + U_{22}(0,\lambda))(U_{12}(0,\lambda) - U_{22}(0,\lambda))^{-1} \\ &= i(I_n - 2c(\lambda I_p - a^\times)^{-1}(b + i\Omega c^*)). \end{aligned} \tag{2.14}$$

Proof. We first look for a $\mathbb{C}^{n \times 2n}$-valued function $P(\lambda)$ such that $x \mapsto P(\lambda)\Theta(x,\lambda)^*$ has square summable entries for $\lambda \in \mathbb{C}^+$. Let $U(\lambda,x)$ be the solution of the differential system (1.1) subject to the asymptotic condition (2.6). Then, $U(x,\lambda) = \Theta(x,\lambda)U(0,\lambda)$. We thus require the entries of the function

$$x \mapsto P(\lambda)U(0,\lambda)^{-*}U(x,\lambda) \tag{2.15}$$

to be square summable. By definition of U, it is necessary for $P(\lambda)U(0,\lambda)^{-*}$ to be of the form $(0,\ p(\lambda))$ where $p(\lambda)$ is $\mathbb{C}^{n\times n}$-valued. It follows from the definition of $U(0,\lambda)$ that one can take

$$P(\lambda) = \begin{pmatrix} 0 & I_n \end{pmatrix} U(0,\lambda)^* = \begin{pmatrix} U_{12}(0,\lambda)^* & U_{22}(0,\lambda)^* \end{pmatrix}$$

and hence the necessity condition. Conversely, we have to show that the function (2.15) has indeed summable entries. But this is just doing the above argument backwards.

The realization formula follows then from the realization formulas for the block entries of the asymptotic equivalence matrix function. $\qquad\square$

Any of the functions in the spectral domain determines all the others, as follows from the next theorem:

Theorem 2.20. *Assume given a differential system of the form* (1.1) *with potential* $k(x)$ *of the form* (2.1). *Assume* $W(\lambda)$, $V(\lambda)$, $R(\lambda)$, $S(\lambda)$ *and* $N(\lambda)$ *are the characteristic spectral functions of* (1.1), *and let* $S = S_- S_+$ *be the spectral factorization of the scattering matrix function* S, *where* S_- *and its inverse are invertible in the closed lower half-plane and* S_+ *and its inverse are invertible in the closed upper half-plane. Then, the connections between these functions are:*

$$\begin{aligned}
W(\lambda) &= S_-(\lambda)^{-1}S_-(\lambda)^{-*} = S_+(\lambda)S_+(\lambda)^*, \\
W(\lambda) &= \operatorname{Im} N(\lambda), \\
S(\lambda) &= S_-(\lambda)S_+(\lambda), \\
R(\lambda) &= (iN(\lambda)^* - I_n)(iN(\lambda)^* + I_n)^{-1}, \\
N(\lambda) &= i(I_n + R(\lambda)^*)(I_n - R(\lambda)^*)^{-1}, \\
V(\lambda) &= \frac{1}{2}\begin{pmatrix} (iN(\lambda)^* + I_n)S_-(\lambda)^* & (-iN(\lambda) - I_n)S_+(\lambda)^{-*} \\ (iN(\lambda)^* - I_n)S_-(\lambda)^* & (-iN(\lambda) + I_n)S_+(\lambda)^{-*} \end{pmatrix}
\end{aligned}$$

for $\lambda \in \mathbb{R}$.

See [10, Theorem 3.1].

We note that $R^* = T_V(0)$. We now wish to relate V to a unitary completion of the reflection coefficient function. It is easier to look at

$$\widetilde{V}(\lambda) = \begin{pmatrix} 0 & I_n \\ I_n & 0 \end{pmatrix} V(\lambda) \begin{pmatrix} 0 & I_n \\ I_n & 0 \end{pmatrix}.$$

We set

$$P = \frac{I_{2n} + J}{2} = \begin{pmatrix} I_n & 0 \\ 0 & 0 \end{pmatrix} \quad \text{and} \quad Q = \frac{I_{2n} - J}{2} = \begin{pmatrix} 0 & 0 \\ 0 & I_n \end{pmatrix}.$$

Theorem 2.21. *Let* $\Theta \in \mathbb{C}^{2n\times 2n}$ *be such that* $\det(P+Q\Theta) \neq 0$. *Then* $\det(P-\Theta Q) \neq 0$ *and*

$$\Theta^{\times} \stackrel{\text{def.}}{=} (P\Theta + Q)(P + Q\Theta)^{-1} = (P - \Theta Q)^{-1}(\Theta P - Q) \qquad (2.16)$$

Finally

$$I_{2n} - \Theta^{\times}\Theta^{\times *} = (P - \Theta Q)^{-1}(J - \Theta J \Theta^*)(P - \Theta Q)^{-*} \tag{2.17}$$

$$I_{2n} - \Theta^{\times *}\Theta^{\times} = (P + Q\Theta)^{-*}(J - \Theta^* J \Theta)(P + Q\Theta)^{-1}. \tag{2.18}$$

Proof. We set $\Theta = \begin{pmatrix} A & B \\ C & C \end{pmatrix}$ where $A \in \mathbb{C}^{n \times n}$. We have:

$$P + Q\Theta = \begin{pmatrix} I_n & 0 \\ C & D \end{pmatrix}, \qquad P - \Theta Q = \begin{pmatrix} I_n & -B \\ 0 & -D \end{pmatrix}.$$

Thus either of these matrices is invertible if and only if D is invertible. Thus both equalities in (2.16) make sense. To prove that they define the same object is equivalent to prove that

$$(P - \Theta Q)(P\Theta + Q) = (\Theta P - Q)(P + Q\Theta),$$

i.e., since $PQ = QP = 0$,

$$P\Theta - \Theta Q = \Theta P - Q\Theta.$$

This in turn clearly holds since $P + Q = I_{2n}$.

We now prove (2.17). The proof of (2.18) is similar and will be omitted. We have

$$
\begin{aligned}
I_{2n} - \Theta^{\times}\Theta^{\times *} &= I_{2n} - (P - \Theta Q)^{-1}(\Theta P - Q)(\Theta P - Q)^*(P - \Theta Q)^{-*} \\
&= (P - \Theta Q)^{-1}\{(P - \Theta Q)(P - \Theta Q)^* - (\Theta P - Q)(\Theta P - Q)^*\} \\
&\quad \times (P - \Theta Q)^{-*} \\
&= (P - \Theta Q)^{-1}\{P - Q + \Theta Q \Theta^* - \Theta P \Theta^*\}(P - \Theta Q)^{-*}
\end{aligned}
$$

and hence the result since $J = P - Q$. $\square$

The function defined by (2.16) is called the Potapov–Ginzburg transform of Θ. We have

$$\Theta^{\times} = \begin{pmatrix} A - BD^{-1}C & BD^{-1} \\ -D^{-1}C & D^{-1} \end{pmatrix}. \tag{2.19}$$

Theorem 2.22. *The Potapov–Ginzburg transform of $\widetilde{V}$ is a unitary completion of the reflection coefficient function.*

Indeed, from (2.19) the 22 block of the Potapov–Ginzburg transform of $\widetilde{V}$ is exactly R. It is not a minimal completion (in particular it has n poles in $\mathbb{C}_-$). See [20] for more information on this transform. Minimal unitary completions of a proper contraction are studied in [33, Theorem 4.1 p. 236].

2.3. The continuous orthogonal polynomials

As already mentioned, for every $x \geq 0$ the function $\lambda \mapsto \Theta(x, \lambda) = U(x, \lambda)U(0, \lambda)^{-1}$ is entire. Albeit their name, the continuous orthogonal polynomials are entire functions, first introduced by M.G. Kreĭn (see [37]) and in terms of which one can compute the matrix function $\Theta(x, \lambda)$. To define these functions we start with a function W of the form

$$W(\lambda) = I_n - \int_{\mathbb{R}} e^{it\lambda} \omega(t) dt, \quad \lambda \in \mathbb{R}, \tag{2.20}$$

with $\omega \in \mathbf{L}_1^{n \times n}(\mathbb{R})$ and such that $W(\lambda) > 0$ for all $\lambda \in \mathbb{R}$. This last condition insures that the integral equation

$$\Gamma_T(t, s) - \int_0^T \omega(t - u) \Gamma_T(u, s) du = \omega(t - s), \quad t, s \in [0, T]$$

has a unique solution for every $T > 0$.

Definition 2.23. *The continuous orthogonal polynomial is given by:*

$$P(t, \lambda) = e^{it\lambda} \left(I_n + \int_0^{2t} \Gamma_{2t}(u, 0) e^{-i\lambda u} du \right).$$

Theorem 2.24. *It holds that*

$$\begin{pmatrix} I_n & I_n \end{pmatrix} \Theta(x, \lambda) = \begin{pmatrix} P(t, -\lambda) & R(t, \lambda) \end{pmatrix}$$

where $R(t, \lambda) = e^{it\lambda} \left(I_n + \int_0^{2t} \Gamma_{2t}(2t - u, 2t) e^{-i\lambda u} du \right).$

In view of Theorem 2.20, note that every rational function analytic at infinity, such that $W(\infty) = I_n$, with no poles and strictly positive on the real line, is the spectral function of a canonical differential expression of the form (1.1) with potential of the form (2.1). Furthermore, let $W(\lambda) = I_n + C(\lambda I_p - A)^{-1}B$ be a minimal realization of W. Then, W is of the form (2.20) with

$$\omega(u) = \begin{cases} iCe^{-iuA}(I_p - P)B, & u > 0, \\ -iCe^{-iuA}PB, & u < 0, \end{cases}$$

where P is the Riesz projection of A in $\mathbb{C}_+$. We recall that

$$P = \int_\gamma (\zeta I_p - A)^{-1} d\zeta$$

where γ is a positively oriented contour which encloses only the eigenvalues of A in $\mathbb{C}_+$.

Theorem 2.25. *Let W be a rational $\mathbb{C}^{n \times n}$-valued function analytic and invertible on $\mathbb{R}$ and at infinity. Assume moreover that $W(\lambda) > 0$ for real λ and that $W(\infty) = I_n$. Let $W(\lambda) = I_n + C(\lambda I_p - A)^{-1}B$ be a minimal realization of W. Let P (resp. $P^\times$) denote the Riesz projection corresponding to the eigenvalues of A (resp. of $A^\times = A - BC$) in $\mathbb{C}_+$. Then, the continuous orthogonal polynomials $P(t, \lambda)$ are given by the formula*

$$P(t, \lambda) = e^{i\lambda t} \left\{ I_n + C(\lambda I_p + A^\times)^{-1}(I_p - e^{-2i\lambda t} e^{-2itA^\times}) \pi_{2t} B \right\}$$

where

$$\pi_t = (I_p - P + Pe^{-itA^\times})^{-1}(I_p - P).$$

Furthermore,

$$\lim_{t\to\infty} e^{-it\lambda} P(t,\lambda) = S_-(-\lambda)^*. \tag{2.21}$$

See [7, Theorem 3.3 p 10]. The computations in [7] use exact formulas for the function $\Gamma_T(t,s)$ in terms of the realization of W which have been developed in [15].

We note that the potential $k(x)$ can be written as

$$k(x) = 2C \left(Pe^{-2ixA^\times} |_{\mathrm{Im}\ P} \right)^{-1} PB \tag{2.22}$$

in terms of the realization of the spectral function W.

2.4. Perturbations

In this subsection we address the following question: assume that $k(x)$ is a strictly pseudo-exponential potential. Is $-k(x)$ also such a potential? This is not quite clear from formulas (2.1) or (2.22). One could attack this problem using the results in [11], where we studied a trace formula for a pair of self-adjoint operators corresponding to the potentials $k(x)$ and $-k(x)$. Here we present a direct argument in the rational case. More precisely, if N is a Nevanlinna function so are the three functions

$$\begin{aligned}
\lambda &\rightarrow -N^{-1}(\lambda), \\
\lambda &\rightarrow -N^{-1}(-\lambda^*)^*, \\
\lambda &\rightarrow N(-\lambda^*)^*,
\end{aligned}$$

and we have three associated weight functions

$$\begin{aligned}
W_-(\lambda) &= \mathrm{Im}\ -N(\lambda)^{-1}, \\
W_1(\lambda) &= \mathrm{Im}\ -N(-\lambda^*)^{-*}, \\
W_2(\lambda) &= \mathrm{Im}\ N(-\lambda^*)^*.
\end{aligned}$$

The relationships between these three weight functions and the original weight function W and the associated potential have been reviewed in the thesis [36] and we recall the results in form of a table:

The potential	The weight function
$v(x) = \begin{pmatrix} 0 & k(x) \\ k(x)^* & 0 \end{pmatrix}$	$W(\lambda) = \mathrm{Im}\ N(\lambda)$
$-v(x) = -\begin{pmatrix} 0 & k(x) \\ k(x)^* & 0 \end{pmatrix}$	$W_-(\lambda) = \mathrm{Im}\ -N(\lambda)^{-1}$
$-\begin{pmatrix} 0 & k(x)^* \\ k(x) & 0 \end{pmatrix}$	$W_1(\lambda) = \mathrm{Im}\ N(-\lambda^*)^*$
$\begin{pmatrix} 0 & k(x)^* \\ k(x) & 0 \end{pmatrix}$	$W_2(\lambda) = \mathrm{Im}\ -N(-\lambda^*)^{-*}$

Let

$$N(\lambda) = i(I + c(\lambda I - a)^{-1}b)$$

be a minimal realization of N. Then,

$$W(\lambda) = I + C(\lambda I - A)^{-1}B$$

is a minimal realization of the weight function W, where

$$A = \begin{pmatrix} a & 0 \\ 0 & a^* \end{pmatrix}, \quad B = \begin{pmatrix} b \\ c^* \end{pmatrix}, \quad C = \frac{1}{2}\begin{pmatrix} c & b^* \end{pmatrix}, \tag{2.23}$$

and the Riesz projection corresponding to the spectrum of A in the open upper half-plane $\mathbb{C}_+$ is

$$P = \begin{pmatrix} I & 0 \\ 0 & 0 \end{pmatrix}. \tag{2.24}$$

Furthermore, the potential associated to the weight function W is given by (2.22) where A, B, C and P are given by (2.23) and (2.24), and

$$A^\times = A - BC = \begin{pmatrix} a - \frac{bc}{2} & -\frac{bb^*}{2} \\ -\frac{c^*c}{2} & (a - \frac{bc}{2})^* \end{pmatrix}.$$

Consider now the weight function W_-. A minimal realization of $-N(\lambda)^{-1}$ is given by

$$-N(\lambda)^{-1} = i(I - c(\lambda I - a^\times)^{-1}b), \quad a^\times = a - bc,$$

and a minimal realization of W_- is given by

$$W_-(\lambda) = I + C_-(\lambda I - A_-)^{-1}B_-,$$

where

$$A_- = \begin{pmatrix} a^\times & 0 \\ 0 & a^{\times *} \end{pmatrix}, \quad B_- = B = \begin{pmatrix} b \\ c^* \end{pmatrix}, \quad C_- = -C = -\frac{1}{2}\begin{pmatrix} c & b^* \end{pmatrix},$$

and the Riesz projection corresponding to the spectrum of A_- in the open upper half-plane $\mathbb{C}_+$ is $P_- = P$ given by (2.24).

The potential associated to the weight function W_- is given by

$$k_-(x) = -2C \left(Pe^{-2itA^\times}\big|_{\operatorname{Im} P}\right)^{-1} PB,$$

where

$$A_-^\times = A_- - B_-C_- = \begin{pmatrix} a - \frac{bc}{2} & \frac{bb^*}{2} \\ \frac{c^*c}{2} & (a - \frac{bc}{2})^* \end{pmatrix}.$$

Setting

$$D = \begin{pmatrix} a - \frac{bc}{2} & 0 \\ 0 & (a - \frac{bc}{2})^* \end{pmatrix}, \quad Z = \begin{pmatrix} 0 & \frac{b^*b}{2} \\ \frac{cc^*}{2} & 0 \end{pmatrix},$$

we have

$$A^\times = D - Z \quad \text{and} \quad A_-^\times = D + Z.$$

We are now in a position to prove the following result:

Theorem 2.26. *Let $k(x)$ be a strictly pseudo-exponential potential with associated Weyl function $N(\lambda)$. The potential associated to $\mathrm{Im} - N^{-1}$ is equal to $k_-(x) = -k(x)$.*

Proof. To prove that $k_-(x) = -k(x)$, it is enough to prove that

$$Pe^{-itA^\times}\big|_{\mathrm{Im}\ P} = Pe^{-it(A_- - B_- C_-)}\big|_{\mathrm{Im}\ P}.$$

To prove this equality, it is enough in turn to prove that for all positive integers ℓ, it holds that

$$PA^{\times \ell}\big|_{\mathrm{Im}\ P} = P(A_- - B_- C_-)^\ell\big|_{\mathrm{Im}\ P},$$

i.e., that

$$\begin{pmatrix} I & 0 \\ 0 & 0 \end{pmatrix}(D+Z)^\ell\begin{pmatrix} I & 0 \\ 0 & 0 \end{pmatrix} = \begin{pmatrix} I & 0 \\ 0 & 0 \end{pmatrix}(D-Z)^\ell\begin{pmatrix} I & 0 \\ 0 & 0 \end{pmatrix}$$

for all positive integers ℓ. Let $\epsilon = \pm 1$. The expression $(D + \epsilon Z)^\ell$ consists of a sum of terms of the form

$$D^{\alpha_1}(\epsilon Z)^{\beta_1} D^{\alpha_2}(\epsilon Z)^{\beta_2} \cdots ,$$

where the α_i and the β_i are equal to 1 or 0 and $\sum_i(\alpha_i + \beta_i) = \ell$. Each factor $D^{\alpha_i} Z^{\beta_i}$ for which $\beta_i \neq 0$ is anti block diagonal. We consider two cases, namely $\sum_i \beta_i$ being odd or even. When $\sum_i \beta_i$ is odd, we have the product of an odd number of anti block diagonal matrices, and the result is anti block diagonal, and so, premultiplying and postmultiplying this product by $\begin{pmatrix} I & 0 \\ 0 & 0 \end{pmatrix}$ we obtain the zero matrix. When $\sum_i \beta_i$ is even, the product is an even function of ϵ and have the same value at $\epsilon = 1$ and at $\epsilon = -1$.

The case of the other two weight functions is treated in much the same way. We focus on $W_1(\lambda) = \mathrm{Im}\ N(-\lambda^*)^*$. A minimal realization of $N(-\lambda^*)^*$ is given by

$$N(-\lambda^*)^* = i(I - b^*(\lambda I + a^*)^{-1}c^*),$$

and a minimal realization of the weight function W_1 is therefore given by

$$W_1(\lambda) = I + C_1(\lambda I - A_1)^{-1}B_1,$$

where

$$A_1 = \begin{pmatrix} -a^* & 0 \\ 0 & -a \end{pmatrix}, \quad B_1 = \begin{pmatrix} c^* \\ b \end{pmatrix}, \quad C_1 = -\frac{1}{2}\begin{pmatrix} b^* & c \end{pmatrix},$$

and the Riesz projection corresponding to the spectrum of A_1 in the open upper half-plane $\mathbb{C}_+$ is $P_1 = P$ given by (2.24). The potential associated to the weight function W_1 is given by

$$k_1(x) = 2C_1 \left(P_1 e^{-2itA_1^\times}\big|_{\mathrm{Im}\ P_1} \right)^{-1} P_1 B_1.$$

We claim that $k_1(x) = -k(x)^*$. Indeed,

$$k_1(x)^* = 2B_1^* P_1^* \left(P_1 e^{2itA_1^{* \times}}\big|_{\mathrm{Im}\ P_1} \right)^{-1} P_1^* C_1^*.$$

But we have that

$$B_1^* P_1^* = 2CP = \begin{pmatrix} c & 0 \end{pmatrix}, \quad P_1 C_1^* = -PB = -\frac{1}{2}\begin{pmatrix} b \\ 0 \end{pmatrix}, \quad A_1^{*\times} = -A^\times,$$

which allows to conclude. $\qquad\qquad\qquad\qquad\qquad\qquad\qquad\qquad\qquad\qquad\qquad\square$

3. The discrete case

3.1. First-order discrete system

In our previous work [6] we studied inverse problems for difference operators associated to Jacobi matrices. Such operators are the discrete counterparts of Sturm–Liouville differential operators, and one can associate to them a number of functions analytic in the open unit disk similar to the characteristic spectral functions of a canonical differential expression. In the present paper we chose a different avenue to define discrete systems, which has more analogy to the continuous case and is more natural. The analogies between the two cases are gathered in form of two tables at the end of the paper.

We note that another type of discrete systems has been considered by L. Sakhnovich in [42, Section 2 p. 389].

Our starting point is the telegraphers' equations (1.2). We now assume that the local impedance function $Z(x)$ defined in (1.2) is equal to a constant, say Z_n, on the interval $[nh, (n+1)h)$ for $n = 0, 1, \ldots$ In particular, $Z(x)$ may have discontinuities at the points nh. On the open interval $(nh, (n+1)h)$, we have $k(x) = 0$ and equation (1.3) becomes

$$\begin{pmatrix} (\frac{\partial}{\partial x} + \frac{\partial}{\partial t}) & 0 \\ 0 & (\frac{\partial}{\partial x} - \frac{\partial}{\partial t}) \end{pmatrix} W(x,t) = 0.$$

Hence one can write

$$W(x,t) = \begin{pmatrix} v_{1n}(x-t) \\ v_{2n}(x+t) \end{pmatrix}$$

on the interval $(nh, (n+1)h)$. Voltage and current are continuous at the points nh. Let us set

$$\alpha(n,t) = \lim_{\substack{x \to nh \\ x > nh}} W(x,t).$$

Taking into account (1.3) one gets to:

$$\alpha(n,t) = \frac{1}{2}\begin{pmatrix} Z_n^{-1/2} & Z_n^{1/2} \\ Z_n^{-1/2} & -Z_n^{1/2} \end{pmatrix}\begin{pmatrix} v(nh,t) \\ i(nh,t) \end{pmatrix}$$

$$\lim_{\substack{x \to nh \\ x < nh}} W(x,t) = \frac{1}{2}\begin{pmatrix} Z_{n-1}^{-1/2} & Z_{n-1}^{1/2} \\ Z_{n-1}^{-1/2} & -Z_{n-1}^{1/2} \end{pmatrix}\begin{pmatrix} v(nh,t) \\ i(nh,t) \end{pmatrix}.$$

We define the backward shift operator on functions of the variable t

$$\Delta f(t) = f(t-h).$$

We have:

$$\lim_{\substack{x \to nh \\ x < nh}} W(x,t) = \begin{pmatrix} v_{1,n-1}(nh - t) \\ v_{2,n-1}(nh + t) \end{pmatrix}$$

$$= \begin{pmatrix} v_{1,n-1}((n-1)h - (t-h)) \\ v_{2,n-1}((n-1)h + t + h) \end{pmatrix}$$

$$= \begin{pmatrix} \Delta & 0 \\ 0 & \Delta^{-1} \end{pmatrix} \begin{pmatrix} v_{1,n-1}((n-1)h - t) \\ v_{2,n-1}((n-1)h + t) \end{pmatrix}$$

$$= \begin{pmatrix} \Delta & 0 \\ 0 & \Delta^{-1} \end{pmatrix} \alpha(n-1,t).$$

Thus,

$$\begin{pmatrix} \Delta & 0 \\ 0 & \Delta^{-1} \end{pmatrix} \alpha(n-1,t) = \frac{1}{2} \begin{pmatrix} Z_{n-1}^{-1/2} & Z_{n-1}^{1/2} \\ Z_{n-1}^{-1/2} & -Z_{n-1}^{1/2} \end{pmatrix} \begin{pmatrix} v(nh,t) \\ i(nh,t) \end{pmatrix},$$

and we have:

$$\alpha(n,t) = \begin{pmatrix} Z_{n+1}^{-1/2} & Z_{n+1}^{1/2} \\ Z_{n+1}^{-1/2} & -Z_{n+1}^{1/2} \end{pmatrix} \begin{pmatrix} Z_n^{-1/2} & Z_n^{1/2} \\ Z_n^{-1/2} & -Z_n^{1/2} \end{pmatrix}^{-1} \begin{pmatrix} \Delta & 0 \\ 0 & \Delta^{-1} \end{pmatrix} \alpha(n-1,t)$$

$$= H(\rho_n) \begin{pmatrix} \Delta & 0 \\ 0 & \Delta^{-1} \end{pmatrix} \alpha(n-1,t)$$

where

$$\rho_n = \frac{Z_{n+1} - Z_n}{Z_{n+1} + Z_n} \quad \text{and} \quad H(\rho) = \frac{1}{\sqrt{1 - |\rho|^2}} \begin{pmatrix} 1 & -\rho \\ -\rho^* & 1 \end{pmatrix}$$

for $|\rho| < 1$. See [19, p. 111].

Replacing Δ by the complex variable and removing the scalar constant factor $\frac{1}{\sqrt{1-|\rho|^2}}$ we see that the discretization of the telegraphers' equations leads to systems of the form

$$Y_{n+1}(z) = \begin{pmatrix} 1 & -\rho_n \\ -\rho_n^* & 1 \end{pmatrix} \begin{pmatrix} z & 0 \\ 0 & z^{-1} \end{pmatrix} Y_n(z), \tag{3.1}$$

which we will call *two-sided first-order discrete systems*.

The solution corresponding to $\rho_n \equiv 0$ is

$$Y_n(z) = \begin{pmatrix} z^n & 0 \\ 0 & z^{-n} \end{pmatrix} Y_0(z),$$

that is, we are in a *two-sided* situation (the negative powers of z corresponding to signals coming from $-\infty$).

Recursions of the related forms

$$X_{n+1}(z) = \begin{pmatrix} 1 & -\rho_n \\ -\rho_n^* & 1 \end{pmatrix} \begin{pmatrix} z & 0 \\ 0 & 1 \end{pmatrix} X_n(z) \tag{3.2}$$

and

$$Z_{n+1}(z) = Z_n(z) \begin{pmatrix} 1 & -\rho_n \\ -\rho_n^* & 1 \end{pmatrix} \begin{pmatrix} z & 0 \\ 0 & 1 \end{pmatrix}$$

are one-sided (in the sense that solutions corresponding to $\rho_n \equiv 0$ involve only positive powers of z) and appear in the covariance extension problem. We here consider equations of the form (3.2). These are sometimes called a *first-order discrete system*. See [1]. Here we will call them *one-sided first-order discrete system*. Connections between the systems (3.1) and (3.2) are studied in the sequel.

Sometimes appears a factor $1/\sqrt{1-|\rho_n|^2}$ on the right side of these equations. For the situation considered here, where the ρ_n are of a special form (and in particular the sequence ρ_n belongs to ℓ_1) this factor can be ignored.

As in the case of canonical differential systems a number of functions of z are associated to such systems: we mention in particular the spectral function, the scattering function and the Weyl function. As in [9] we focus on the scalar case and postpone the matrix-valued case to a later publication. We refer to [21] for more information on discrete systems.

The potential $k(x)$ in (1.1) is now replaced by a sequence of numbers $\rho_n, n = 0, 1, 2 \ldots$ in the open unit disk $\mathbb{D}$. We will call such sequences *Schur sequences*. Strictly pseudo-exponential potentials are now replaced by sequences of the form

$$\rho_n = -ca^n(I_p - \Delta a^{*(n+1)}\Omega a^{n+1})^{-1}b. \tag{3.3}$$

In this equation $(a, b, c) \in \mathbb{C}^{p\times p} \times \mathbb{C}^{p\times 1} \times \mathbb{C}^{1\times p}$ is a minimal triple of matrices, the spectrum of a is in the open unit disk and Δ and Ω are the solutions of the Stein equations

$$\Delta - a\Delta a^* = bb^* \tag{3.4}$$

$$\Omega - a^*\Omega a = c^*c. \tag{3.5}$$

Furthermore, one requires that a is invertible and that it holds that

$$\Omega^{-1} > \Delta. \tag{3.6}$$

One recognizes in (3.3) the counterpart of (2.1). Moreover, as $n \to \infty$,

$$\rho_n \sim -ca^n b, \tag{3.7}$$

which is the analogue of (2.4). These sequences were introduced in our previous work [9] and called *strictly pseudo-exponential sequences*. The form of the ρ_n and the condition (3.6) call for some explanations, which we now give. In [9] the starting point was the Nehari extension problem associated to a sequence $\gamma_j, j = 0, -1, \ldots$: *find all elements $f \in \mathcal{W}$ such that*

$$f_j = \gamma_j, \qquad j = 0, -1, \ldots$$

$$\sup_{|z|=1} |f(z)| < 1.$$

In this problem an important role is played by the Hankel operator

$$\Gamma = \begin{pmatrix} \gamma_0 & \gamma_{-1} & \cdots \\ \gamma_{-1} & \gamma_{-2} & \cdots \\ \vdots & \vdots & \\ \vdots & \vdots & \end{pmatrix} \quad : \quad \ell_2 \to \ell_2.$$

A necessary and sufficient condition for the problem to have a solution is that $\|\Gamma\| < 1$. In [9] we took $\gamma_{-j} = ca^j b$. With this choice of γ_j the norm condition on Γ is equivalent to (3.6). This follows for instance from the formula

$$(I_{\ell_2} - \Gamma^*\Gamma)^{-1} = I_{\ell_2} + B^*(I_p - \Omega\Delta)^{-1}\Omega B$$
$$= I_{\ell_2} + B^*\Omega^{1/2}(I_p - \Omega^{1/2}\Delta\Omega^{1/2})^{-1}\Omega^{1/2}B$$

where $B = \begin{pmatrix} b & ab & a^2 b & \cdots \end{pmatrix}$.

As a direct consequence of the results in [9] one can get a formula for the solution of the systems (3.2) and (3.1) with various boundary conditions in terms of the matrices a, b and c.

The analogue of Theorem 2.26 is:

Remark 3.1. *If ρ_n is a strictly pseudo-exponential sequence so is $-\rho_n$.*

Indeed, it suffices to replace c by $-c$. This does not affect the matrices Y and Ω.

In the remaining of this section we study the spectral functions associated to a one-sided discrete first-order system. The two-sided case is considered in the next section.

3.2. The asymptotic equivalence matrix function

As for the continuous case there are two distinguished solutions to the systems (3.1) and (3.2); the first, related to the *inverse spectral problem*, fixes the value of the solution for $n = 0$ while the second, related to the *inverse scattering problem*, fixes the asymptotic value as $n \to \infty$. For (3.1) the asymptotic behavior at ∞ is

$$\lim_{n\to\infty} \begin{pmatrix} z^{-n} & 0 \\ 0 & z^n \end{pmatrix} Y_n(z) = I_2$$

while for (3.2) it is

$$\lim_{n\to\infty} \begin{pmatrix} z^{-n} & 0 \\ 0 & 1 \end{pmatrix} X_n(z) = I_2.$$

We begin with the analogue of Theorem 2.1.

Theorem 3.2. *Let $\rho_0, \rho_1, \ldots$ be a strictly pseudo-exponential sequence of Schur coefficients. Every solution of the first-order discrete system (3.2) is of the form*

$$X_n(z) = \prod_{\ell=0}^{n-1}(1 - |\rho_\ell|^2) \begin{pmatrix} 1 & 0 \\ 0 & z \end{pmatrix} H_n(z)^{-1} \begin{pmatrix} z^n & 0 \\ 0 & 1 \end{pmatrix} H_0(z) \begin{pmatrix} 1 & 0 \\ 0 & z^{-1} \end{pmatrix} X_0(z), \quad (3.8)$$

where

$$H_n(z) = \begin{pmatrix} \alpha_n(z) & \beta_n(z) \\ \gamma_n(z) & \delta_n(z) \end{pmatrix}$$

and

$$
\begin{aligned}
\alpha_n(z) &= 1 + ca^n z(zI_p - a)^{-1}(I_p - \Delta\Omega_n)^{-1}\Delta a^{*n}c^* & (3.9)\\
\beta_n(z) &= ca^n z(zI_p - a)^{-1}(I_p - \Delta\Omega_n)^{-1}b & (3.10)\\
\gamma_n(z) &= b^*(I_p - za^*)^{-1}(I_p - \Omega_n\Delta)^{-1}a^{*n}c^* & (3.11)\\
\delta_n(z) &= 1 + b^*(I_p - za^*)^{-1}(I_p - \Omega_n\Delta)^{-1}\Omega_n b, & (3.12)
\end{aligned}
$$

where $\Omega_n = a^{*n}\Omega a^n$. The solution K_n with the asymptotic

$$\lim_{n\to\infty} \begin{pmatrix} z^{-n} & 0 \\ 0 & 1 \end{pmatrix} X_n(z) = I_2$$

corresponds to

$$X_0(z) = \frac{1}{\prod_{\ell=0}^{\infty}(1 - |\rho_\ell|^2)} \begin{pmatrix} 1 & 0 \\ 0 & z \end{pmatrix} H_0(z)^{-1} \begin{pmatrix} 1 & 0 \\ 0 & z^{-1} \end{pmatrix}, \qquad (3.13)$$

that is,

$$K_n(z) = \frac{\prod_{\ell=0}^{n-1}(1 - |\rho_\ell|^2)}{\prod_{\ell=0}^{\infty}(1 - |\rho_\ell|^2)} \begin{pmatrix} 1 & 0 \\ 0 & z \end{pmatrix} H_n(z)^{-1} \begin{pmatrix} z^n & 0 \\ 0 & z^{-1} \end{pmatrix}, \qquad (3.14)$$

while the solution for which the initial value is identity at $n = 0$ corresponds to $X_0(z) = I_2$. In particular we have

$$
\begin{aligned}
\overset{\ell=n-1}{\underset{\ell=0}{\prod}} & \begin{pmatrix} 1 & -\rho_\ell \\ -\rho_\ell^* & 1 \end{pmatrix} \begin{pmatrix} z & 0 \\ 0 & 1 \end{pmatrix} \\
&= \prod_{\ell=0}^{n-1}(1 - |\rho_\ell|^2) \begin{pmatrix} 1 & 0 \\ 0 & z \end{pmatrix} H_n(z)^{-1} \begin{pmatrix} z^n & 0 \\ 0 & 1 \end{pmatrix} H_0(z) \begin{pmatrix} 1 & 0 \\ 0 & z^{-1} \end{pmatrix},
\end{aligned}
\qquad (3.15)
$$

where we denote

$$\overset{\ell=n-1}{\underset{\ell=0}{\prod}} A_\ell = A_{n-1}\cdots A_0.$$

Proof. We first recall the following results, proved in [9]. It holds that

$$\delta_n(z)^* = \alpha_n(1/z^*), \qquad \beta_n(z)^* = \gamma_n(1/z^*), \qquad (3.16)$$

and

$$\det H_0(z) = \frac{1}{\prod_{\ell=0}^{\infty}(1 - |\rho_\ell|^2)}. \qquad (3.17)$$

Furthermore, the matrix functions H_n satisfy the recurrence equation

$$H_{n+1}(z) = \begin{pmatrix} 1 & 0 \\ 0 & \frac{1}{z} \end{pmatrix} H_n(z) \begin{pmatrix} 1 & \rho_n \\ \rho_n^* & 1 \end{pmatrix} \begin{pmatrix} 1 & 0 \\ 0 & z \end{pmatrix}, \qquad n = 0, 1, 2, \ldots \qquad (3.18)$$

We rewrite (3.18) as

$$H_{n+1}(z)\begin{pmatrix} 1 & 0 \\ 0 & z^{-1} \end{pmatrix} = \begin{pmatrix} 1 & 0 \\ 0 & z^{-1} \end{pmatrix} H_n(z)\begin{pmatrix} 1 & \rho_n \\ \rho_n^* & 1 \end{pmatrix},$$

and we multiply this equation and equation (3.2) side by side. We obtain:

$$H_{n+1}(z)\begin{pmatrix} 1 & 0 \\ 0 & z^{-1} \end{pmatrix} X_{n+1}(z) = (1-|\rho_n|^2)\begin{pmatrix} 1 & 0 \\ 0 & z^{-1} \end{pmatrix} H_n(z)\begin{pmatrix} z & 0 \\ 0 & 1 \end{pmatrix} X_n(z)$$

$$= (1-|\rho_n|^2)\begin{pmatrix} z & 0 \\ 0 & 1 \end{pmatrix} H_n(z)\begin{pmatrix} 1 & 0 \\ 0 & z^{-1} \end{pmatrix} X_n(z).$$

Reiterating, we obtain that

$$H_{n+1}(z)\begin{pmatrix} 1 & 0 \\ 0 & z^{-1} \end{pmatrix} X_{n+1}(z) = \left(\prod_{\ell=0}^{n} 1-|\rho_\ell|^2\right)\begin{pmatrix} z^{n+1} & 0 \\ 0 & 1 \end{pmatrix} H_0(z)\begin{pmatrix} 1 & 0 \\ 0 & z^{-1} \end{pmatrix} X_0(z)$$

and hence we obtain formula (3.8) for $X_n(z)$. The other claims are easily verified. $\qquad\square$

We note that the solution X_n to (3.2) corresponding to $X_0 = I_2$ is a polynomial for every n (in the continuous case, the solution is an entire function). X_n can be expressed in terms of the orthogonal polynomials. We also remark that $\begin{pmatrix} z^n & 0 \\ 0 & 1 \end{pmatrix}$ is the solution of (3.2) corresponding to $\rho_n \equiv 0$.

Definition 3.3. *The function*

$$V(z) = \begin{pmatrix} \delta_0(z) & -\dfrac{\beta_0(z)}{z} \\ -z\gamma_0(z) & \alpha_0(z) \end{pmatrix} \tag{3.19}$$

is called the asymptotic equivalence matrix function of the one-sided first-order discrete system (3.2).

The terminology is explained in the next theorem:

Theorem 3.4. *Let c_1 and c_2 be in $\mathbb{C}^2$, and let $X_n^{(1)}$ and $X_n^{(2)}$ be the $\mathbb{C}^2$-valued solutions of* (3.2), *corresponding to the case of zero potential and to a potential ρ_n respectively and with initial conditions $X_0^{(1)}(z) = c_1$ and $X_0^{(2)}(z) = c_2$. Then, for every z on the unit circle,*

$$\lim_{n\to\infty} \|X_n^{(1)}(z) - X_n^{(2)}(z)\| = 0 \quad\Longleftrightarrow\quad c_2 = V(z)c_1.$$

Proof. By definition, $X_n^{(1)}(z) = \begin{pmatrix} z^n & 0 \\ 0 & 1 \end{pmatrix} c_1$. On the other hand,

$$X_n^{(2)}(z) = \prod_{\ell=0}^{n-1}(1-|\rho_\ell|^2)\begin{pmatrix} 1 & 0 \\ 0 & z \end{pmatrix} H_n(z)^{-1}\begin{pmatrix} z^n & 0 \\ 0 & 1 \end{pmatrix} H_0(z)\begin{pmatrix} 1 & 0 \\ 0 & z^{-1} \end{pmatrix} c_2.$$

The result follows since $\lim_{n\to\infty} H_n(z)^{-1} = I_2$ for z on the unit circle and since $\det H_0(z) = 1/\prod_{\ell=0}^{\infty}(1-|\rho_\ell|^2)$. $\qquad\square$

We note that the function $X_0(z)$ given by (3.13) is equal to $V(z)$.

The asymptotic equivalence matrix function takes (up to a constant) J-unitary values on the unit circle, with

$$J = \begin{pmatrix} 1 & 0 \\ 0 & -1 \end{pmatrix}.$$

Minimal realizations of rational functions which take J-unitary values on the unit circle $\mathbb{T}$ were studied in [5], where the following theorem is proved (in the more general setting of matrix-valued functions, i.e., where J is an arbitrary $m \times m$ matrix both unitary and self-adjoint; see [5, Theorem 3.1 p. 197]).

Theorem 3.5. *Let R be a $\mathbb{C}^{2\times 2}$-valued rational function analytic and invertible both at infinity and at the origin. Let $R(z) = D + C(zI - A)^{-1}B$ be a minimal realization of H. Then, R takes J-unitary values on $\mathbb{T}$ if and only if there is an Hermitian invertible matrix such that*

$$\begin{pmatrix} A & B \\ C & D \end{pmatrix}^* \begin{pmatrix} H & 0 \\ 0 & -J \end{pmatrix} \begin{pmatrix} A & B \\ C & D \end{pmatrix} = \begin{pmatrix} H & 0 \\ 0 & -J \end{pmatrix}. \qquad (3.20)$$

Note that (3.20) can be rewritten as

$$\begin{aligned} H - A^*HA &= -C^*JC, \\ C^*JD &= A^*HB, \\ J - D^*JD &= -B^*HB. \end{aligned}$$

The matrix H is uniquely defined from the given minimal realization and is called the *associated Hermitian matrix* to the given realization.

Theorem 3.6. *A minimal realization of the matrix function H_0 is given by $H_0(z) = D + C(zI - A)^{-1}B$ where*

$$A = \begin{pmatrix} a & 0 \\ 0 & a^{-*} \end{pmatrix},$$

$$B = \begin{pmatrix} a & 0 \\ 0 & a^{-*} \end{pmatrix} \begin{pmatrix} (I_p - \Delta\Omega)^{-1}\Delta & (I_p - \Delta\Omega)^{-1}, \\ -(I_p - \Omega\Delta)^{-1} & -(I_p - \Omega\Delta)^{-1}\Omega \end{pmatrix} \begin{pmatrix} a^*c^* & 0 \\ 0 & b \end{pmatrix},$$

$$C = \begin{pmatrix} c & 0 \\ 0 & b^* \end{pmatrix},$$

$$D = \begin{pmatrix} 1 + c(I_p - \Delta\Omega)^{-1}\Delta c^* & c(I_p - \Delta\Omega)^{-1}b \\ 0 & 1 \end{pmatrix}.$$

Let

$$t = 1 + c(I_p - \Delta\Omega)^{-1}\Delta c^*.$$

Then, $t > 0$ and the function $\frac{1}{\sqrt{t}}H_0(z)$ is J-unitary on the unit circle, with minimal realization

$$\frac{1}{\sqrt{t}}H_0(z) = \frac{1}{\sqrt{t}}D + C(zI - A)^{-1}\frac{1}{\sqrt{t}}B.$$

The associated Hermitian matrix to this realization is given by

$$X = \begin{pmatrix} -\Omega & -I_p \\ -I_p & -a\Delta a^* \end{pmatrix}.$$

We now recall the analogue of Theorem 2.5 for minimal J-unitary factorizations on the unit circle (see [5, Theorem 3.7 p. 205]):

Theorem 3.7. *Let R be a rational function J-unitary on the unit circle and analytic and invertible at ∞. Let $R(z) = D + C(zI - A)^{-1}B$ be a minimal realization of R, with associated Hermitian matrix H. Let $\mathcal{M}$ be a A-invariant subspace non-degenerate in the metric $[\cdot, \cdot]_H$ induced by H. Finally, let π denote the orthogonal projection defined by*

$$\ker \pi = \mathcal{M}, \quad \operatorname{Im} \pi = \mathcal{M}^{[\perp]}.$$

Then $R = R_1 R_2$ with

$$R_1(z) = (I + C(zI - A)^{-1}(I - \pi)BD^{-1})D_1^{-1}$$
$$R_2(z) = D_2(I + D^{-1}C\pi(zI - A)^{-1}B)D$$

with

$$D_1 = I + C_1 H_1^{-1}(I - \alpha A_1^*)^{-1}C_1^* J, \quad D_2 = DD_1^{-1}$$

where $|\alpha| = 1$ and

$$C_1 = C|_{\mathcal{M}}, \quad A_1 = A|_{\mathcal{M}}, \quad H_1 = \pi H|_{\mathcal{M}}$$

is a minimal J-unitary factorization of R, and every minimal J-unitary factorization of R is obtained in such a way.

Using this result we obtain:

Theorem 3.8. *The matrix function H_0 admits a minimal J-unitary factorization*

$$H_0(z) = U_1(z)^{-1}U_2(z)$$

where U_1 and U_2 are J-inner. The asymptotic equivalence matrix function admits a minimal J-unitary factorization

$$\frac{1}{\det H_0(z)}V(z) = V_1(z)^{-1}V_2(z)$$

where V_1 and V_2 are J-inner.

Indeed, the space $\begin{pmatrix} \mathbb{C}^p \\ 0 \end{pmatrix}$ is A invariant and H-negative. Furthermore,

$$\begin{pmatrix} -\Omega & -I_p \\ -I_p & -a\Delta a^* \end{pmatrix} = \begin{pmatrix} I_p & 0 \\ \Omega^{-1} & I_p \end{pmatrix} \begin{pmatrix} -\Omega & 0 \\ 0 & \Omega^{-1} - a\Delta a^* \end{pmatrix} \begin{pmatrix} I_p & 0 \\ \Omega^{-1} & I_p \end{pmatrix}^*,$$

and by (3.6) and (3.4), $\Omega^{-1} - a\Delta a^* > 0$. This insures that U_2 is J-inner.

To prove the second claim, we remark that the function $\begin{pmatrix} 1 & 0 \\ 0 & z^{-1} \end{pmatrix}$ is J-inner and set $V_1(z) = U_2(z)\begin{pmatrix} 1 & 0 \\ 0 & z^{-1} \end{pmatrix}$ and $V_2(z) = U_1(z)\begin{pmatrix} 1 & 0 \\ 0 & z^{-1} \end{pmatrix}.$ $\qquad \square$

belongs to $\mathbf{H}_{2,J}$ and is such that

$$(\Theta_\ell^{-1}\Theta_k)(0)\begin{pmatrix}1\\1\end{pmatrix} = \begin{pmatrix}0\\0\end{pmatrix}.$$

Thus,

$$\left\langle \Theta_\ell\begin{pmatrix}1\\1\end{pmatrix}, \Theta_k\begin{pmatrix}1\\1\end{pmatrix}\right\rangle_{\mathcal{H}(\Theta_n)} = \left\langle \begin{pmatrix}1\\1\end{pmatrix}, \Theta_\ell^{-1}\Theta_k\begin{pmatrix}1\\1\end{pmatrix}\right\rangle_{\mathcal{H}(\Theta_n)} = 0$$

The proof that the inner product is equal to 2 when $\ell = k$ is proved in the same way. The last claim follows from (3.37). $\qquad\square$

4. Two-sided systems and an example

4.1. Two-sided discrete first-order systems

We now turn to the systems of the form (3.1), that is,

$$Y_{n+1}(z) = \begin{pmatrix}1 & -\rho_n\\ -\rho_n^* & 1\end{pmatrix}\begin{pmatrix}z & 0\\ 0 & z^{-1}\end{pmatrix} Y_n(z),$$

and begin with the definition of the asymptotic equivalence matrix function.

Theorem 4.1. *Let ρ_n be a strictly pseudo-exponential sequence. Every solution of the system (3.1) is of the form*

$$Y_n(z) = \prod_{\ell=0}^{n-1}(1 - |\rho_\ell|^2)\begin{pmatrix}1 & 0\\ 0 & z^2\end{pmatrix} H_n(z^2)^{-1}\begin{pmatrix}z^n & 0\\ 0 & z^{-n}\end{pmatrix} H_0(z^2)\begin{pmatrix}1 & 0\\ 0 & \frac{1}{z^2}\end{pmatrix} Y_0(z).$$

The solution such that

$$\lim_{n\to\infty}\begin{pmatrix}z^{-n} & 0\\ 0 & z^n\end{pmatrix} Y_n(z) = I_2$$

corresponds to

$$Y_0(z) = \frac{1}{\prod_{\ell=0}^{\infty}(1 - |\rho_\ell|^2)}\begin{pmatrix}1 & 0\\ 0 & z^2\end{pmatrix} H_0(z^2)^{-1}\begin{pmatrix}1 & 0\\ 0 & z^{-2}\end{pmatrix},$$

while the solution with value I_2 at $n = 0$ corresponds to $Y_0(z) = I_2$.

Proof. Replacing z by z^2 in the recursion (3.18) we obtain:

$$H_{n+1}(z^2)\begin{pmatrix}1 & 0\\ 0 & \frac{1}{z^2}\end{pmatrix} = \begin{pmatrix}1 & 0\\ 0 & \frac{1}{z^2}\end{pmatrix} H_n(z^2)\begin{pmatrix}1 & \rho_n\\ \rho_n^* & 1\end{pmatrix}. \tag{4.1}$$

Note that

$$\begin{pmatrix}1 & -\rho_n\\ -\rho_n^* & 1\end{pmatrix}\begin{pmatrix}1 & \rho_n\\ \rho_n^* & 1\end{pmatrix} = (1 - |\rho_n|^2)I_2.$$

Thus, multiplying side by side (4.1) and (3.1) we obtain:

$$
H_{n+1}(z^2) \begin{pmatrix} 1 & 0 \\ 0 & \frac{1}{z^2} \end{pmatrix} Y_{n+1}(z) = (1 - |\rho_n|^2) \begin{pmatrix} 1 & 0 \\ 0 & \frac{1}{z^2} \end{pmatrix} H_n(z^2) \begin{pmatrix} z & 0 \\ 0 & z^{-1} \end{pmatrix} Y_n(z)
$$

$$
= (1 - |\rho_n|^2) \begin{pmatrix} z & 0 \\ 0 & z^{-1} \end{pmatrix} H_n(z^2) \begin{pmatrix} 1 & 0 \\ 0 & \frac{1}{z^2} \end{pmatrix} Y_n(z)
$$

from which we obtain:

$$
H_{n+1}(z^2) \begin{pmatrix} 1 & 0 \\ 0 & \frac{1}{z^2} \end{pmatrix} Y_{n+1}(z) =
$$

$$
= \begin{pmatrix} z^{n+1} & 0 \\ 0 & z^{-(n+1)} \end{pmatrix} H_0(z^2) \begin{pmatrix} 1 & 0 \\ 0 & \frac{1}{z^2} \end{pmatrix} Y_0(z) \left(\prod_{\ell=0}^{n} 1 - |\rho_\ell|^2 \right)
$$

and hence the formula for $Y_n(z)$. $\square$

Definition 4.2. *The function*

$$
\widetilde{V}(z) = \frac{1}{\prod_{\ell=0}^{n-1} (1 - |\rho_\ell|^2)} \begin{pmatrix} 1 & 0 \\ 0 & z^2 \end{pmatrix} H_0(z^2)^{-1} \begin{pmatrix} 1 & 0 \\ 0 & z^{-2} \end{pmatrix}
$$

is called the asymptotic equivalence matrix of the two-sided first-order discrete system (3.1).

We note that it is related to the asymptotic equivalence matrix (3.19) of the discrete system (3.2) by the transformation $z \mapsto z^2$. The proof of the following result is similar to the proof of Theorem 3.4.

Theorem 4.3. *Let c_1 and c_2 be in $\mathbb{C}^2$, and let $Y^{(1)}$ and $Y^{(2)}$ be the $\mathbb{C}^2$-valued solutions of (3.1), corresponding to the case of $\rho_n \equiv 0$ and to the strictly pseudo-exponential sequence ρ_n respectively and with initial conditions $Y_0^{(1)}(z) = c_1$ and $Y_0^{(2)}(z) = c_2$. Then, for every z on the unit circle it holds that*

$$
\lim_{n \to \infty} \|Y_n^{(1)}(z)c_1 - Y_n^{(2)}(z)c_2\| = 0 \quad \Longleftrightarrow \quad c_2 = \widetilde{V}(z)c_1.
$$

Proof. By definition, $Y_n^{(1)}(z) = \begin{pmatrix} z^n & 0 \\ 0 & z^{-n} \end{pmatrix} c_1$. On the other hand,

$$
Y_n^{(2)}(z) = \prod_{\ell=0}^{n-1} (1 - |\rho_\ell|^2) \begin{pmatrix} 1 & 0 \\ 0 & z^2 \end{pmatrix} H_n(z^2)^{-1} \begin{pmatrix} z^n & 0 \\ 0 & z^{-n} \end{pmatrix} H_0(z^2) \begin{pmatrix} 1 & 0 \\ 0 & z^{-2} \end{pmatrix} c_2.
$$

The result follows since $\lim_{n \to \infty} H_n(z^2)^{-1} = I_2$ for z on the unit circle. $\square$

The other spectral functions of the systems (3.2) and (3.1) are also related by the transformation $z \mapsto z^2$. The definitions and results are identical to the one-sided case.

Theorem 4.4. *Let $\rho_n, n = 0, 1, \ldots$ be a strictly pseudo-exponential sequence of the form (3.3). The reflection coefficient function of the associated discrete system (3.1) is given by the formula:*

$$R(z) = c\left\{(I - \Delta a^*\Omega a) - z^2(I - \Delta\Omega)a\right\}^{-1}b. \tag{4.2}$$

The scattering function is defined as follows. We look for the $\mathbb{C}^2$-valued solution of the system (3.2), with the boundary conditions

$$\begin{pmatrix} 1 & -1 \end{pmatrix} Y_0(z) = 0,$$
$$\begin{pmatrix} 0 & 1 \end{pmatrix} Y_n(z) = z^{-n} + o(n).$$

Then the limit

$$\lim_{n \to \infty} \begin{pmatrix} 1 & 0 \end{pmatrix} Y_n(z) z^{-n}$$

exists and is called the scattering function of the system (3.1). It is related to the scattering function of the system (3.2) by the map $z \mapsto z^2$.

We also mention that J-inner polynomials are now replaced by J-unitary functions with possibly poles at the origin and at infinity, but with constant determinant.

4.2. An illustrative example

As a simple example we take $a = \alpha \in (0, 1)$, $b = 1$ and $c = c^*$. Then

$$\Delta = \frac{1}{1 - \alpha^2}, \qquad \Omega = \frac{c^2}{1 - \alpha^2},$$

and

$$\rho_n = -\alpha^n \frac{c}{1 - \frac{c^2\alpha^{2n+2}}{(1-\alpha^2)^2}}. \tag{4.3}$$

The numbers c and α need to satisfy (3.6), that is $(1 - \alpha^2)^2 > c^2$. Note that this condition implies that

$$|\rho_0| = \frac{c}{1 - \alpha^2\left(\frac{c^2}{(1-\alpha^2)^2}\right)} < \frac{c}{1 - \alpha^2} < 1,$$

and more generally,

$$\begin{aligned}
|\rho_n| &= \frac{\alpha^n c}{1 - \frac{c^2\alpha^{2n+2}}{(1-\alpha^2)^2}} \\
&\leq \frac{\alpha^n c}{1 - \alpha^{2n+2}} \\
&= \frac{\alpha^n(1 - \alpha^2)}{1 - \alpha^{2n+2}} \frac{c}{1 - \alpha^2} < \frac{\alpha^n}{1 + \alpha^2 + \cdots + \alpha^{2n}} < 1,
\end{aligned}$$

as it should be.

	Continuous case
The system	$iJf' - Vf = zf$
Special solutions	Entire J-inner functions
Potential	$v(x) = \begin{pmatrix} 0 & k(x) \\ k(x)^* & 0 \end{pmatrix}$ $k(x) = -2ce^{ita}\left(I_p + \Omega\left(Y - e^{-2ixa^*}Ye^{2ixa}\right)\right)^{-1}$
Solution asymptotic to the solution with $k \equiv 0$	Theorem 2.1
$-k$ is also a potential	Theorem 2.26
Asymptotic property	Formula (2.4)
Reflection coefficient	Formulas (2.11) and (2.10)
Weyl function	Formula (2.14)
Weyl function for $-k(x)$	Theorem 2.26
Factorization of the asymptotic equivalence matrix	Theorem 2.6
Asymptotic behavior of the orthogonal polynomial	Equation (2.21)

TABLE 1

The reflection coefficient is equal to:

$$R(z) = \frac{c}{1 - \frac{\alpha^2 c^2}{(1-\alpha^2)^2} - z\alpha(1 - \frac{c^2}{(1-\alpha^2)^2})}.$$

We check directly that it is indeed a Schur function as follows: we have for $|z| \leq 1$

$$|R(z)| \leq \frac{c}{1 - \frac{\alpha^2 c^2}{(1-\alpha^2)^2} - \alpha(1 - \frac{c^2}{(1-\alpha^2)^2})}.$$

We thus need to check that

$$c \leq 1 - \frac{\alpha^2 c^2}{(1-\alpha^2)^2} - \alpha(1 - \frac{c^2}{(1-\alpha^2)^2}),$$

that is, with $T = \frac{c}{(1-\alpha^2)}$,

$$c \leq 1 - \alpha^2 T^2 - \alpha(1 - T^2) = (1-\alpha)(1 + T^2\alpha),$$

	Discrete case (one-sided case)
The system	$Y_{n+1}(z) = \begin{pmatrix} z & -\rho_n \\ -z\rho_n^* & 1 \end{pmatrix} Y_n(z)$
Special solutions	J-inner polynomials
Potential: the Schur coefficients ρ_n	$\rho_n = -ca^n(I - \Delta a^{*(n+1)}\Omega a^{n+1})^{-1}b$
Solution asymptotic to the solution with $\rho_n \equiv 0$	Formula (3.14)
$-\rho_n$ is also pseudo-exponential	Remark 3.1
Asymptotic property	Formula (3.7)
Reflection coefficient	Formulas (3.23) and (3.22)
Weyl function	Formula (3.30)
Weyl function for $-\rho_n$	Remark 3.18
Factorization of the asymptotic equivalence matrix	Theorem 3.8
Asymptotic behavior of the orthogonal polynomial	Equation (3.33)

TABLE 2

that is, $T \leq \frac{1}{1+\alpha}(1 + T^2\alpha)$. This last inequality in turn holds since T and α are in $(0,1)$.

Finally, from (3.27) we obtain the expression for the Weyl function:

$$N(z) = i\frac{1 - \frac{\alpha^2 c^2}{(1-\alpha^2)^2} - z\alpha(1 - \frac{c^2}{(1-\alpha^2)^2}) - zc}{1 - \frac{\alpha^2 c^2}{(1-\alpha^2)^2} - z\alpha(1 - \frac{c^2}{(1-\alpha^2)^2}) + zc}.$$

We summarize the parallels between the continuous case and the one-sided discrete case in Tables 1 and 2.

References

[1] V.M. Adamyan and S.E. Nechayev. Nuclear Hankel matrices and orthogonal trigonometric polynomials. *Contemporary Mathematics*, 189:1–15, 1995.

[2] D. Alpay, T. Azizov, A. Dijksma, and H. Langer. The Schur algorithm for generalized Schur functions. III. *J*-unitary matrix polynomials on the circle. *Linear Algebra Appl.*, 369:113–144, 2003.

[3] D. Alpay and H. Dym. Hilbert spaces of analytic functions, inverse scattering and operator models, I. *Integral Equation and Operator Theory*, 7:589–641, 1984.

[4] D. Alpay and H. Dym. On applications of reproducing kernel spaces to the Schur algorithm and rational *J*-unitary factorization. In I. Gohberg, editor, *I. Schur methods in operator theory and signal processing*, volume 18 of *Operator Theory: Advances and Applications*, pages 89–159. Birkhäuser Verlag, Basel, 1986.

[5] D. Alpay and I. Gohberg. Unitary rational matrix functions. In I. Gohberg, editor, *Topics in interpolation theory of rational matrix-valued functions*, volume 33 of *Operator Theory: Advances and Applications*, pages 175–222. Birkhäuser Verlag, Basel, 1988.

[6] D. Alpay and I. Gohberg. Inverse spectral problems for difference operators with rational scattering matrix function. *Integral Equations Operator Theory*, 20(2):125–170, 1994.

[7] D. Alpay and I. Gohberg. Inverse spectral problem for differential operators with rational scattering matrix functions. *Journal of differential equations*, 118:1–19, 1995.

[8] D. Alpay and I. Gohberg. Inverse scattering problem for differential operators with rational scattering matrix functions. In I. Böttcher and I. Gohberg, editors, *Singular integral operators and related topics (Tel Aviv, 1995)*, volume 90 of *Operator Theory: Advances and Applications*, pages 1–18. Birkhäuser Verlag, Basel, 1996.

[9] D. Alpay and I. Gohberg. Connections between the Carathéodory-Toeplitz and the Nehari extension problems: the discrete scalar case. *Integral Equations Operator Theory*, 37(2):125–142, 2000.

[10] D. Alpay and I. Gohberg. Inverse problems associated to a canonical differential system. In L. Kérchy, C. Foias, I. Gohberg, and H. Langer, editors, *Recent advances in operator theory and related topics (Szeged, 1999)*, Operator theory: Advances and Applications, pages 1–27. Birkhäuser, Basel, 2001.

[11] D. Alpay and I. Gohberg. A trace formula for canonical differential expressions. *J. Funct. Anal.*, 197(2):489–525, 2003.

[12] D. Alpay, I. Gohberg, M.A. Kaashoek, and A.L. Sakhnovich. Direct and inverse scattering problem for canonical systems with a strictly pseudo-exponential potential. *Math. Nachr.*, 215:5–31, 2000.

[13] D. Alpay, I. Gohberg, and L. Sakhnovich. Inverse scattering for continuous transmission lines with rational reflection coefficient function. In I. Gohberg, P. Lancaster, and P.N. Shivakumar, editors, *Proceedings of the International Conference on Applications of Operator Theory held in Winnipeg, Manitoba, October 2–6, 1994*, volume 87 of *Operator theory: Advances and Applications*, pages 1–16. Birkhäuser Verlag, Basel, 1996.

[14] H. Bart, I. Gohberg, and M.A. Kaashoek. *Minimal factorization of matrix and operator functions*, volume 1 of *Operator Theory: Advances and Applications*. Birkhäuser Verlag, Basel, 1979.

[15] H. Bart, I. Gohberg, and M.A. Kaashoek. Convolution equations and linear systems. *Integral Equations Operator Theory*, 5:283–340, 1982.

[16] A.M. Bruckstein and T. Kailath. Inverse scattering for discrete transmission-line models. *SIAM Rev.*, 29(3):359–389, 1987.

[17] K. Clancey and I. Gohberg. *Factorization of matrix functions and singular integral operators*, volume 3 of *Operator Theory: Advances and Applications*. Birkhäuser Verlag, Basel, 1981.

[18] D de Cogan. *Transmission line matrix (LTM) techniques for diffusion applications*. Gordon and Breach Science Publishers, 1998.

[19] T. Constantinescu. *Schur parameters, factorization and dilation problems*, volume 82 of *Operator Theory: Advances and Applications*. Birkhäuser Verlag, Basel, 1996.

[20] H. Dym. *J-contractive matrix functions, reproducing kernel Hilbert spaces and interpolation*. Published for the Conference Board of the Mathematical Sciences, Washington, DC, 1989.

[21] H. Dym and A. Iacob. Applications of factorization and Toeplitz operators to inverse problems. In I. Gohberg, editor, *Toeplitz centennial (Tel Aviv, 1981)*, volume 4 of *Operator Theory: Adv. Appl.*, pages 233–260. Birkhäuser, Basel, 1982.

[22] H. Dym and A. Iacob. Positive definite extensions, canonical equations and inverse problems. In H. Dym and I. Gohberg, editors, *Proceedings of the workshop on applications of linear operator theory to systems and networks held at Rehovot, June 13–16, 1983*, volume 12 of *Operator Theory: Advances and Applications*, pages 141–240. Birkhäuser Verlag, Basel, 1984.

[23] B. Fritzsche and B. Kirstein, editors. *Ausgewählte Arbeiten zu den Ursprüngen der Schur-Analysis*, volume 16 of *Teubner-Archiv zur Mathematik*. B.G. Teubner Verlagsgesellschaft, Stuttgart–Leipzig, 1991.

[24] I. Gohberg, S. Goldberg, and M.A. Kaashoek. *Classes of linear operators. Vol. II*, volume 63 of *Operator Theory: Advances and Applications*. Birkhäuser Verlag, Basel, 1993.

[25] I. Gohberg and M.A. Kaashoek. Block Toeplitz operators with rational symbols. In I. Gohberg, J.W. Helton, and L. Rodman, editors, *Contributions to operator theory and its applications (Mesa, AZ, 1987)*, volume 35 of *Oper. Theory Adv. Appl.*, pages 385–440. Birkhäuser, Basel, 1988.

[26] I. Gohberg, M.A. Kaashoek, and A.L. Sakhnovich. Canonical systems with rational spectral densities: explicit formulas and applications. *Math. Nachr.*, 194:93–125, 1998.

[27] I. Gohberg, M.A. Kaashoek, and A.L. Sakhnovich. Pseudo-canonical systems with rational Weyl functions: explicit formulas and applications. *Journal of differential equations*, 146:375–398, 1998.

[28] I. Gohberg, M.A. Kaashoek, and F. van Schagen. Szegő–Kac–Achiezer formulas in terms of realizations of the symbol. *J. Funct. Anal.*, 74:24–51, 1987.

[29] I. Gohberg, P. Lancaster, and L. Rodman. *Matrices and indefinite scalar products*, volume 8 of *Operator Theory: Advances and Applications*. Birkhäuser Verlag, Basel, 1983.

[30] I. Gohberg, P. Lancaster, and L. Rodman. *Invariant subspaces of matrices with applications*. Canadian Mathematical Society Series of Monographs and Advanced Texts. John Wiley & Sons Inc., New York, 1986. A Wiley-Interscience Publication.

[31] I. Gohberg and Ju. Leiterer. General theorems on the factorization of operator-valued functions with respect to a contour. I. Holomorphic functions. *Acta Sci. Math. (Szeged)*, 34:103–120, 1973.

[32] I. Gohberg and Ju. Leiterer. General theorems on the factorization of operator-valued functions with respect to a contour. II. Generalizations. *Acta Sci. Math. (Szeged)*, 35:39–59, 1973.

[33] I. Gohberg and S. Rubinstein. Proper contractions and their unitary minimal completions. In I. Gohberg, editor, *Topics in interpolation theory of rational matrix-valued functions*, volume 33 of *Operator Theory: Advances and Applications*, pages 223–247. Birkhäuser Verlag, Basel, 1988.

[34] I.C. Gohberg and I.A. Fel'dman. *Convolution equations and projection methods for their solution*. American Mathematical Society, Providence, R.I., 1974. Translated from the Russian by F.M. Goldware, Translations of Mathematical Monographs, Vol. 41.

[35] G.J. Groenewald. Toeplitz operators with rational symbols and realizations: an alternative version. Technical Report WS:–362, Vrije Universiteit Amsterdam, 1990.

[36] A. Iacob. *On the spectral theory of a class of canonical systems of differential equations*. PhD thesis, The Weizmann Institute of Sciences, 1986.

[37] M.G. Kreĭn. Continuous analogues of propositions for polynomials orthogonal on the unit circle. *Dokl. Akad. Nauk. SSSR*, 105:637–640, 1955.

[38] M.G. Kreĭn. *Topics in differential and integral equations and operator theory*, volume 7 of *Operator theory: Advances and Applications*. Birkhäuser Verlag, Basel, 1983. Edited by I. Gohberg, Translated from the Russian by A. Iacob.

[39] M.G. Kreĭn and H. Langer. Über die verallgemeinerten Resolventen und die charakteristische Funktion eines isometrischen Operators im Raume Π_k. In *Hilbert space operators and operator algebras (Proc. Int. Conf. Tihany, 1970)*, pages 353–399. North-Holland, Amsterdam, 1972. Colloquia Math. Soc. János Bolyai.

[40] L.Golinskii and P. Nevai. Szegő difference equations, transfer matrices and orthogonal polynomials on the unit circle. *Comm. Math. Phys.*, 223(2):223–259, 2001.

[41] F.E. Melik-Adamyan. On a class of canonical differential operators. *Izvestya Akademii Nauk. Armyanskoi SSR Matematica*, 24:570–592, 1989. English translation in: Soviet Journal of Contemporary Mathematics, vol. 24, pages 48–69 (1989).

[42] L. Sakhnovich. Dual discrete canonical systems and dual orthogonal polynomials. In D. Alpay, I. Gohberg, and V. Vinnikov, editors, *Interpolation theory, systems theory and related topics (Tel Aviv/Rehovot, 1999)*, volume 134 of *Oper. Theory Adv. Appl.*, pages 385–401. Birkhäuser, Basel, 2002.

[43] I. Schur. Über die Potenzreihen, die im Innern des Einheitkreises beschränkten sind, I. *Journal für die Reine und Angewandte Mathematik*, 147:205–232, 1917. English

translation in: I. Schur methods in operator theory and signal processing. (Operator theory: Advances and Applications OT 18 (1986), Birkhäuser Verlag), Basel.

[44] B. Simon. Analogs of the m-function in the theory of orthogonal polynomials on the unit circle. *J. Comput. Appl. Math.*, 171(1-2):411–424, 2004.

[45] F. Wenger, T. Gustafsson, and L. Svensson. Perturbation theory for inhomogeneous transmission lines. *IEEE Trans. Circuits Systems I Fund. Theory Appl.*, 49(3):289–297, 2002.

[46] A. Yagle and B. Levy. The Schur algorithm and its applications. *Acta Applicandae Mathematicae*, 3:255–284, 1985.

Daniel Alpay
Department of Mathematics
Ben–Gurion University of the Negev
Beer-Sheva 84105
Israel
e-mail: `dany@math.bgu.ac.il`

Israel Gohberg
School of Mathematical Sciences
The Raymond and Beverly Sackler Faculty of Exact Sciences
Tel–Aviv University
Tel–Aviv, Ramat–Aviv 69989
Israel
e-mail: `gohberg@post.tau.ac.il`

Operator Theory:
Advances and Applications, Vol. 161, 49–113
© 2005 Birkhäuser Verlag Basel/Switzerland

Matrix-J-unitary Non-commutative Rational Formal Power Series

D. Alpay and D.S. Kalyuzhnyĭ-Verbovetzkiĭ

Abstract. Formal power series in N non-commuting indeterminates can be considered as a counterpart of functions of one variable holomorphic at 0, and some of their properties are described in terms of coefficients. However, really fruitful analysis begins when one considers for them evaluations on N-tuples of $n \times n$ matrices (with $n = 1, 2, \ldots$) or operators on an infinite-dimensional separable Hilbert space. Moreover, such evaluations appear in control, optimization and stabilization problems of modern system engineering.

In this paper, a theory of realization and minimal factorization of rational matrix-valued functions which are J-unitary on the imaginary line or on the unit circle is extended to the setting of non-commutative rational formal power series. The property of J-unitarity holds on N-tuples of $n \times n$ skew-Hermitian versus unitary matrices ($n = 1, 2, \ldots$), and a rational formal power series is called *matrix-J-unitary* in this case. The close relationship between minimal realizations and structured Hermitian solutions H of the Lyapunov or Stein equations is established. The results are specialized for the case of *matrix-J-inner* rational formal power series. In this case $H > 0$, however the proof of that is more elaborated than in the one-variable case and involves a new technique. For the rational *matrix-inner* case, i.e., when $J = I$, the theorem of Ball, Groenewald and Malakorn on unitary realization of a formal power series from the non-commutative Schur–Agler class admits an improvement: the existence of a minimal (thus, finite-dimensional) such unitary realization and its uniqueness up to a unitary similarity is proved. A version of the theory for *matrix-selfadjoint* rational formal power series is also presented. The concept of non-commutative formal reproducing kernel Pontryagin spaces is introduced, and in this framework the backward shift realization of a matrix-J-unitary rational formal power series in a finite-dimensional non-commutative de Branges–Rovnyak space is described.

Mathematics Subject Classification (2000). Primary 47A48; Secondary 13F25, 46C20, 46E22, 93B20, 93D05.

The second author was supported by the Center for Advanced Studies in Mathematics, Ben-Gurion University of the Negev.

Keywords. J-unitary matrix functions, non-commutative, rational, formal power series, minimal realizations, Lyapunov equation, Stein equation, minimal factorizations, Schur–Agler class, reproducing kernel Pontryagin spaces, backward shift, de Branges–Rovnyak space.

Contents

1. Introduction

In the present paper we study a non-commutative analogue of rational matrix-valued functions which are J-unitary on the imaginary line or on the unit circle and, as a special case, J-inner ones. Let $J \in \mathbb{C}^{q \times q}$ be a signature matrix, i.e., a matrix which is both self-adjoint and unitary. A $\mathbb{C}^{q \times q}$-valued rational function F is *J-unitary on the imaginary line* if

$$F(z)JF(z)^* = J \tag{1.1}$$

at every point of holomorphy of F on the imaginary line. It is called *J-inner* if moreover

$$F(z)JF(z)^* \leq J \tag{1.2}$$

at every point of holomorphy of F in the open right half-plane Π. Replacing the imaginary line by the unit circle $\mathbb{T}$ in (1.1) and the open right half-plane Π by the open unit disk $\mathbb{D}$ in (1.2), one defines J-unitary functions on the unit circle (resp., J-inner functions in the open unit disk). These classes of rational functions were studied in [7] and [6] using the theory of realizations of rational matrix-valued functions, and in [4] using the theory of reproducing kernel Pontryagin spaces. The circle and line cases were studied in a unified way in [5]. We mention also the earlier papers [36, 23] that inspired much of investigation of these and other classes of rational matrix-valued functions with symmetries.

We now recall some of the arguments in [7], then explain the difficulties appearing in the several complex variables setting, and why the arguments of [7] extend to the non-commutative framework. So let F be a rational function which is J-unitary on the imaginary line, and assume that F is holomorphic in a neighborhood of the origin. It then admits a minimal realization

$$F(z) = D + C(I_\gamma - zA)^{-1}zB$$

where $D = F(0)$, and A, B, C are matrices of appropriate sizes (the size $\gamma \times \gamma$ of the square matrix A is minimal possible for such a realization). Rewrite (1.1) as

$$F(z) = JF(-\bar{z})^{-*}J, \tag{1.3}$$

where z is in the domain of holomorphy of both $F(z)$ and $F(-\bar{z})^{-*}$. We can rewrite (1.3) as

$$D + C(I_\gamma - zA)^{-1}zB = J\left(D^{-*} + D^{-*}B^*(I_\gamma + z(A - BD^{-1}C)^*)^{-1}zC^*D^{-*}\right)J.$$

The above equality gives two minimal realizations of a given rational matrix-valued function. These realizations are therefore similar, and there is a uniquely defined matrix (which, for convenience, we denote by $-H$) such that

$$\begin{pmatrix} -H & 0 \\ 0 & I_q \end{pmatrix} \begin{pmatrix} A & B \\ C & D \end{pmatrix} = \begin{pmatrix} -(A^* - C^*D^{-*}B^*) & C^*D^{-*}J \\ JD^{-*}B^* & JD^{-*}J \end{pmatrix} \begin{pmatrix} -H & 0 \\ 0 & I_q \end{pmatrix}. \tag{1.4}$$

The matrix $-H^*$ in the place of $-H$ also satisfies (1.4), and by uniqueness of the similarity matrix we have $H = H^*$, which leads to the following theorem.

Theorem 1.1. *Let F be a rational matrix-valued function holomorphic in a neighborhood of the origin and let $F(z) = D + C(I_\gamma - zA)^{-1}zB$ be a minimal realization of F. Then F is J-unitary on the imaginary line if and only if the following conditions hold:*

(1) D is J-unitary, that is, $DJD^* = J$;

(2) there exists an Hermitian invertible matrix H such that

$$A^*H + HA = -C^*JC, \tag{1.5}$$

$$B = -H^{-1}C^*JD. \tag{1.6}$$

The matrix H is uniquely determined by a given minimal realization (it is called the associated Hermitian matrix to this realization). It holds that

$$\frac{J - F(z)JF(z')^*}{z + \overline{z'}} = C(I_\gamma - zA)^{-1}H^{-1}(I_\gamma - z'A)^{-*}C^*. \tag{1.7}$$

In particular, F is J-inner if and only if $H > 0$.

The finite-dimensional reproducing kernel Pontryagin space $\mathcal{K}(F)$ with reproducing kernel

$$K^F(z, z') = \frac{J - F(z)JF(z')^*}{(z + \overline{z'})}$$

provides a minimal state space realization for F: more precisely (see [4]),

$$F(z) = D + C(I_\gamma - zA)^{-1}zB,$$

where

$$\begin{pmatrix} A & B \\ C & D \end{pmatrix} : \begin{pmatrix} \mathcal{K}(F) \\ \mathbb{C}^q \end{pmatrix} \rightarrow \begin{pmatrix} \mathcal{K}(F) \\ \mathbb{C}^q \end{pmatrix}$$

is defined by

$$(Af)(z) = (R_0 f)(z) := \frac{f(z) - f(0)}{z}, \quad Bu = \frac{F(z) - F(0)}{z}u, \quad Cf = f(0), \quad Dx = F(0)x.$$

Another topic considered in [7] and [4] is J-unitary factorization. Given a matrix-valued function F which is J-unitary on the imaginary line one looks for all minimal factorizations of F (see [15]) into factors which are themselves J-unitary on the imaginary line. There are two equivalent characterizations of these factorizations: the first one uses the theory of realization and the second one uses the theory of reproducing kernel Pontryagin spaces.

Theorem 1.2. *Let F be a rational matrix-valued function which is J-unitary on the imaginary line and holomorphic in a neighborhood of the origin, and let $F(z) = D + C(I_\gamma - zA)^{-1}zB$ be a minimal realization of F, with the associated Hermitian matrix H. There is a one-to-one correspondence between minimal J-unitary factorizations of F (up to a multiplicative J-unitary constant) and A-invariant subspaces which are non-degenerate in the (possibly, indefinite) metric induced by H.*

In general, F may fail to have non-trivial J-unitary factorizations.

Theorem 1.3. *Let F be a rational matrix-valued function which is J-unitary on the imaginary line and holomorphic in a neighborhood of the origin. There is a one-to-one correspondence between minimal J-unitary factorizations of F (up to a multiplicative J-unitary constant) and R_0-invariant non-degenerate subspaces of $\mathcal{K}(F)$.*

The arguments in the proof of Theorem 1.1 do not go through in the several complex variables context. Indeed, uniqueness, up to a similarity, of minimal realizations doesn't hold anymore (see, e.g., [27, 25, 33]). On the other hand, the notion of realization still makes sense in the non-commutative setting, namely for non-commutative rational *formal power series* (*FPSs* in short), and there is a uniqueness result for minimal realizations in this case (see [16, 39, 11]). The latter allows us to extend the notion and study of J-unitary matrix-valued functions to the non-commutative case. We introduce the notion of a *matrix-J-unitary* rational FPS as a formal power series in N non-commuting indeterminates which is $J \otimes I_n$-unitary on N-tuples of $n \times n$ skew-Hermitian versus unitary matrices for $n = 1, 2, \ldots$. We extend to this case the theory of minimal realizations, minimal J-unitary factorizations, and backward shift models in finite-dimensional de Branges–Rovnyak spaces. We also introduce, in a similar way, the notion of matrix-selfadjoint rational formal power series, and show how to deduce the related theory for them from the theory of matrix-J-unitary ones.

We now turn to the outline of this paper. It consists of eight sections. Section 1 is this introduction. In Section 2 we review various results in the theory of FPSs. Let us note that the theorem on null spaces for matrix substitutions and its corollary, from our paper [8], which are recollected in the end of Section 2, become an important tool in our present work on FPSs. In Section 3 we study the properties of observability, controllability and minimality of Givone-Roesser nodes in the non-commutative setting and give the corresponding criteria in terms of matrix evaluations for their "formal transfer functions". We also formulate a theorem on minimal factorizations of a rational FPS. In Section 4 we define the non-commutative analogue of the imaginary line and study matrix-J-unitary FPSs for this case. We in particular obtain a non-commutative version of Theorem 1.1. We obtain a counterpart of the Lyapunov equation (1.5) and of Theorem 1.2 on minimal J-unitary factorizations. The unique solution of the Lyapunov equation has in this case a block diagonal structure: $H = \mathrm{diag}(H_1, \ldots, H_N)$, and is said to be *the associated structured Hermitian matrix* (associated with a given minimal realization of a matrix-J-unitary FPS). Section 5 contains the analogue of the previous section for the case of a non-commutative counterpart of the unit circle. These two sections do not take into account a counterpart of condition (1.2), which is considered in Section 6 where we study matrix-J-inner rational FPSs. In particular, we show that the associated structured Hermitian matrix $H = \mathrm{diag}(H_1, \ldots, H_N)$ is strictly positive in this case, which generalizes the statement in Theorem 1.1 on J-inner functions. We define non-commutative counterparts of the right half-plane and the unit disk, and formulate our results for both of these domains. The second one is the disjoint union of the products of N copies of $n \times n$ matrix unit disks, $n = 1, 2, \ldots$, and plays a role of a "non-commutative polydisk". In Theorem 6.6 we show that any (not necessarily rational) FPS with operator coefficients, which takes contractive values in this domain, belongs to the non-commutative Schur–Agler class, defined by J.A. Ball, G. Groenewald and T. Malakorn in [12]. (The opposite is trivial: any function from this class has the above-mentioned property.) In other words, the contractivity of values of a FPS on N-tuples of strictly contractive $n \times n$ matrices, $n = 1, 2, \ldots$, is sufficient for the contractivity of its values on N-tuples of strictly contractive operators in an infinite-dimensional separable Hilbert space. Thus, matrix-inner rational FPSs (i.e., matrix-J-inner ones for the case $J = I_q$) belong to the non-commutative Schur–Agler class. For this case, we recover the theorem on unitary realizations for FPSs from the latter class which was obtain in [12]. Moreover, our Theorem 6.4 establishes the existence of a *minimal*, thus *finite-dimensional*, unitary Givone–Roesser realization of a rational matrix-inner FPS and *the uniqueness of such a realization up to a unitary similarity*. This implies, in particular, non-commutative Lossless Bounded Real Lemma (see [41, 7] for its one-variable counterpart). A non-commutative version of standard Bounded Real Lemma (see [47]) has been presented recently in [13]. In Section 7 we study matrix-selfadjoint rational FPSs. In Section 8 we introduce non-commutative formal reproducing kernel Pontryagin spaces in a way which extends one that J.A. Ball and V. Vinnikov have introduced in [14] non-commutative formal reproducing kernel Hilbert spaces. We describe minimal backward shift realizations in non-commutative formal reproducing kernel Pontryagin spaces which serve as a counterpart of finite-dimensional de Branges–Rovnyak spaces. Let us note that we derive an explicit formula (8.12) for the corresponding reproducing kernels. In the last subsection of Section 8 we present examples of matrix-inner rational FPSs with scalar coefficients, in two non-commuting indeterminates, and the corresponding reproducing kernels computed by formula (8.12).

2. Preliminaries

In this section we introduce the notations which will be used throughout this paper and review some definitions from the theory of formal power series. The symbol $\mathbb{C}^{p \times q}$ denotes the set of $p \times q$ matrices with complex entries, and $(\mathbb{C}^{r \times s})^{p \times q}$ is the space of $p \times q$ block matrices with block entries in $\mathbb{C}^{r \times s}$. The tensor product $A \otimes B$ of matrices $A \in \mathbb{C}^{r \times s}$ and $B \in \mathbb{C}^{p \times q}$ is the element of $(\mathbb{C}^{r \times s})^{p \times q}$ with (i, j)th block entry equal to Ab_{ij}. The tensor product $\mathbb{C}^{r \times s} \otimes \mathbb{C}^{p \times q}$ is the linear span of finite sums of the form $C = \sum_{k=1}^{n} A_k \otimes B_k$ where $A_k \in \mathbb{C}^{r \times s}$ and $B_k \in \mathbb{C}^{p \times q}$. One identifies $\mathbb{C}^{r \times s} \otimes \mathbb{C}^{p \times q}$ with $(\mathbb{C}^{r \times s})^{p \times q}$. Different representations for an element $C \in \mathbb{C}^{r \times s} \otimes \mathbb{C}^{p \times q}$ can be reduced to a unique one:

$$C = \sum_{\mu=1}^{r} \sum_{\nu=1}^{s} \sum_{\tau=1}^{p} \sum_{\sigma=1}^{q} c_{\mu\nu\tau\sigma} E'_{\mu\nu} \otimes E''_{\tau\sigma},$$

where the matrices $E'_{\mu\nu} \in \mathbb{C}^{r\times s}$ and $E''_{\tau\sigma} \in \mathbb{C}^{p\times q}$ are given by

$$\left(E'_{\mu\nu}\right)_{ij} = \begin{cases} 1 & \text{if} \quad (i,j) = (\mu,\nu) \\ 0 & \text{if} \quad (i,j) \neq (\mu,\nu) \end{cases}, \qquad \mu, i = 1,\ldots,r \quad \text{and} \quad \nu, j = 1,\ldots s,$$

$$\left(E''_{\tau\sigma}\right)_{k\ell} = \begin{cases} 1 & \text{if} \quad (k,\ell) = (\tau,\sigma) \\ 0 & \text{if} \quad (k,\ell) \neq (\tau,\sigma) \end{cases}, \qquad \tau, k = 1,\ldots,p \quad \text{and} \quad \sigma, \ell = 1,\ldots q.$$

We denote by $\mathcal{F}_N$ the free semigroup with N generators $g_1, \ldots, g_N$ and the identity element $\emptyset$ with respect to the concatenation product. This means that the generic element of $\mathcal{F}_N$ is a word $w = g_{i_1}\cdots g_{i_n}$, where $i_\nu \in \{1,\ldots,N\}$ for $\nu = 1,\ldots,n$, the identity element $\emptyset$ corresponds to the empty word, and for another word $w' = g_{j_1}\cdots g_{j_m}$, one defines the product as

$$ww' = g_{i_1}\cdots g_{i_n} g_{j_1}\cdots g_{j_m}, \quad w\emptyset = \emptyset w = w.$$

We denote by $w^T = g_{i_n}\cdots g_{i_1} \in \mathcal{F}_N$ the *transpose* of $w = g_{i_1}\cdots g_{i_n} \in \mathcal{F}_N$ and by $|w| = n$ the *length* of the word w. Correspondingly, $\emptyset^T = \emptyset$, and $|\emptyset| = 0$.

A *formal power series* (*FPS* in short) in non-commuting indeterminates $z_1,\ldots,z_N$ with coefficients in a linear space $\mathcal{E}$ is given by

$$f(z) = \sum_{w \in \mathcal{F}_N} f_w z^w, \quad f_w \in \mathcal{E}, \tag{2.1}$$

where for $w = g_{i_1}\cdots g_{i_n}$ and $z = (z_1,\ldots,z_N)$ we set $z^w = z_{i_1}\cdots z_{i_n}$, and $z^\emptyset = 1$. We denote by $\mathcal{E}\langle\langle z_1,\ldots,z_N\rangle\rangle$ the linear space of FPSs in non-commuting indeterminates $z_1,\ldots,z_N$ with coefficients in $\mathcal{E}$. A series $f \in \mathbb{C}^{p\times q}\langle\langle z_1,\ldots,z_N\rangle\rangle$ of the form (2.1) can also be viewed as a $p \times q$ matrix whose entries are formal power series with coefficients in $\mathbb{C}$, i.e., belong to the space $\mathbb{C}\langle\langle z_1,\ldots,z_N\rangle\rangle$, which has an additional structure of non-commutative ring (we assume that the indeterminates z_j formally commute with the coefficients f_w). The *support* of a FPS f given by (2.1) is the set

$$\operatorname{supp} f = \{w \in \mathcal{F}_N : f_w \neq 0\}.$$

Non-commutative polynomials are formal power series with finite support. We denote by $\mathcal{E}\langle z_1,\ldots,z_N\rangle$ the subspace in the space $\mathcal{E}\langle\langle z_1,\ldots,z_N\rangle\rangle$ consisting of non-commutative polynomials. Clearly, a FPS is determined by its coefficients f_w. Sums and products of two FPSs f and g with matrix coefficients of compatible sizes (or with operator coefficients) are given by

$$(f+g)_w = f_w + g_w, \qquad (fg)_w = \sum_{w'w''=w} f_{w'} g_{w''}. \tag{2.2}$$

A FPS f with coefficients in $\mathbb{C}$ is invertible if and only if $f_\emptyset \neq 0$. Indeed, assume that f is invertible. From the definition of the product of two FPSs in (2.2) we get $f_\emptyset (f^{-1})_\emptyset = 1$, and hence $f_\emptyset \neq 0$. On the other hand, if $f_\emptyset \neq 0$ then f^{-1} is given by

$$f^{-1}(z) = \sum_{k=0}^{\infty} \left(1 - f_\emptyset^{-1} f(z)\right)^k f_\emptyset^{-1}.$$

The formal power series in the right-hand side is well defined since the expansion of $\left(1 - f_\emptyset^{-1} f\right)^k$ contains words of length at least k, and thus the coefficients $(f^{-1})_w$ are finite sums.

A FPS with coefficients in $\mathbb{C}$ is called *rational* if it can be expressed as a finite number of sums, products and inversions of non-commutative polynomials. A formal power series with coefficients in $\mathbb{C}^{p \times q}$ is called *rational* if it is a $p \times q$ matrix whose all entries are rational FPSs with coefficients in $\mathbb{C}$. We will denote by $\mathbb{C}^{p \times q} \langle\langle z_1, \ldots, z_N \rangle\rangle_{\mathrm{rat}}$ the linear space of rational FPSs with coefficients in $\mathbb{C}^{p \times q}$. Define the product of $f \in \mathbb{C}^{p \times q} \langle\langle z_1, \ldots, z_N \rangle\rangle_{\mathrm{rat}}$ and $p \in \mathbb{C} \langle z_1, \ldots, z_N \rangle$ as follows:

1. $f \cdot 1 = f$ for every $f \in \mathbb{C}^{p \times q} \langle\langle z_1, \ldots, z_N \rangle\rangle_{\mathrm{rat}}$;
2. For every word $w' \in \mathcal{F}_N$ and every $f \in \mathbb{C}^{p \times q} \langle\langle z_1, \ldots, z_N \rangle\rangle_{\mathrm{rat}}$,

$$f \cdot z^{w'} = \sum_{w \in \mathcal{F}_N} f_w z^{ww'} = \sum_w f_v z^w$$

    where the last sum is taken over all w which can be written as $w = vw'$ for some $v \in \mathcal{F}_N$;
3. For every $f \in \mathbb{C}^{p \times q} \langle\langle z_1, \ldots, z_N \rangle\rangle_{\mathrm{rat}}$, $p_1, p_2 \in \mathbb{C} \langle z_1, \ldots, z_N \rangle$ and $\alpha_1, \alpha_2 \in \mathbb{C}$,

$$f \cdot (\alpha_1 p_1 + \alpha_2 p_2) = \alpha_1 (f \cdot p_1) + \alpha_2 (f \cdot p_2).$$

The space $\mathbb{C}^{p \times q} \langle\langle z_1, \ldots, z_N \rangle\rangle_{\mathrm{rat}}$ is a right module over the ring $\mathbb{C} \langle z_1, \ldots, z_N \rangle$ with respect to this product. A structure of left $\mathbb{C} \langle z_1, \ldots, z_N \rangle$-module can be defined in a similar way since the indeterminates commute with coefficients.

Formal power series are used in various branches of mathematics, e.g., in abstract algebra, enumeration problems and combinatorics; rational formal power series have been extensively used in theoretical computer science, mostly in automata theory and language theory (see [18]). The Kleene–Schützenberger theorem [35, 44] (see also [24]) says that a FPS f with coefficients in $\mathbb{C}^{p \times q}$ is rational if and only if it is *recognizable*, i.e., there exist $r \in \mathbb{N}$ and matrices $C \in \mathbb{C}^{p \times r}$, $A_1, \ldots, A_N \in \mathbb{C}^{r \times r}$ and $B \in \mathbb{C}^{r \times q}$ such that for every word $w = g_{i_1} \cdots g_{i_n} \in \mathcal{F}_N$ one has

$$f_w = C A^w B, \quad \text{where} \quad A^w = A_{i_1} \ldots A_{i_n}. \tag{2.3}$$

Let $\mathcal{H}_f$ be the *Hankel matrix* whose rows and columns are indexed by the words of $\mathcal{F}_N$ and defined by

$$(\mathcal{H}_f)_{w,w'} = f_{ww'^T}, \quad w, w' \in \mathcal{F}_N.$$

It follows from (2.3) that if the FPS f is recognizable then $(\mathcal{H}_f)_{w,w'} = C A^{ww'^T} B$ for all $w, w' \in \mathcal{F}_N$. M. Fliess has shown in [24] that a FPS f is rational (that is, recognizable) if and only if

$$\gamma := \mathrm{rank}\, \mathcal{H}_f < \infty.$$

In this case the number γ is the smallest possible r for a representation (2.3).

In control theory, rational FPSs appear as the input/output mappings of linear systems with structured uncertainties. For instance, in [17] a system matrix is given by

$$M = \begin{pmatrix} A & B \\ C & D \end{pmatrix} \in \mathbb{C}^{(r+p)\times(r+q)},$$

and the uncertainty operator is given by

$$\Delta(\delta) = \mathrm{diag}(\delta_1 I_{r_1}, \ldots, \delta_N I_{r_N}),$$

where $r_1 + \cdots + r_N = r$. The uncertainties δ_k are linear operators on ℓ^2 representing disturbances or small perturbation parameters which enter the system at different locations. Mathematically, they can be interpreted as non-commuting indeterminates. The input/output map is a linear fractional transformation

$$LFT(M, \Delta(\delta)) = D + C(I_r - \Delta(\delta)A)^{-1}\Delta(\delta)B, \tag{2.4}$$

which can be interpreted as a non-commutative transfer function T_α^{nc} of a linear system α with evolution on $\mathcal{F}_N$:

$$\alpha : \begin{cases} x_j(g_j w) & = A_{j1}x_1(w) + \cdots + A_{jN}x_N(w) + B_j u(w), \quad j = 1,\ldots,N, \\ y(w) & = C_1 x_1(w) + \cdots + C_N x_N(w) + Du(w), \end{cases} \tag{2.5}$$

where $x_j(w) \in \mathbb{C}^{r_j}$ $(j = 1,\ldots,N)$, $u(w) \in \mathbb{C}^q$, $y(w) \in \mathbb{C}^p$, and the matrices A_{jk}, B and C are of appropriate sizes along the decomposition $\mathbb{C}^r = \mathbb{C}^{r_1} \oplus \cdots \oplus \mathbb{C}^{r_N}$. Such a system appears in [39, 11, 12, 13] and is known as the *non-commutative Givone–Roesser model* of multidimensional linear system; see [26, 27, 42] for its commutative counterpart.

In this paper we do not consider system evolutions (i.e., equations (2.5)). We will use the terminology *N-dimensional Givone–Roesser operator node* (for brevity, *GR-node*) for the collection of data

$$\alpha = (N; A, B, C, D; \mathbb{C}^r = \bigoplus_{j=1}^{N} \mathbb{C}^{r_j}, \mathbb{C}^q, \mathbb{C}^p). \tag{2.6}$$

Sometimes instead of spaces $\mathbb{C}^r, \mathbb{C}^{r_j}$ $(j = 1,\ldots,N), \mathbb{C}^q$ and $\mathbb{C}^p$ we shall consider abstract finite-dimensional linear spaces $\mathcal{X}$ (the *state space*), $\mathcal{X}_j$ $(j = 1,\ldots,N), \mathcal{U}$ (the *input space*) and $\mathcal{Y}$ (the *output space*), respectively, and a node

$$\alpha = (N; A, B, C, D; \mathcal{X} = \bigoplus_{j=1}^{N} \mathcal{X}_j, \mathcal{U}, \mathcal{Y}),$$

where A, B, C, D are linear operators in the corresponding pairs of spaces. The *non-commutative transfer function of a GR-node* α is a rational FPS

$$T_\alpha^{\mathrm{nc}}(z) = D + C(I_r - \Delta(z)A)^{-1}\Delta(z)B. \tag{2.7}$$

Minimal GR-realizations (2.6) of non-commutative rational FPSs, that is, representations of them in the form (2.7), with minimal possible r_k for $k = 1,\ldots,N$ were studied in [17, 16, 39, 11]. For $k = 1,\ldots,N$, the *kth observability matrix* is

$$\mathcal{O}_k = \mathrm{col}(C_k, C_1 A_{1k}, \ldots, C_N A_{Nk}, C_1 A_{11} A_{1k}, \ldots C_1 A_{1N} A_{Nk}, \ldots)$$

and the *kth controllability matrix* is

$$\mathcal{C}_k = \mathrm{row}(B_k, A_{k1}B_1, \dots, A_{kN}B_N, A_{k1}A_{11}B_1, \dots A_{kN}A_{N1}B_1, \dots)$$

(note that these are infinite block matrices). A GR-node α is called *observable* (resp., *controllable*) if $\mathrm{rank}\,\mathcal{O}_k = r_k$ (resp., $\mathrm{rank}\,\mathcal{C}_k = r_k$) for $k = 1, \dots, N$. A GR-node $\alpha = (N; A, B, C, D; \mathbb{C}^r = \bigoplus_{j=1}^{N} \mathbb{C}^{r_j}, \mathbb{C}^q, \mathbb{C}^p)$ is observable if and only if its *adjoint GR-node* $\alpha^* = (N; A^*, C^*, B^*, D^*; \mathbb{C}^r = \bigoplus_{j=1}^{N} \mathbb{C}^{r_j}, \mathbb{C}^p, \mathbb{C}^q)$ is controllable. (Clearly, $(\alpha^*)^* = \alpha$.)

In view of the sequel, we introduce some notations. We set:

$$A^{wg_\nu} = A_{j_1 j_2} A_{j_2 j_3} \cdots A_{j_{k-1} j_k} A_{j_k \nu},$$
$$(C\flat A)^{g_\nu w} = C_\nu A_{\nu j_1} A_{j_1 j_2} \cdots A_{j_{k-1} j_k},$$
$$(A\sharp B)^{wg_\nu} = A_{j_1 j_2} \cdots A_{j_{k-1} j_k} A_{j_k \nu} B_\nu,$$
$$(C\flat A\sharp B)^{g_\mu wg_\nu} = C_\mu A_{\mu j_1} A_{j_1 j_2} \cdots A_{j_{k-1} j_k} A_{j_k \nu} B_\nu,$$

where $w = g_{j_1} \cdots g_{j_k} \in \mathcal{F}_N$ and $\mu, \nu \in \{1, \dots, N\}$. We also define:

$$A^{g_\nu} = A^{\emptyset} = I_\gamma$$
$$(C\flat A)^{g_\nu} = C_\nu,$$
$$(A\sharp B)^{g_\nu} = B_\nu,$$
$$(C\flat A\sharp B)^{g_\nu} = C_\nu B_\nu,$$
$$(C\flat A\sharp B)^{g_\mu g_\nu} = C_\mu A_{\mu\nu} B_\nu,$$

and hence, with the lexicographic order of words in $\mathcal{F}_N$,

$$\mathcal{O}_k = \mathrm{col}_{w \in \mathcal{F}_N}(C\flat A)^{wg_k} \quad \text{and} \quad \mathcal{C}_k = \mathrm{row}_{w \in \mathcal{F}_N}(A\sharp B)^{g_k w^T},$$

and the coefficients of the FPS T_α^{nc} (defined by (2.7)) are given by

$$(T_\alpha^{\mathrm{nc}})_\emptyset = D, \qquad (T_\alpha^{\mathrm{nc}})_w = (C\flat A\sharp B)^w \quad \text{for} \quad w = g_{j_1} \cdots g_{j_n} \in \mathcal{F}_N.$$

The *kth Hankel matrix associated with a FPS f* is defined in [39] (see also [11]) as

$$(\mathcal{H}_{f,k})_{w,w'g_k} = f_{wg_k w'^T} \quad \text{with} \quad w, w' \in \mathcal{F}_N,$$

that is, the rows of $\mathcal{H}_{f,k}$ are indexed by all the words of $\mathcal{F}_N$ and the columns of $\mathcal{H}_{f,k}$ are indexed by all the words of $\mathcal{F}_N$ ending by g_k, provided the lexicographic order is used. If a GR-node α defines a realization of f, that is, $f = T_\alpha^{\mathrm{nc}}$, then

$$(\mathcal{H}_{f,k})_{w,w'g_k} = (C\flat A\sharp B)^{wg_k w'^T} = (C\flat A)^{wg_k}(A\sharp B)^{g_k w'^T},$$

i.e., $\mathcal{H}_{f,k} = \mathcal{O}_k \mathcal{C}_k$. Hence, the node α is minimal if and only if α is both observable and controllable, i.e.,

$$\gamma_k := \mathrm{rank}\,\mathcal{H}_{f,k} = r_k \quad \text{for all} \quad k \in \{1, \dots, N\}.$$

This last set of conditions is an analogue of the above mentioned result of Fliess on minimal recognizable representations of rational formal power series. Every non-commutative rational FPS has a minimal GR-realization.

Finally, we note (see [17, 39]) that two minimal GR-realizations of a given rational FPS are *similar*: if $\alpha^{(i)} = (N; A^{(i)}, B^{(i)}, C^{(i)}, D; \mathbb{C}^\gamma = \bigoplus_{k=1}^N \mathbb{C}^{\gamma_k}, \mathbb{C}^q, \mathbb{C}^p)$ (i=1,2) are minimal GR-nodes such that $T_{\alpha^{(1)}}^{\mathrm{nc}} = T_{\alpha^{(2)}}^{\mathrm{nc}}$ then there exists a block diagonal invertible matrix $T = \mathrm{diag}(T_1, \ldots, T_N)$ (with $T_k \in \mathbb{C}^{\gamma_k \times \gamma_k}$) such that

$$A^{(1)} = T^{-1} A^{(2)} T, \quad B^{(1)} = T^{-1} B^{(2)}, \quad C^{(1)} = C^{(2)} T. \tag{2.8}$$

Of course, the converse is also true, moreover, any two similar (not necessarily minimal) GR-nodes have the same transfer functions.

Now we turn to the discussion on substitutions of matrices for indeterminates in formal power series. Many properties of non-commutative FPSs or non-commutative polynomials are described in terms of matrix substitutions, e.g., matrix-positivity of non-commutative polynomials (non-commutative Positivstellensatz) [29, 40, 31, 32], matrix-positivity of FPS kernels [34], matrix-convexity [21, 30]. The non-commutative Schur–Agler class, i.e., the class of FPSs with operator coefficients, which take contractive values on all N-tuples of strictly contractive operators on ℓ^2, was studied in [12] [1]; we will show in Section 6 that in order that a FPS belongs to this class it suffices to check its contractivity on N-tuples of strictly contractive $n \times n$ matrices, for all $n \in \mathbb{N}$. The notions of matrix-J-unitary (in particular, matrix-J-inner) and matrix-selfadjoint rational FPS, which will be introduced and studied in the present paper, are also defined in terms of substitutions of matrices (of a certain class) for indeterminates.

Let $p(z) = \sum_{|w| \leq m} p_w z^w \in \mathbb{C}\langle z_1, \ldots, z_N \rangle$. For $n \in \mathbb{N}$ and an N-tuple of matrices $Z = (Z_1, \ldots, Z_N) \in (\mathbb{C}^{n \times n})^N$, set

$$p(Z) = \sum_{|w| \leq m} p_w Z^w,$$

where $Z^w = Z_{i_1} \cdots Z_{i_{|w|}}$ for $w = g_{i_1} \cdots g_{i_{|w|}} \in \mathcal{F}_N$, and $Z^\emptyset = I_n$. Then for any rational expression for a FPS $f \in \mathbb{C}\langle\langle z_1, \ldots, z_N \rangle\rangle_{\mathrm{rat}}$ its value at $Z \in (\mathbb{C}^{n \times n})^N$ is well defined provided all of the inversions of polynomials $p^{(j)} \in \mathbb{C}\langle z_1, \ldots, z_N \rangle$ in this expression are well defined at Z. The latter is the case at least in some neighborhood of $Z = 0$, since $p_\emptyset^{(j)} \neq 0$.

Now, if $f \in \mathbb{C}^{p \times q}\langle\langle z_1, \ldots, z_N \rangle\rangle_{\mathrm{rat}}$ then the value $f(Z)$ at some $Z \in (\mathbb{C}^{n \times n})^N$ is well defined whenever the values of matrix entries $(f_{ij}(Z))$ ($i = 1, \ldots, p; j = 1, \ldots, q$) are well defined at Z. As a function of matrix entries $(Z_k)_{ij}$ ($k = 1, \ldots, N; i, j = 1, \ldots, n$), $f(Z)$ is rational $\mathbb{C}^{p \times q} \otimes \mathbb{C}^{n \times n}$-valued function, which is holomorphic on an open and dense set in $\mathbb{C}^{n \times n}$. The latter set contains some neighborhood

$$\Gamma_n(\varepsilon) := \{ Z \in (\mathbb{C}^{n \times n})^N : \|Z_k\| < \varepsilon, \ k = 1, \ldots, N \} \tag{2.9}$$

---

[1]In fact, a more general class was studied in [12], however for our purposes it is enough to consider here only the case mentioned above.

of $Z = 0$, where $f(Z)$ is given by

$$f(Z) = \sum_{w \in \mathcal{F}_N} f_w \otimes Z^w.$$

The following results from [8] on matrix substitutions are used in the sequel.

Theorem 2.1. *Let $f \in \mathbb{C}^{p \times q} \langle\langle z_1, \ldots, z_N \rangle\rangle_{\mathrm{rat}}$, and $m \in \mathbb{Z}_+$ be such that*

$$\bigcap_{w \in \mathcal{F}_N : |w| \leq m} \ker f_w = \bigcap_{w \in \mathcal{F}_N} \ker f_w.$$

Then there exists $\varepsilon > 0$ such that for every $n \in \mathbb{N} : n \geq m^m$ (in the case $m = 0$, for every $n \in \mathbb{N}$),

$$\bigcap_{Z \in \Gamma_n(\varepsilon)} \ker f(Z) = \left(\bigcap_{w \in \mathcal{F}_N : |w| \leq m} \ker f_w \right) \otimes \mathbb{C}^n, \tag{2.10}$$

and moreover, there exist $l \in \mathbb{N} : l \leq qn$, and N-tuples of matrices $Z^{(1)}, \ldots, Z^{(l)}$ from $\Gamma_n(\varepsilon)$ such that

$$\bigcap_{j=1}^{l} \ker f(Z^{(j)}) = \left(\bigcap_{w \in \mathcal{F}_N : |w| \leq m} \ker f_w \right) \otimes \mathbb{C}^n.$$

Corollary 2.2. *In conditions of Theorem 2.1, if for some $n \in \mathbb{N} : n \geq m^m$ (in the case $m = 0$, for some $n \in \mathbb{N}$) one has $f(Z) = 0$, $\forall Z \in \Gamma_n(\varepsilon)$, then $f = 0$.*

3. More on observability, controllability, and minimality in the non-commutative setting

In this section we prove a number of results on observable, controllable and minimal GR-nodes in the multivariable non-commutative setting, which generalize some well-known statements for one-variable nodes (see [15]).

Let us introduce the *kth truncated observability matrix* $\widetilde{\mathcal{O}}_k$ and the *kth truncated controllability matrix* $\widetilde{\mathcal{C}}_k$ of a GR-node (2.6) by

$$\widetilde{\mathcal{O}}_k = \mathrm{col}_{|w| < pr} (C\flat A)^{w g_k}, \quad \widetilde{\mathcal{C}}_k = \mathrm{row}_{|w| < rq} (A \sharp B)^{g_k w^T},$$

with the lexicographic order of words in $\mathcal{F}_N$.

Theorem 3.1. *For each $k \in \{1, \ldots, N\} : \mathrm{rank}\, \widetilde{\mathcal{O}}_k = \mathrm{rank}\, \mathcal{O}_k$ and $\mathrm{rank}\, \widetilde{\mathcal{C}}_k = \mathrm{rank}\, \mathcal{C}_k$.*

Proof. Let us show that for every fixed $k \in \{1, \ldots, N\}$ matrices of the form $(C\flat A)^{w g_k}$ with $|w| \geq pr$ are representable as linear combinations of matrices $(C\flat A)^{\widetilde{w} g_k}$ with $|\widetilde{w}| < pr$. First we remark that if for each fixed $k \in \{1, \ldots, N\}$ and $j \in \mathbb{N}$ all matrices of the form $(C\flat A)^{w g_k}$ with $|w| = j$ are representable as linear combinations of matrices of the form $(C\flat A)^{w' g_k}$ with $|w'| < j$ then the same holds for matrices of the form $(C\flat A)^{w g_k}$ with $|w| = j+1$. Indeed, if $w = i_1 \cdots i_j i_{j+1}$

then there exist words $w'_1, \ldots, w'_s$ with $|w'_1| < j, \ldots, |w'_s| < j$ and $a_1, \ldots, a_s \in \mathbb{C}$ such that

$$(C \flat A)^w = \sum_{\nu=1}^{s} a_\nu (C \flat A)^{w'_\nu g_{i_{j+1}}}.$$

Then for every $k \in \{1, \ldots, N\}$,

$$(C \flat A)^{w g_k} = (C \flat A)^w A_{i_{j+1},k} = \sum_{\nu=1}^{s} a_\nu (C \flat A)^{w'_\nu g_{i_{j+1}}} A_{i_{j+1},k}$$

$$= \sum_{\nu: |w'_\nu| < j-1} a_\nu (C \flat A)^{w'_\nu g_{i_{j+1}}} A_{i_{j+1},k} + \sum_{\nu: |w'_\nu| = j-1} a_\nu (C \flat A)^{w'_\nu g_{i_{j+1}}} A_{i_{j+1},k}$$

$$= \sum_{\nu: |w'_\nu| < j-1} a_\nu (C \flat A)^{w'_\nu g_{i_{j+1}} g_k} + \sum_{\nu: |w'_\nu| = j-1} a_\nu (C \flat A)^{w'_\nu g_{i_{j+1}} g_k}.$$

Consider these two sums separately. All the terms in the first sum are of the form $a_\nu (C \flat A)^{(w'_\nu g_{i_{j+1}}) g_k}$ with $|w'_\nu g_{i_{j+1}}| < j$. In the second sum, by the assumption, for each matrix $(C \flat A)^{w'_\nu g_{i_{j+1}} g_k}$ there exist words $w''_{1\nu}, \ldots, w''_{t\nu}$ of length strictly less than j and complex numbers $b_{1\nu}, \ldots, b_{t\nu}$ such that

$$(C \flat A)^{w'_\nu g_{i_{j+1}} g_k} = \sum_{\mu=1}^{t} b_{\mu\nu} (C \flat A)^{w''_{\mu\nu} g_k}.$$

Hence $(C \flat A)^{w g_k}$ is a linear combination of matrices of the form $(C \flat A)^{\widetilde{w} g_k}$ with $|\widetilde{w}| < j$. Reiterating this argument we obtain that any matrix of the form $(C \flat A)^{w g_k}$ with $|w| \geq j$ and fixed $k \in \{1, \ldots, N\}$ can be represented as a linear combination of matrices of the form $(C \flat A)^{\widetilde{w} g_k}$ with $|\widetilde{w}| < j$. In particular,

$$\operatorname{rank} \operatorname{col}_{|w| < j} (C \flat A)^{w g_k} = \operatorname{rank} \mathcal{O}_k, \quad k = 1, \ldots, N. \tag{3.1}$$

Since for any $k \in \{1, \ldots, N\}$ one has $(C \flat A)^{w g_k} \in \mathbb{C}^{p \times r_k}$ and $\dim \mathbb{C}^{p \times r_k} = p r_k$, we obtain that for some $j \leq pr$, and moreover for $j = pr$ (3.1) is true, i.e., $\operatorname{rank} \widetilde{\mathcal{O}}_k = \operatorname{rank} \mathcal{O}_k$.

The second equality is proved analogously. $\qquad\square$

Remark 3.2. The sizes of the truncated matrices $\widetilde{\mathcal{O}}_k$ and $\widetilde{\mathcal{C}}_k$ depend only on the sizes of matrices A, B and C, and do not depend on these matrices themselves. Our estimate for the size of $\widetilde{\mathcal{O}}_k$ is rough, and one could probably improve it. For our present purposes, only the finiteness of the matrices $\widetilde{\mathcal{O}}_k$ and $\widetilde{\mathcal{C}}_k$ is important, and not their actual sizes.

Corollary 3.3. *A GR-node* (2.6) *is observable (resp., controllable) if and only if for every $k \in \{1, \ldots, N\}$:*

$$\operatorname{rank} \widetilde{\mathcal{O}}_k = r_k \quad (resp, \ \operatorname{rank} \widetilde{\mathcal{C}}_k = r_k),$$

or equivalently, the matrix $\mathcal{O}_k$ (resp., $\mathcal{C}_k$) is left (resp., right) invertible.

Remark 3.4. Corollary 3.3 is comparable with Theorems 7.4 and 7.7 in [39], however we note again that the matrices $\widetilde{\mathcal{O}}_k$ and $\widetilde{\mathcal{C}}_k$ here are finite.

Theorem 3.5. *Let* $\alpha^{(i)} = (N; A^{(i)}, B^{(i)}, C^{(i)}, D, \mathbb{C}^\gamma = \oplus_{k=1}^N \mathbb{C}^{\gamma_k}, \mathbb{C}^q, \mathbb{C}^p)$, $i = 1, 2$, *be minimal GR-nodes with the same transfer function. Then they are similar, the similarity transform is unique and given by* $T = \operatorname{diag}(T_1, \ldots, T_N)$ *where*

$$T_k = \left(\widetilde{\mathcal{O}}_k^{(2)}\right)^+ \widetilde{\mathcal{O}}_k^{(1)} = \widetilde{\mathcal{C}}_k^{(2)} \left(\widetilde{\mathcal{C}}_k^{(1)}\right)^\dagger \tag{3.2}$$

(here "$+$" denotes a left inverse, while "$\dagger$" denotes a right inverse).

Proof. We already mentioned in Section 2 that two minimal nodes with the same transfer function are similar. Let $T' = \operatorname{diag}(T'_1, \ldots, T'_N)$ and $T'' = \operatorname{diag}(T''_1, \ldots, T''_N)$ be two similarity transforms. Let $x \in \mathbb{C}^{\gamma_k}$. Then, for every $w \in \mathcal{F}_N$,

$$(C^{(2)}\flat A^{(2)})^{wg_k} (T''_k - T'_k) x = (C^{(1)}\flat A^{(1)})^{wg_k} x - (C^{(1)}\flat A^{(1)})^{wg_k} x = 0.$$

Since x is arbitrary, from the observability of $\alpha^{(2)}$ we get $T'_k = T''_k$ for $k = 1, \ldots, N$, hence the similarity transform is unique. Comparing the coefficients in the two FPS representations of the transfer function, we obtain

$$(C^{(1)}\flat A^{(1)}\sharp B^{(1)})^w = (C^{(2)}\flat A^{(2)}\sharp B^{(2)})^w$$

for all of $w \in \mathcal{F}_N \setminus \{\emptyset\}$, and therefore

$$\widetilde{\mathcal{O}_k^{(1)}}\,\widetilde{\mathcal{C}_k^{(1)}} = \widetilde{\mathcal{O}_k^{(2)}}\,\widetilde{\mathcal{C}_k^{(2)}}, \quad k = 1, \ldots, N.$$

Thus we obtain

$$\left(\widetilde{\mathcal{O}_k^{(2)}}\right)^+ \widetilde{\mathcal{O}_k^{(1)}} = \widetilde{\mathcal{C}_k^{(2)}} \left(\widetilde{\mathcal{C}_k^{(1)}}\right)^\dagger, \quad k = 1, \ldots, N.$$

Denote the operators which appear in these equalities by T_k, $k = 1, \ldots, N$. A direct computation shows that T_k are invertible with

$$T_k^{-1} = \left(\widetilde{\mathcal{O}_k^{(1)}}\right)^+ \widetilde{\mathcal{O}_k^{(2)}} = \widetilde{\mathcal{C}_k^{(1)}} \left(\widetilde{\mathcal{C}_k^{(2)}}\right)^\dagger.$$

Let us verify that $T = \operatorname{diag}(T_1, \ldots, T_N) \in \mathbb{C}^{\gamma \times \gamma}$ is a similarity transform between $\alpha^{(1)}$ and $\alpha^{(2)}$. It follows from the controllability of $\alpha^{(1)}$ that for arbitrary $k \in \{1, \ldots, N\}$ and $x \in \mathbb{C}^{\gamma_k}$ there exist words $w_j \in \mathcal{F}_N$, with $|w_j| < \gamma q$, scalars $a_j \in \mathbb{C}$ and vectors $u_j \in \mathbb{C}^q$, $j = 1, \ldots, s$, such that

$$x = \sum_{\nu=1}^s a_\nu (A^{(1)}\sharp B^{(1)})^{g_k w_\nu^T} u_\nu.$$

Then

$$T_k x = \left(\widetilde{\mathcal{O}^{(2)}_k}\right)^+ \widetilde{\mathcal{O}^{(1)}_k} x = \sum_{\nu=1}^{s} a_\nu \left(\widetilde{\mathcal{O}^{(2)}_k}\right)^+ \widetilde{\mathcal{O}^{(1)}_k} (A^{(1)} \natural B^{(1)})^{g_k} w_\nu^T u_\nu$$

$$= \sum_{\nu=1}^{s} a_\nu \left(\widetilde{\mathcal{O}^{(2)}_k}\right)^+ \widetilde{\mathcal{O}^{(2)}_k} (A^{(2)} \natural B^{(2)})^{g_k} w_\nu^T u_\nu = \sum_{\nu=1}^{s} a_\nu (A^{(2)} \natural B^{(2)})^{g_k} w_\nu^T u_\nu.$$

This explicit formula implies the set of equalities

$$T_k B_k^{(1)} = B_k^{(2)}, \quad T_k A_{kj}^{(1)} = A_{kj}^{(2)} T_j, \quad C_k^{(1)} = C_k^{(2)} T_k, \quad k, j = 1, \dots, N,$$

which is equivalent to (2.8). $\qquad\square$

Remark 3.6. Theorem 3.5 is comparable with Theorem 7.9 in [39]. However, we establish in Theorem 3.5 the uniqueness and an explicit formula for the similarity transform T.

Using Theorem 2.1, we will prove now the following criteria of observability, controllability, and minimality for GR-nodes analogous to the ones proven in [8, Theorem 3.3] for recognizable FPS representations.

Theorem 3.7. *A GR node α of the form* (2.6) *is observable* (*resp., controllable*) *if and only if for every $k \in \{1, \dots, N\}$ and $n \in \mathbb{N}$: $n \geq (pr - 1)^{pr-1}$ (resp, $n \geq (rq - 1)^{rq-1}$), which means in the case of $pr = 1$ (resp., $rq = 1$): "for every $n \in \mathbb{N}$",*

$$\bigcap_{Z \in \Gamma_n(\varepsilon)} \ker \varphi_k(Z) = 0 \tag{3.3}$$

$$(resp., \quad \bigvee_{Z \in \Gamma_n(\varepsilon)} \mathrm{ran}\, \psi_k(Z) = \mathbb{C}^{r_k} \otimes \mathbb{C}^n), \tag{3.4}$$

where the rational FPSs φ_k and ψ_k are defined by

$$\varphi_k(z) = C(I_r - \Delta(z)A)^{-1}\big|_{\mathbb{C}^{r_k}}, \tag{3.5}$$

$$\psi_k(z) = P_k(I_r - A\Delta(z))^{-1}B, \tag{3.6}$$

with P_k standing for the orthogonal projection onto $\mathbb{C}^{r_k}$ (which is naturally identified here with the subspace in $\mathbb{C}^r$), the symbol "$\bigvee$" means linear span, $\varepsilon = \|A\|^{-1}$ ($\varepsilon > 0$ is arbitrary in the case $A = 0$), and $\Gamma_n(\varepsilon)$ is defined by (2.9). *This GR-node is minimal if both of conditions* (3.3) *and* (3.4) *are fulfilled.*

Proof. First, let us remark that for all $k = 1, \dots, N$ the functions φ_k and ψ_k are well defined in $\Gamma_n(\varepsilon)$, and holomorphic as functions of matrix entries $(Z_j)_{\mu\nu}$, $j = 1, \dots, N$, $\mu, \nu = 1, \dots, n$. Second, Theorem 3.1 implies that in Theorem 2.1 applied to φ_k one can choose $m = pr - 1$, and then from (2.10) obtain that observability for a GR-node α is equivalent to condition (3.3). Since α is controllable if and only if α^* is observable, controllability for α is equivalent to condition (3.4). Since minimality for a GR-node α is equivalent to controllability and observability together, it is in turn equivalent to conditions (3.3) and (3.4) together. $\qquad\square$

Let $\alpha' = (N; A', B', C', D'; \mathbb{C}^{r'} = \bigoplus_{j=1}^{N} \mathbb{C}^{r'_j}, \mathbb{C}^s, \mathbb{C}^p)$ and $\alpha'' = (N; A'', B'',$
$C'', D''; \mathbb{C}^{r''} = \bigoplus_{j=1}^{N} \mathbb{C}^{r''_j}, \mathbb{C}^q, \mathbb{C}^s)$ be GR-nodes. For $k, j = 1, \ldots, N$ set $r_j = r'_j + r''_j$, and

$$A_{kj} = \begin{pmatrix} A'_{kj} & B'_k C''_j \\ 0 & A''_{kj} \end{pmatrix} \in \mathbb{C}^{r_k \times r_j}, \quad B_k = \begin{pmatrix} B'_k D'' \\ B''_k \end{pmatrix} \in \mathbb{C}^{r_k \times q},$$
$$C_j = \begin{pmatrix} C'_j & D' C''_j \end{pmatrix} \in \mathbb{C}^{p \times r_j}, \qquad D = D' D'' \in \mathbb{C}^{p \times q}. \tag{3.7}$$

Then $\alpha = (N; A, B, C, D; \mathbb{C}^r = \bigoplus_{j=1}^{N} \mathbb{C}^{r_j}, \mathbb{C}^q, \mathbb{C}^p)$ will be called the *product of GR-nodes* α' and α'' and denoted by $\alpha = \alpha'\alpha''$. A straightforward calculation shows that

$$T_\alpha^{\mathrm{nc}} = T_{\alpha'}^{\mathrm{nc}} T_{\alpha''}^{\mathrm{nc}}.$$

Consider a GR-node

$$\alpha = (N; A, B, C, D; \mathbb{C}^r = \bigoplus_{j=1}^{N} \mathbb{C}^{r_j}, \mathbb{C}^q) := (N; A, B, C, D; \mathbb{C}^r = \bigoplus_{j=1}^{N} \mathbb{C}^{r_j}, \mathbb{C}^q, \mathbb{C}^q)$$
$$\tag{3.8}$$

with invertible operator D. Then

$$\alpha^\times = (N; A^\times, B^\times, C^\times, D^\times; \mathbb{C}^r = \bigoplus_{j=1}^{N} \mathbb{C}^{r_j}, \mathbb{C}^q),$$

with

$$A^\times = A - BD^{-1}C, \quad B^\times = BD^{-1}, \quad C^\times = -D^{-1}C, \quad D^\times = D^{-1}, \tag{3.9}$$

will be called the *associated GR-node*, and $A^\times$ the *associated main operator*, of α. It is easy to see that, as well as in the one-variable case, $(T_\alpha^{\mathrm{nc}})^{-1} = T_{\alpha^\times}^{\mathrm{nc}}$. Moreover, $(\alpha^\times)^\times = \alpha$ (in particular, $(A^\times)^\times = A$), and $(\alpha'\alpha'')^\times = \alpha''^\times \alpha'^\times$ up to the natural identification of $\mathbb{C}^{r'_j} \oplus \mathbb{C}^{r''_j}$ with $\mathbb{C}^{r''_j} \oplus \mathbb{C}^{r'_j}$, $j = 1, \ldots, N$, which is a similarity transform.

Theorem 3.8. *A GR-node* (3.8) *with invertible operator D is minimal if and only if its associated GR-node $\alpha^\times$ is minimal.*

Proof. Let a GR-node α of the form (3.8) with invertible operator D be minimal, and $x \in \ker \mathcal{O}_k^\times$ for some $k \in \{1, \ldots, N\}$, where $\mathcal{O}_k^\times$ is the kth observability matrix for the GR-node $\alpha^\times$. Then $x \in \ker(C^\times \flat A^\times)^{wg_k}$ for every $w \in \mathcal{F}_N$. Let us show that $x \in \ker \mathcal{O}_k = \bigcap_{w \in \mathcal{F}_N} \ker(C \flat A)^{wg_k}$, i.e, $x = 0$.

For $w = \emptyset$, $C_k^\times x = 0$ means $-D^{-1}C_k x = 0$ (see (3.9)), which is equivalent to $C_k x = 0$. For $|w| > 0$, $w = g_{i_1} \cdots g_{i_{|w|}}$,

$$\begin{aligned}
(C \flat A)^{wg_k} &= C_{i_1} A_{i_1 i_2} \cdots A_{i_{|w|} k} \\
&= -D C_{i_1}^\times (A_{i_1 i_2}^\times + B_{i_1} D^{-1} C_{i_2}) \cdots (A_{i_{|w|} k}^\times + B_{i_{|w|}} D^{-1} C_k) \\
&= L_0 C_k^\times + \sum_{j=1}^{|w|} L_j C_{i_j}^\times A_{i_j i_{j+1}}^\times \cdots A_{i_{|w|} k}^\times,
\end{aligned}$$

with some matrices $L_j \in \mathbb{C}^{q \times q}$, $j = 0, 1, \ldots, |w|$. Thus, $x \in \ker(C\flat A)^{wg_k}$ for every $w \in \mathcal{F}_N$, i.e., $x = 0$, which means that $\alpha^\times$ is observable.

Since α is controllable if and only if α^* is observable (see Section 2), and D^* is invertible whenever D is invertible, the same is true for $\alpha^\times$ and $(\alpha^\times)^* = (\alpha^*)^\times$. Thus, the controllability of $\alpha^\times$ follows from the controllability of α. Finally, the minimality of $\alpha^\times$ follows from the minimality of α. Since $(\alpha^\times)^\times = \alpha$, the minimality of α follows from the minimality of $\alpha^\times$. $\qquad\square$

Suppose that for a GR-node (3.8), projections Π_k on $\mathbb{C}^{r_k}$ are defined such that

$$A_{kj} \ker \Pi_j \subset \ker \Pi_k, \quad (A^\times)_{kj} \operatorname{ran} \Pi_j \subset \operatorname{ran} \Pi_k, \quad k, j = 1, \ldots, N.$$

We do not assume that Π_k are orthogonal. We shall call Π_k a *kth supporting projection* for α. Clearly, the map $\Pi = \operatorname{diag}(\Pi_1, \ldots, \Pi_N) : \; \mathbb{C}^r \to \mathbb{C}^r$ satisfies

$$A \ker \Pi \subset \ker \Pi, \quad A^\times \operatorname{ran} \Pi \subset \operatorname{ran} \Pi,$$

i.e., it is a *supporting projection* for the one-variable node $(1; A, B, C, D; \mathbb{C}^r, \mathbb{C}^q)$ in the sense of [15]. If Π is a supporting projection for α, then $I_r - \Pi$ is a supporting projection for $\alpha^\times$.

The following theorem and corollary are analogous to, and are proved in the same way as Theorem 1.1 and its corollary in [15, pp. 7–9] (see also [43, Theorem 2.1]).

Theorem 3.9. *Let* (3.8) *be a GR-node with invertible operator D. Let Π_k be a projection on $\mathbb{C}^{r_k}$, and let*

$$A = \begin{pmatrix} A_{kj}^{(11)} & A_{kj}^{(12)} \\ A_{kj}^{(21)} & A_{kj}^{(22)} \end{pmatrix}, \quad B_j = \begin{pmatrix} B_j^{(1)} \\ B_j^{(2)} \end{pmatrix}, \quad C_k = \begin{pmatrix} C_k^{(1)} & C_k^{(2)} \end{pmatrix}$$

be the block matrix representations of the operators A_{kj}, B_j and C_k with respect to the decompositions $\mathbb{C}^{r_k} = \ker \Pi_k \dotplus \operatorname{ran} \Pi_k$, for $k, j \in \{1, \ldots, N\}$. Assume that $D = D'D''$, where D' and D'' are invertible operators on $\mathbb{C}^q$, and set

$$\alpha' = (N; A^{(11)}, B^{(1)}(D'')^{-1}, C^{(1)}, D'; \ker \Pi = \bigoplus_{k=1}^{N} \ker \Pi_k, \mathbb{C}^q),$$

$$\alpha'' = (N; A^{(22)}, B^{(2)}, (D')^{-1}C^{(2)}, D''; \operatorname{ran} \Pi = \bigoplus_{k=1}^{N} \operatorname{ran} \Pi_k, \mathbb{C}^q).$$

Then $\alpha = \alpha'\alpha''$ (up to a similarity which maps $\mathbb{C}^{r_k} = \ker \Pi_k \dotplus \operatorname{ran} \Pi_k$ onto $\mathbb{C}^{\dim(\ker \Pi_k)} \oplus \mathbb{C}^{\dim(\operatorname{ran}\Pi_k)}$ $(k = 1, \ldots, N)$ such that $\ker \Pi_k \dotplus \{0\}$ is mapped onto $\mathbb{C}^{\dim(\ker \Pi_k)} \oplus \{0\}$ and $\{0\} \dotplus \operatorname{ran}\Pi_k$ is mapped onto $\{0\} \oplus \mathbb{C}^{\dim(\operatorname{ran}\Pi_k)})$ if and only if Π is a supporting projection for α.

Corollary 3.10. *In the assumptions of Theorem 3.9,*

$$T_\alpha^{\mathrm{nc}} = F'F'',$$

where

$$F'(z) = D' + C(I_r - \Delta(z)A)^{-1}(I_r - \Pi)\Delta(z)B(D'')^{-1},$$

$$F''(z) = D'' + (D')^{-1}C\Pi(I_r - \Delta(z)A)^{-1}\Delta(z)B.$$

We assume now that the external operator of the GR-node (3.8) is equal to $D = I_q$ and that we also take $D' = D'' = I_q$. Then, the GR-nodes α' and α'' of Theorem 3.9 are called *projections of α with respect to the supporting projections $I_r - \Pi$ and Π*, respectively, and we use the notations

$$\alpha' = \mathrm{pr}_{I_r - \Pi}(\alpha) = \left(N; A^{(11)}, B^{(1)}, C^{(1)}, D'; \ker \Pi = \bigoplus_{k=1}^{N} \ker \Pi_k, \mathbb{C}^q \right),$$

$$\alpha'' = \mathrm{pr}_\Pi(\alpha) = \left(N; A^{(22)}, B^{(2)}, C^{(2)}, D''; \mathrm{ran}\,\Pi = \bigoplus_{k=1}^{N} \mathrm{ran}\,\Pi_k, \mathbb{C}^q \right).$$

Let F', F'' and F be rational FPSs with coefficients in $\mathbb{C}^{q \times q}$ such that

$$F = F'F''. \tag{3.10}$$

The factorization (3.10) will be said to be *minimal* if whenever α' and α'' are minimal GR-realizations of F' and F'', respectively, $\alpha'\alpha''$ is a minimal GR-realization of F.

In the sequel, we will use the notation

$$\alpha = \left(N; A, B, C, D; \mathbb{C}^\gamma = \bigoplus_{k=1}^{N} \mathbb{C}^{\gamma_k \times \gamma_k}, \mathbb{C}^q \right) \tag{3.11}$$

for a minimal GR-realization (i.e., $r_k = \gamma_k$ for $k = 1, \dots, N$) of a rational FPS F in the case when $p = q$.

The following theorem is the multivariable non-commutative version of [15, Theorem 4.8]. It gives a complete description of all minimal factorizations in terms of supporting projections.

Theorem 3.11. *Let F be a rational FPS with a minimal GR-realization (3.11). Then the following statements hold:*

(i) *if $\Pi = \mathrm{diag}(\Pi_1, \dots, \Pi_N)$ is a supporting projection for α, then F' is the transfer function of $\mathrm{pr}_{I_\gamma - \Pi}(\alpha)$, F'' is the transfer function of $\mathrm{pr}_\Pi(\alpha)$, and $F = F'F''$ is a minimal factorization of F;*

(ii) *if $F = F'F''$ is a minimal factorization of F, then there exists a uniquely defined supporting projection $\Pi = \mathrm{diag}(\Pi_1, \dots, \Pi_N)$ for the GR-node α such that F' and F'' are the transfer functions of $\mathrm{pr}_{I_\gamma - \Pi}(\alpha)$ and $\mathrm{pr}_\Pi(\alpha)$, respectively.*

Proof. (i). Let Π be a supporting projection for α. Then, by Theorem 3.9,

$$\alpha = \mathrm{pr}_{I_\gamma - \Pi}(\alpha)\mathrm{pr}_\Pi(\alpha).$$

By the assumption, α is minimal. We now show that the GR-nodes $\alpha' = \mathrm{pr}_{I_\gamma - \Pi}(\alpha)$ and $\alpha'' = \mathrm{pr}_\Pi(\alpha)$ are also minimal. To this end, let $x \in \mathrm{ran}\,\Pi_k$. Then

$$\left(C^{(2)}\flat A^{(22)} \right)^{w g_k} x = (C\flat A)^{w g_k} \Pi_k x = (C\flat A)^{w g_k} x.$$

Thus, if $\mathcal{O}_k''$ denotes the kth observability matrix of α'', then $x \in \ker \mathcal{O}_k''$ implies $x \in \ker \mathcal{O}_k$, and the observability of α implies that α'' is also observable. Since

$$\left(A^{(22)}\sharp B^{(2)} \right)^{g_k w^T} = \Pi_k \left(A\sharp B \right)^{g_k w^T},$$

one has $\mathcal{C}_k'' = \Pi_k \mathcal{C}_k$, where $\mathcal{C}_k''$ is the kth controllability matrix of α''. Thus, the controllability of α implies the controllability of α''. Hence, we have proved the minimality of α''. Note that we have used that $\ker \Pi = \mathrm{ran}\,(I_\gamma - \Pi)$ is A-invariant. Since $\mathrm{ran}\,\Pi = \ker(I_\gamma - \Pi)$ is $A^\times$-invariant, by Theorem 3.8 $\alpha^\times$ is minimal. Using

$$\alpha^\times = (\alpha'\alpha'')^\times = (\alpha'')^\times (\alpha')^\times,$$

we prove the minimality of $(\alpha')^\times$ in the same way as that of α''. Applying once again Theorem 3.8, we obtain the minimality of α'. The dimensions of the state spaces of the minimal GR-nodes α', α'' and α are related by

$$\gamma_k = \gamma_k' + \gamma_k'', \qquad k = 1, \ldots, N.$$

Therefore, given any minimal GR-realizations β' and β'' of F' and F'', respectively, the same equalities hold for the state space dimensions of β', β'' and β. Thus, $\beta'\beta''$ is a minimal GR-node, and the factorization $F = F'F''$ is minimal.

(ii). Assume that the factorization $F = F'F''$ is minimal. Let β' and β'' be minimal GR-realizations of F' and F'' with k-th state space dimensions equal to γ_k' and γ_k'', respectively ($k = 1, \ldots, N$). Then $\beta'\beta''$ is a minimal GR-realization of F and its kth state space dimension is equal to $\gamma_k = \gamma_k' + \gamma_k''$ ($k = 1, \ldots, N$). Hence $\beta'\beta''$ is similar to α. We denote the corresponding GR-node similarity by $T = \mathrm{diag}(T_1, \ldots, T_N)$, where

$$T_k : \mathbb{C}^{\gamma'} \oplus \mathbb{C}^{\gamma''} \to \mathbb{C}^\gamma, \quad k = 1, \ldots N,$$

is the canonical isomorphism. Let Π_k be the projection of $\mathbb{C}^{\gamma_k}$ along $T_k \mathbb{C}^{\gamma_k'}$ onto $T_k \mathbb{C}^{\gamma_k''}$, $k = 1, \ldots, N$, and set $\Pi = \mathrm{diag}(\Pi_1, \ldots, \Pi_k)$. Then Π is a supporting projection for α. Moreover $\mathrm{pr}_{I_\gamma - \Pi}(\alpha)$ is similar to β', and $\mathrm{pr}_\Pi(\alpha)$ is similar to β''. The uniqueness of Π is proved in the same way as in [15, Theorem 4.8]. The uniqueness of the GR-node similarity follows from Theorem 3.5. $\qquad\square$

4. Matrix-J-unitary formal power series: A multivariable non-commutative analogue of the line case

In this section we study a multivariable non-commutative analogue of rational $q \times q$ matrix-valued functions which are J-unitary on the imaginary line $i\mathbb{R}$ of the complex plane $\mathbb{C}$.

4.1. Minimal Givone–Roesser realizations and the Lyapunov equation

Denote by $\mathbb{H}^{n\times n}$ the set of Hermitian $n \times n$ matrices. Then $(i\mathbb{H}^{n\times n})^N$ will denote the set of N-tuples of skew-Hermitian matrices. In our paper, the set

$$\mathcal{J}_N = \coprod_{n\in\mathbb{N}} \left(i\mathbb{H}^{n\times n}\right)^N,$$

where "$\coprod$" stands for a disjoint union, will be a counterpart of the imaginary line $i\mathbb{R}$.

Let $J \in \mathbb{C}^{q\times q}$ be a signature matrix. We will call a rational FPS $F \in \mathbb{C}^{q\times q} \langle\langle z_1,\ldots,z_N\rangle\rangle_{\mathrm{rat}}$ *matrix-J-unitary* on $\mathcal{J}_N$ if for every $n \in \mathbb{N}$,

$$F(Z)(J \otimes I_n)F(Z)^* = J \otimes I_n \tag{4.1}$$

at all points $Z \in (i\mathbb{H}^{n\times n})^N$ where it is defined. For a fixed $n \in \mathbb{N}$, $F(Z)$ as a function of matrix entries is rational and holomorphic on some open neighborhood $\Gamma_n(\varepsilon)$ of $Z = 0$, e.g., of the form (2.9), and $\Gamma_n(\varepsilon) \cap (i\mathbb{H}^{n\times n})^N$ is a uniqueness set in $(\mathbb{C}^{n\times n})^N$ (see [45] for the uniqueness theorem in several complex variables). Thus, (4.1) implies that

$$F(Z)(J \otimes I_n)F(-Z^*)^* = J \otimes I_n \tag{4.2}$$

at all points $Z \in (\mathbb{C}^{n\times n})^N$ where $F(Z)$ is holomorphic and invertible (the set of such points is open and dense, since $\det F(Z) \not\equiv 0$).

The following theorem is a counterpart of Theorem 2.1 in [7].

Theorem 4.1. *Let F be a rational FPS with a minimal GR-realization (3.11). Then F is matrix-J-unitary on $\mathcal{J}_N$ if and only if the following conditions are fulfilled:*

a) *D is J-unitary, i.e., $DJD^* = J$;*

b) *there exists an invertible Hermitian solution $H = \mathrm{diag}(H_1,\ldots,H_N)$, with $H_k \in \mathbb{C}^{\gamma_k\times\gamma_k}$, $k = 1,\ldots,N$, of the Lyapunov equation*

$$A^*H + HA = -C^*JC, \tag{4.3}$$

and

$$B = -H^{-1}C^*JD. \tag{4.4}$$

The property b) is equivalent to

b') *there exists an invertible Hermitian matrix $H = \mathrm{diag}(H_1,\ldots,H_N)$, with $H_k \in \mathbb{C}^{\gamma_k\times\gamma_k}$, $k = 1,\ldots,N$, such that*

$$H^{-1}A^* + AH^{-1} = -BJB^*, \tag{4.5}$$

and

$$C = -DJB^*H. \tag{4.6}$$

Proof. Let F be matrix-J-unitary. Then F is holomorphic at the point $Z = 0$ in $\mathbb{C}^N$, hence $D = F(0)$ is J-unitary (in particular, invertible). Equality (4.2) may be rewritten as

$$F(Z)^{-1} = (J \otimes I_n)F(-Z^*)^*(J \otimes I_n). \tag{4.7}$$

Since (4.7) holds for all $n \in \mathbb{N}$, it follows from Corollary 2.2 that the FPSs corresponding to the left and the right sides of equality (4.7) coincide. Due to Theorem 3.8, $\alpha^\times = (N; A^\times, B^\times, C^\times, D^\times; \mathbb{C}^\gamma = \bigoplus_{k=1}^{N} \mathbb{C}^{\gamma_k}, \mathbb{C}^q)$ with $A^\times, B^\times, C^\times, D^\times$ given by (3.9) is a minimal GR-realization of F^{-1}. Due to (4.7), another minimal GR-realization of F^{-1} is $\tilde{\alpha} = (N; \tilde{A}, \tilde{B}, \tilde{C}, \tilde{D}; \mathbb{C}^\gamma = \bigoplus_{k=1}^{N} \mathbb{C}^{\gamma_k}, \mathbb{C}^q)$, where

$$\tilde{A} = -A^*, \quad \tilde{B} = C^* J, \quad \tilde{C} = -JB^*, \quad \tilde{D} = JD^* J.$$

By Theorem 3.5, there exists unique similarity transform $T = \mathrm{diag}(T_1, \ldots, T_N)$ which relates $\alpha^\times$ and $\tilde{\alpha}$, where $T_k \in \mathbb{C}^{\gamma_k \times \gamma_k}$ are invertible for $k = 1, \ldots, N$, and

$$T(A - BD^{-1}C) = -A^*T, \quad TBD^{-1} = C^*J, \quad D^{-1}C = JB^*T. \tag{4.8}$$

Note that the relation $D^{-1} = JD^*J$, which means J-unitarity of D, has been already established above. It is easy to check that relations (4.8) are also valid for T^* in the place of T. Hence, by the uniqueness of similarity matrix, $T = T^*$. Setting $H = -T$, we obtain from (4.8) the equalities (4.3) and (4.4), as well as (4.5) and (4.6), by a straightforward calculation.

Let us prove now a slightly more general statement than the converse. Let α be a (not necessarily minimal) GR-realization of F of the form (3.8), where D is J-unitary, and let $H = \mathrm{diag}(H_1, \ldots, H_N)$ with $H_k \in \mathbb{C}^{r_k \times r_k}$, $k = 1, \ldots, N$, be a Hermitian invertible matrix satisfying (4.3) and (4.4). Then in the same way as in [7, Theorem 2.1] for the one-variable case, we obtain for $Z, Z' \in \mathbb{C}^{n \times n}$:

$$F(Z)(J \otimes I_n)F(Z')^* = J \otimes I_n - (C \otimes I_n)\left(I_r \otimes I_n - \Delta(Z)(A \otimes I_n)\right)^{-1}$$
$$\times \Delta(Z + Z'^*)(H^{-1} \otimes I_n)\left(I_r \otimes I_n - (A^* \otimes I_n)\Delta(Z'^*)\right)^{-1}(C^* \otimes I_n) \tag{4.9}$$

(note that $\Delta(Z)$ commutes with $H^{-1} \otimes I_n$). It follows from (4.9) that $F(Z)$ is $(J \otimes I_n)$-unitary on $(i\mathbb{H}^{n \times n})^N$ at all points Z where it is defined. Since $n \in \mathbb{N}$ is arbitrary, F is matrix-J-unitary on $\mathcal{J}_N$. Clearly, conditions a) and b') also imply the matrix-J-unitarity of F on $\mathcal{J}_N$. $\qquad\square$

Let us make some remarks. First, it follows from the proof of Theorem 4.1 that the structured solution $H = \mathrm{diag}(H_1, \ldots, H_N)$ of the Lyapunov equation (4.3) is uniquely determined by a given minimal GR-realization of F. The matrix $H = \mathrm{diag}(H_1, \ldots, H_N)$ is called the *associated structured Hermitian matrix* (associated with this minimal GR-realization of F). The matrix H_k will be called the *kth component of the associated Hermitian matrix* ($k = 1, \ldots, N$). The explicit formulas for H_k follow from (3.2):

$$H_k = -\left[\mathrm{col}_{|w| \leq qr-1}\left((JB^*)\flat(-A^*)\right)^{wg_k}\right]^+ \mathrm{col}_{|w| \leq qr-1}\left((D^{-1}C)\flat A^\times\right)^{wg_k}$$

$$= -\mathrm{row}_{|w| \leq qr-1}\left((-A^*)\sharp(C^*J)\right)^{g_k w^T}\left[\mathrm{row}_{|w| \leq qr-1}\left(A^\times\sharp(BD^{-1})\right)^{g_k w^T}\right]^\dagger.$$

Second, let α be a (not necessarily minimal) GR-realization of F of the form (3.8), where D is J-unitary, and let $H = \mathrm{diag}(H_1, \ldots, H_N)$ with $H_k \in \mathbb{C}^{r_k \times r_k}$, $k = 1, \ldots, N$, be an Hermitian, not necessarily invertible, matrix satisfying (4.3) and

(4.6). Then in the same way as in [7, Theorem 2.1] for the one-variable case, we obtain for $Z, Z' \in \mathbb{C}^{n \times n}$:

$$F(Z')^*(J \otimes I_n)F(Z) = J \otimes I_n - (B^* \otimes I_n)\left(I_r \otimes I_n - \Delta(Z'^*)(A^* \otimes I_n)\right)^{-1}$$
$$\times (H \otimes I_n)\Delta(Z'^* + Z)\left(I_r \otimes I_n - (A \otimes I_n)\Delta(Z)\right)^{-1}(B \otimes I_n) \qquad (4.10)$$

(note that $\Delta(Z)$ commutes with $H \otimes I_n$). It follows from (4.10) that $F(Z)$ is $(J \otimes I_n)$-unitary on $(i\mathbb{H}^{n \times n})^N$ at all points Z where it is defined. Since $n \in \mathbb{N}$ is arbitrary, F is matrix-J-unitary on $\mathcal{J}_N$.

Third, if α is a (not necessarily minimal) GR-realization of F of the form (3.8), where D is J-unitary, and equalities (4.5) and (4.6) are valid with H^{-1} replaced by some, possibly not invertible, Hermitian matrix $Y = \mathrm{diag}(Y_1, \ldots, Y_N)$ with $Y_k \in \mathbb{C}^{r_k \times r_k}$, $k = 1, \ldots, N$, then F is matrix-J-unitary on $\mathcal{J}_N$. This follows from the fact that (4.9) is valid with H^{-1} replaced by Y.

Theorem 4.2. *Let (C, A) be an observable pair of matrices $C \in \mathbb{C}^{q \times r}, A \in \mathbb{C}^{r \times r}$ in the sense that $\mathbb{C}^r = \bigoplus_{k=1}^N \mathbb{C}^{r_k}$ and $\mathcal{O}_k$ has full column rank for each $k \in \{1, \ldots, N\}$, and let $J \in \mathbb{C}^{q \times q}$ be a signature matrix. Then there exists a matrix-J-unitary on $\mathcal{J}_N$ rational FPS F with a minimal GR-realization $\alpha = (N; A, B, C, D; \mathbb{C}^r = \bigoplus_{k=1}^N \mathbb{C}^{r_k}, \mathbb{C}^q)$ if and only if the Lyapunov equation (4.3) has a structured solution $H = \mathrm{diag}(H_1, \ldots, H_N)$ which is both Hermitian and invertible. If such a solution H exists, possible choices of D and B are*

$$D_0 = I_q, \quad B_0 = -H^{-1}C^*J. \qquad (4.11)$$

Finally, for a given such H, all other choices of D and B differ from D_0 and B_0 by a right multiplicative J-unitary constant matrix.

Proof. Let $H = \mathrm{diag}(H_1, \ldots, H_N)$ be a structured solution of the Lyapunov equation (4.3) which is both Hermitian and invertible. We first check that the pair $(A, -H^{-1}C^*J)$ is controllable, or equivalently, that the pair $(-JCH^{-1}, A^*)$ is observable. Using the Lyapunov equation (4.3), one can see that for any $k \in \{1, \ldots, N\}$ and $w = g_{i_1} \cdots g_{i_{|w|}} \in \mathcal{F}_N$ there exist matrices $K_0, \ldots, K_{|w|-1}$ such that

$$\begin{aligned} (C\flat A)^{wg_k} &= (-1)^{|w|-1}J((-JCH^{-1})\flat A^*)^{wg_k}H_k \\ &+ K_0 J(-JC_{i_2}H_{i_2}^{-1}(A^*)_{i_2 i_3} \cdots (A^*)_{i_{|w|}k})H_k + \cdots \\ &+ K_{|w|-2}J(-JC_{i_{|w|}}(A^*)_{i_{|w|}k})H_k + K_{|w|-1}J(-JC_kH_k^{-1})H_k. \end{aligned}$$

Thus, if $x \in \ker((-JCH^{-1})\flat A^*)^{wg_k}$ for all of $w \in \mathcal{F}_N$ then $H_k^{-1}x \in \ker \mathcal{O}_k$, and the observability of the pair (C, A) implies that $x = 0$. Therefore, the pair $(-JCH^{-1}, A^*)$ is observable, and the pair $(A, -H^{-1}C^*J)$ is controllable. By Theorem 4.1 we obtain that

$$F_0(z) = I_q - C(I_r - \Delta(z)A)^{-1}\Delta(z)H^{-1}C^*J \qquad (4.12)$$

is a matrix-J-unitary on $\mathcal{J}_N$ rational FPS, which has a minimal GR-realization $\alpha_0 = (N : A, -H^{-1}C^*J, C, I_q; \mathbb{C}^r = \bigoplus_{k=1}^N \mathbb{C}^{r_k}, \mathbb{C}^q)$ with the associated structured Hermitian matrix H.

Conversely, let $\alpha = (N; A, B, C, D; \mathbb{C}^r = \bigoplus_{k=1}^{N} \mathbb{C}^{r_k}, \mathbb{C}^q)$ be a minimal GR-node. Then by Theorem 4.1 there exists an Hermitian and invertible matrix $H = \mathrm{diag}(H_1, \ldots, H_N)$ which solves (4.3).

Given $H = \mathrm{diag}(H_1, \ldots, H_N)$, let B, D be any solution of the inverse problem, i.e., $\alpha = (N; A, B, C, D; \mathbb{C}^r = \bigoplus_{k=1}^{N} \mathbb{C}^{r_k}, \mathbb{C}^q)$ is a minimal GR-node with the associated structured Hermitian matrix H. Then for $F_0 = T_{\alpha_0}^{\mathrm{nc}}$ and $F = T_{\alpha}^{\mathrm{nc}}$ we obtain from (4.9) that

$$F(Z)(J \otimes I_n)F(Z')^* = F_0(Z)(J \otimes I_n)F_0(Z')^*$$

for any $n \in \mathbb{N}$, at all points $Z, Z' \in (\mathbb{C}^{n \times n})^N$ where both F and F_0 are defined. By the uniqueness theorem in several complex variables (matrix entries for Z_k's and Z'^{*}_k's, $k = 1, \ldots, N$), we obtain that $F(Z)$ and $F_0(Z)$ differ by a right multiplicative $(J \otimes I_n)$-unitary constant, which clearly has to be $D \otimes I_n$, i.e.,

$$F(Z) = F_0(Z)(D \otimes I_n).$$

Since $n \in \mathbb{N}$ is arbitrary, by Corollary 2.2 we obtain

$$F(z) = F_0(z)D.$$

Equating the coefficients of these two FPSs, we easily deduce using the observability of the pair (C, A) that $B = -H^{-1}C^*JD$. $\qquad\square$

The following dual theorem is proved analogously.

Theorem 4.3. *Let (A, B) be a controllable pair of matrices $A \in \mathbb{C}^{r \times r}, B \in \mathbb{C}^{r \times q}$ in the sense that $\mathbb{C}^r = \bigoplus_{k=1}^{N} \mathbb{C}^{r_k}$ and C_k has full row rank for each $k \in \{1, \ldots, N\}$, and let $J \in \mathbb{C}^{q \times q}$ be a signature matrix. Then there exists a matrix-J-unitary on $\mathcal{J}_N$ rational FPS F with a minimal GR-realization $\alpha = (N; A, B, C, D; \mathbb{C}^r = \bigoplus_{k=1}^{N} \mathbb{C}^{r_k}, \mathbb{C}^q)$ if and only if the Lyapunov equation*

$$GA^* + AG = -BJB^*$$

has a structured solution $G = \mathrm{diag}(G_1, \ldots, G_N)$ which is both Hermitian and invertible. If such a solution G exists, possible choices of D and C are

$$D_0 = I_q, \quad C_0 = -JB^*G^{-1}. \tag{4.13}$$

Finally, for a given such G, all other choices of D and C differ from D_0 and C_0 by a left multiplicative J-unitary constant matrix.

Theorem 4.4. *Let F be a matrix-J-unitary on $\mathcal{J}_N$ rational FPS, and α be its GR-realization. Let $H = \mathrm{diag}(H_1, \ldots, H_N)$ with $H_k \in \mathbb{C}^{r_k \times r_k}$, $k = 1, \ldots, N$, be an Hermitian invertible matrix satisfying (4.3) and (4.4), or equivalently, (4.5) and (4.6). Then α is observable if and only if α is controllable.*

Proof. Suppose that α is observable. Since by Theorem 4.1 $D = F_{\emptyset}$ is J-unitary, by Theorem 4.2 α is a minimal GR-node. In particular, α is controllable.

Suppose that α is controllable. Then by Theorem 4.3 α is minimal, and in particular, observable. $\qquad\square$

4.2. The associated structured Hermitian matrix

Lemma 4.5. *Let F be a matrix-J-unitary on $\mathcal{J}_N$ rational FPS, and let $\alpha^{(i)} = (N; A^{(i)}, B^{(i)}, C^{(i)}, D; \mathbb{C}^\gamma = \bigoplus_{k=1}^{N} \mathbb{C}^{\gamma_k}, \mathbb{C}^q)$ be minimal GR-realizations of F, with the associated structured Hermitian matrices $H^{(i)} = \mathrm{diag}(H_1^{(i)}, \ldots, H_N^{(i)})$, $i = 1, 2$. Then $\alpha^{(1)}$ and $\alpha^{(2)}$ are similar, i.e., (2.8) holds with a uniquely defined invertible matrix $T = \mathrm{diag}(T_1, \ldots, T_N)$, and*

$$H_k^{(1)} = T_k^* H_k^{(2)} T_k, \qquad k = 1, \ldots, N. \tag{4.14}$$

In particular, the matrices $H_k^{(1)}$ and $H_k^{(2)}$ have the same signature.

The proof is easy and analogous to the proof of Lemma 2.1 in [7].

Remark 4.6. The similarity matrix $T = \mathrm{diag}(T_1, \ldots, T_N)$ is a unitary mapping from $\mathbb{C}^\gamma = \bigoplus_{k=1}^{N} \mathbb{C}^{\gamma_k}$ endowed with the inner product $[\,\cdot\,, \cdot\,]_{H^{(1)}}$ onto $\mathbb{C}^\gamma = \bigoplus_{k=1}^{N} \mathbb{C}^{\gamma_k}$ endowed with the inner product $[\,\cdot\,, \cdot\,]_{H^{(2)}}$, where

$$[x, y]_{H^{(i)}} = \langle H^{(i)} x, y \rangle_{\mathbb{C}^\gamma}, \qquad x, y \in \mathbb{C}^\gamma, \ i = 1, 2,$$

that is,

$$[x, y]_{H^{(i)}} = \sum_{k=1}^{N} [x_k, y_k]_{H_k^{(i)}}, \qquad i = 1, 2,$$

where $x_k, y_k \in \mathbb{C}^{\gamma_k}$, $x = \mathrm{col}_{k=1,\ldots,N}(x_k)$, $y = \mathrm{col}_{k=1,\ldots,N}(y_k)$, and

$$[x_k, y_k]_{H_k^{(i)}} = \langle H_k^{(i)} x_k, y_k \rangle_{\mathbb{C}^{\gamma_k}}, \qquad k = 1, \ldots, N, \ i = 1, 2.$$

Recall the following definition [37]. Let $K_{w,w'}$ be a $\mathbb{C}^{q \times q}$-valued function defined for w and w' in some set E and such that $(K_{w,w'})^* = K_{w',w}$. Then $K_{w,w'}$ is called a *kernel with κ negative squares* if for any $m \in \mathbb{N}$, any points $w_1, \ldots, w_m$ in E, and any vectors $c_1, \ldots, c_m$ in $\mathbb{C}^q$ the matrix $(c_j^* K_{w_j, w_i} c_i)_{i,j=1,\ldots,m} \in \mathbb{H}^{m \times m}$ has at most κ negative eigenvalues, and has exactly κ negative eigenvalues for some choice of $m, w_1, \ldots, w_m, c_1, \ldots, c_m$. We will use this definition to give a characterization of the number of negative eigenvalues of the kth component H_k, $k = 1, \ldots, N$, of the associated structured Hermitian matrix H.

Theorem 4.7. *Let F be a matrix-J-unitary on $\mathcal{J}_N$ rational FPS, and let α be its minimal GR-realization of the form (3.11), with the associated structured Hermitian matrix $H = \mathrm{diag}(H_1, \ldots, H_N)$. Then for $k = 1, \ldots, N$ the number of negative eigenvalues of the matrix H_k is equal to the number of negative squares of each of the kernels*

$$K_{w,w'}^{F,k} \;=\; (C\flat A)^{wg_k} H_k^{-1} (A^* \natural C^*)^{g_k w'^T}, \qquad w, w' \in \mathcal{F}_N, \tag{4.15}$$

$$K_{w,w'}^{F^*,k} \;=\; (B^* \flat A^*)^{wg_k} H_k (A \natural B)^{g_k w'^T}, \qquad w, w' \in \mathcal{F}_N, \tag{4.16}$$

For $k = 1, \ldots, N$, denote by $\mathcal{K}_k(F)$ (resp., $\mathcal{K}_k(F^*)$) the linear span of the functions $w \mapsto K_{w,w'}^{F,k} c$ (resp., $w \mapsto K_{w,w'}^{F^*,k} c$) where $w' \in \mathcal{F}_N$ and $c \in \mathbb{C}^q$. Then

$$\dim \mathcal{K}_k(F) = \dim \mathcal{K}_k(F^*) = \gamma_k.$$

Proof. Let $m \in \mathbb{N}$, $w_1, \ldots, w_m \in \mathcal{F}_N$, and $c_1, \ldots, c_m \in \mathbb{C}^q$. Then the matrix equality

$$(c_j^* K_{w_j, w_i}^{F,k} c_i)_{i,j=1,\ldots,m} = X^* H_k^{-1} X,$$

with

$$X = \mathrm{row}_{1 \leq i \leq m} \left((A^* \sharp C^*)^{g_k w_i^T} c_i \right),$$

implies that the kernel $K_{w,w'}^{F,k}$ has at most κ_k negative squares, where κ_k denotes the number of negative eigenvalues of H_k. The pair (C, A) is observable, hence we can choose a basis of $\mathbb{C}^q$ of the form $x_i = (A^* \sharp C^*)^{g_k w_i^T} c_i$, $i = 1, \ldots, q$. Since the matrix $X = \mathrm{row}_{i=1,\ldots,q}(x_i)$ is non-degenerate, and therefore the matrix $X^* H_k^{-1} X$ has exactly κ_k negative eigenvalues, the kernel $K_{w,w'}^{F,k}$ has κ_k negative squares. Analogously, from the controllability of the pair (A, B) one can obtain that the kernel $\mathcal{K}_k(F^*)$ has κ_k negative squares.

Since $\mathcal{K}_k(F)$ is the span of functions (of variable $w \in \mathcal{F}_N$) of the form $(C \flat A)^{w g_k} y$, $y \in \mathbb{C}^{\gamma_k}$, it follows that $\dim \mathcal{K}_k(F) \leq \gamma_k$. From the observability of the pair (C, A) we obtain that $(C \flat A)^{w g_k} y \equiv 0$ implies $y = 0$, thus $\dim \mathcal{K}_k(F) = \gamma_k$. In the same way we obtain that the controllability of the pair (A, B) implies that $\dim \mathcal{K}_k(F^*) = \gamma_k$. $\qquad\square$

We will denote by $\nu_k(F)$ the number of negative squares of either the kernel $K_{w,w'}^{F,k}$ or the kernel $K_{w,w'}^{F^*,k}$ defined by (4.15) and (4.16), respectively.

Theorem 4.8. *Let $F^{(i)}$ be matrix-J-unitary on $\mathcal{J}_N$ rational FPSs, with minimal GR-realizations $\alpha^{(i)} = (N; A^{(i)}, B^{(i)}, C^{(i)}, D^{(i)}; \mathbb{C}^{\gamma^{(i)}} = \bigoplus_{k=1}^N \mathbb{C}^{\gamma_k^{(i)}}, \mathbb{C}^q)$ and the associated structured Hermitian matrices $H^{(i)} = \mathrm{diag}(H_1^{(i)}, \ldots, H_N^{(i)})$, respectively, $i = 1, 2$. Suppose that the product $\alpha = \alpha^{(1)} \alpha^{(2)}$ is a minimal GR-node. Then the matrix $H = \mathrm{diag}(H_1, \ldots, H_N)$, with*

$$H_k = \begin{pmatrix} H_k^{(1)} & 0 \\ 0 & H_k^{(2)} \end{pmatrix} \in \mathbb{C}^{(\gamma_k^{(1)} + \gamma_k^{(2)}) \times (\gamma_k^{(1)} + \gamma_k^{(2)})}, \quad k = 1, \ldots, N, \qquad (4.17)$$

is the associated structured Hermitian matrix for $\alpha = \alpha^{(1)} \alpha^{(2)}$.

Proof. It suffices to check that (4.3) and (4.4) hold for the matrices A, B, C, D defined as in (3.7), and $H = \mathrm{diag}(H_1, \ldots, H_N)$ where H_k, $k = 1, \ldots, N$, are defined in (4.17). This is an easy computation which is omitted. $\qquad\square$

Corollary 4.9. *Let F_1 and F_2 be matrix-J-unitary on $\mathcal{J}_N$ rational FPSs, and suppose that the factorization $F = F_1 F_2$ is minimal. Then*

$$\nu_k(F_1 F_2) = \nu_k(F_1) + \nu_k(F_2), \quad k = 1, \ldots, N.$$

4.3. Minimal matrix-J-unitary factorizations

In this subsection we consider minimal factorizations of rational formal power series which are matrix-J-unitary on $\mathcal{J}_N$ into factors both of which are also matrix-J-unitary on $\mathcal{J}_N$. Such factorizations will be called *minimal matrix-J-unitary factorizations*.

Let $H \in \mathbb{C}^{r \times r}$ be an invertible Hermitian matrix. We denote by $[\,\cdot\,,\,\cdot\,]_H$ the Hermitian sesquilinear form

$$[x, y]_H = \langle Hx, y \rangle$$

where $\langle\,\cdot\,,\,\cdot\,\rangle$ denotes the standard inner product of $\mathbb{C}^r$. Two vectors x and y in $\mathbb{C}^r$ are called H-orthogonal if $[x, y]_H = 0$. For any subspace $M \subset \mathbb{C}^r$ denote

$$M^{[\perp]} = \{y \in \mathbb{C}^r : \langle y, m \rangle_H = 0 \quad \forall m \in M\}.$$

The subspace $M \subset \mathbb{C}^r$ is called *non-degenerate* if $M \cap M^{[\perp]} = \{0\}$. In this case,

$$M[\dot{+}]M^{[\perp]} = \mathbb{C}^r$$

where $[\dot{+}]$ denotes the *H-orthogonal direct sum*.

In the case when $H = \mathrm{diag}(H_1, \ldots, H_N)$ is the structured Hermitian matrix associated with a given minimal GR-realization of a matrix-J-unitary on $\mathcal{J}_N$ rational FPS F, we will call $[\,\cdot\,,\,\cdot\,]_H$ the *associated inner product* (associated with the given minimal GR-realization of F). In more details,

$$[x, y]_H = \sum_{k=1}^{N} [x_k, y_k]_{H_k},$$

where $x_k, y_k \in \mathbb{C}^{\gamma_k}$ and $x = \mathrm{col}_{k=1,\ldots,N}(x_k)$, $y = \mathrm{col}_{k=1,\ldots,N}(y_k)$, and

$$[x_k, y_k]_{H_k} = \langle H_k x_k, y_k \rangle_{\mathbb{C}^{\gamma_k}}, \quad k = 1, \ldots, N.$$

The following theorem (as well as its proof) is analogous to its one-variable counterpart, Theorem 2.6 from [7] (see also [43, Chapter II]).

Theorem 4.10. *Let F be a matrix-J-unitary on $\mathcal{J}_N$ rational FPS, and let α be its minimal GR-realization of the form (3.11), with the associated structured Hermitian matrix $H = \mathrm{diag}(H_1, \ldots, H_N)$. Let $\mathcal{M} = \bigoplus_{k=1}^{N} \mathcal{M}_k$ be an A-invariant subspace such that $\mathcal{M}_k \subset \mathbb{C}^{\gamma_k}$, $k = 1, \ldots, N$, and $\mathcal{M}$ is non-degenerate in the associated inner product $[\,\cdot\,,\,\cdot\,]_H$. Let $\Pi = \mathrm{diag}(\Pi_1, \ldots, \Pi_N)$ be the projection defined by*

$$\ker \Pi = \mathcal{M}, \quad \mathrm{ran}\, \Pi = \mathcal{M}^{[\perp]},$$

or in more details,

$$\ker \Pi_k = \mathcal{M}_k, \quad \mathrm{ran}\, \Pi_k = \mathcal{M}_k^{[\perp]}, \quad k = 1, \ldots, N.$$

Let $D = D_1 D_2$ be a factorization of D into two J-unitary factors. Then the factorization $F = F_1 F_2$ where

$$\begin{aligned}
F_1(z) &= D_1 + C(I_\gamma - \Delta(z)A)^{-1}\Delta(z)(I_\gamma - \Pi)BD_2^{-1}, \\
F_2(z) &= D_2 + D_1^{-1}C\Pi(I_\gamma - \Delta(z)A)^{-1}\Delta(z)B,
\end{aligned}$$

is a minimal matrix-J-unitary factorization of F.

Conversely, any minimal matrix-J-unitary factorization of F can be obtained in such a way. For a fixed J-unitary decomposition $D = D_1 D_2$, the correspondence between minimal matrix-J-unitary factorizations of F and nondegenerate A-invariant subspaces of the form $\mathcal{M} = \bigoplus_{k=1}^{N} \mathcal{M}_k$, where $\mathcal{M}_k \subset \mathbb{C}^{\gamma_k}$ for $k = 1, \ldots, N$, is one-to-one.

Remark 4.11. We omit here the proof, which can be easily restored, with making use of Theorem 3.9 and Corollary 3.10.

Remark 4.12. Minimal matrix-J-unitary factorizations do not always exist, even for $N = 1$. Examples of J-unitary on $i\mathbb{R}$ rational functions which have non-trivial minimal factorizations but lack minimal J-unitary factorizations can be found in [4] and [7].

4.4. Matrix-unitary rational formal power series

In this subsection we specialize some of the preceding results to the case $J = I_q$. We call the corresponding rational formal power series *matrix-unitary* on $\mathcal{J}_N$.

Theorem 4.13. *Let F be a rational FPS and α be its minimal GR-realization of the form (3.11). Then F is matrix-unitary on $\mathcal{J}_N$ if and only if the following conditions are fulfilled:*

a) *D is a unitary matrix, i.e., $DD^* = I_q$;*
b) *there exists an Hermitian solution $H = \mathrm{diag}(H_1, \ldots, H_N)$, with $H_k \in \mathbb{C}^{\gamma_k \times \gamma_k}$, $k = 1, \ldots, N$, of the Lyapunov equation*

$$A^*H + HA = -C^*C, \tag{4.18}$$

    and

$$C = -D^{-1}B^*H. \tag{4.19}$$

The property b) is equivalent to

b′) *there exists an Hermitian solution $G = \mathrm{diag}(G_1, \ldots, G_N)$, with $G_k \in \mathbb{C}^{\gamma_k \times \gamma_k}$, $k = 1, \ldots, N$, of the Lyapunov equation*

$$GA^* + AG = -BB^*, \tag{4.20}$$

    and

$$B = -GC^*D^{-1}. \tag{4.21}$$

Proof. To obtain Theorem 4.13 from Theorem 4.1 it suffices to show that any structured Hermitian solution to the Lyapunov equation (4.18) (resp., (4.20)) is invertible. Let $H = \mathrm{diag}(H_1, \ldots, H_N)$ be a structured Hermitian solution to (4.18), and $x \in \ker H$, i.e., $x = \mathrm{col}_{1 \leq k \leq N}(x_k)$ and $x_k \in \ker H_k$, $k = 1, \ldots, N$. Then

$\langle HAx, x \rangle = \langle Ax, Hx \rangle = 0$, and equation (4.18) implies $Cx = 0$. In particular, for every $k \in \{1, \ldots, N\}$ one can define $\tilde{x} = \mathrm{col}(0, \ldots, 0, x_k, 0, \ldots, 0)$ where $x_k \in \ker H_k$ is on the kth block entry of $\tilde{x}$, and from $C\tilde{x} = 0$ get $C_k x_k = 0$. Thus, $\ker H_k \subset \ker C_k$, $k = 1, \ldots, N$. Consider the following block representations with respect to the decompositions $\mathbb{C}^{\gamma_k} = \ker H_k \oplus \mathrm{ran}\, H_k$:

$$A_{ij} = \begin{pmatrix} A_{ij}^{(11)} & A_{ij}^{(12)} \\ A_{ij}^{(21)} & A_{ij}^{(22)} \end{pmatrix}, \quad C_k = \begin{pmatrix} 0 & C_k^{(2)} \end{pmatrix}, \quad H_k = \begin{pmatrix} 0 & 0 \\ 0 & H_k^{(22)} \end{pmatrix},$$

where $i, j, k = 1, \ldots, N$. Then (4.18) implies

$$(A^*H + HA)_{ij}^{(12)} = (A_{ji}^* H_j + H_i A_{ij})^{(12)} = \left(A_{ji}^{(21)} \right)^* H_j^{(22)} = 0,$$

and $A_{ji}^{(21)} = 0$, $i, j = 1, \ldots, N$. Therefore, for any $w \in \mathcal{F}_N$ we have

$$(C\flat A)^{wg_k} = \begin{pmatrix} 0 & (C^{(2)}\flat A^{(22)})^{wg_k} \end{pmatrix}, \quad k = 1, \ldots, N,$$

where $C^{(2)} = \mathrm{row}_{1 \le k \le N}(C_k^{(2)})$, $A^{(22)} = (A_{ij}^{(22)})_{i,j=1,\ldots,N}$. If there exists $k \in \{1, \ldots, N\}$ such that $\ker H_k \ne \{0\}$, then the pair (C, A) is not observable, which contradicts to the assumption on α. Thus, H is invertible.

In a similar way one can show that any structured Hermitian solution $G = \mathrm{diag}(G_1, \ldots, G_N)$ of the Lyapunov equation (4.20) is invertible. $\square$

A counterpart of Theorem 4.2 in the present case is the following theorem.

Theorem 4.14. *Let (C, A) be an observable pair of matrices $C \in \mathbb{C}^{q \times r}, A \in \mathbb{C}^{r \times r}$ in the sense that $\mathbb{C}^r = \bigoplus_{k=1}^{N} \mathbb{C}^{r_k}$ and $\mathcal{O}_k$ has full column rank for each $k \in \{1, \ldots, N\}$. Then there exists a matrix-unitary on $\mathcal{J}_N$ rational FPS F with a minimal GR-realization $\alpha = (N; A, B, C, D; \mathbb{C}^r = \bigoplus_{k=1}^{N} \mathbb{C}^{r_k}, \mathbb{C}^q)$ if and only if the Lyapunov equation (4.18) has a structured Hermitian solution $H = \mathrm{diag}(H_1, \ldots, H_N)$. If such a solution H exists, it is invertible, and possible choices of D and B are*

$$D_0 = I_q, \quad B_0 = -H^{-1}C^*. \tag{4.22}$$

Finally, for a given such H, all other choices of D and B differ from D_0 and B_0 by a right multiplicative unitary constant matrix.

The proof of Theorem 4.14 is a direct application of Theorem 4.2 and Theorem 4.13. One can prove analogously the following theorem which is a counterpart of Theorem 4.3.

Theorem 4.15. *Let (A, B) be a controllable pair of matrices $A \in \mathbb{C}^{r \times r}, B \in \mathbb{C}^{r \times q}$ in the sense that $\mathbb{C}^r = \bigoplus_{k=1}^{N} \mathbb{C}^{r_k}$ and $\mathcal{C}_k$ has full row rank for each $k \in \{1, \ldots, N\}$. Then there exists a matrix-unitary on $\mathcal{J}_N$ rational FPS F with a minimal GR-realization $\alpha = (N; A, B, C, D; \mathbb{C}^r = \bigoplus_{k=1}^{N} \mathbb{C}^{r_k}, \mathbb{C}^q)$ if and only if the Lyapunov equation (4.20) has a structured Hermitian solution $G = \mathrm{diag}(G_1, \ldots, G_N)$. If such a solution G exists, it is invertible, and possible choices of D and C are*

$$D_0 = I_q, \quad C_0 = -B^*G^{-1}. \tag{4.23}$$

Finally, for a given such G, all other choices of D and C differ from D_0 and C_0 by a left multiplicative unitary constant matrix.

Let $\overline{A} = (\overline{A_1}, \ldots, \overline{A_N})$ be an N-tuple of $r \times r$ matrices. A non-zero vector $x \in \mathbb{C}^r$ is called a *common eigenvector for $\overline{A}$* if there exists $\lambda = (\lambda_1, \ldots, \lambda_N) \in \mathbb{C}^N$ (which is called a *common eigenvalue for $\overline{A}$*) such that

$$\overline{A_k} x = \lambda_k x, \quad k = 1, \ldots, N.$$

The following theorem, which is a multivariable non-commutative counterpart of statements a) and b) of Theorem 2.10 in [7], gives a necessary condition on a minimal GR-realization of a matrix-unitary on $\mathcal{J}_N$ rational FPS.

Theorem 4.16. *Let F be a matrix-unitary on $\mathcal{J}_N$ rational FPS and α be its minimal GR-realization, with the associated structured Hermitian matrix $H = \mathrm{diag}(H_1, \ldots, H_N)$ and the associated inner products $[\cdot, \cdot]_{H_k}$, $k = 1, \ldots, N$. Let P_k denote the orthogonal projection in $\mathbb{C}^\gamma$ onto the subspace $\{0\} \oplus \cdots \oplus \{0\} \oplus \mathbb{C}^{\gamma_k} \oplus \{0\} \oplus \cdots \oplus \{0\}$, and $\overline{A_k} = AP_k$, $k = 1, \ldots, N$. If $x \in \mathbb{C}^\gamma$ is a common eigenvector for $\overline{A}$ corresponding to a common eigenvalue $\lambda \in \mathbb{C}^N$ then there exists $j \in \{1, \ldots, N\}$ such that $\mathrm{Re}\,\lambda_j \neq 0$ and $[P_j x, P_j x]_{H_j} \neq 0$. In particular, $\overline{A}$ has no common eigenvalues on $(i\mathbb{R})^N$.*

Proof. By (4.18), we have for every $k \in \{1, \ldots, N\}$,

$$(\overline{\lambda_k} + \lambda_k)[P_k x, P_k x]_{H_k} = -\langle CP_k x, CP_k x \rangle.$$

Suppose that for all $k \in \{1, \ldots, N\}$ the left-hand side of this equality is zero, then $CP_k x = 0$. Since for $\emptyset \neq w = g_{i_1} \cdots g_{i_{|w|}} \in \mathcal{F}_N$,

$$(C\flat A)^{wg_k} P_k x = CP_{i_1}\overline{A_{i_2}} \cdots \overline{A_{i_{|w|}}} \cdot \overline{A_k} x = \lambda_{i_2} \cdots \lambda_{i_{|w|}} \lambda_k CP_{i_1} x = 0,$$

the observability of the pair (C, A) implies $P_k x = 0$, $k = 1, \ldots, N$, i.e., $x = 0$ which contradicts to the assumption that x is a common eigenvector for $\overline{A}$. Thus, there exists $j \in \{1, \ldots, N\}$ such that $(\overline{\lambda_j} + \lambda_j)[P_j x, P_j x]_{H_j} \neq 0$, as desired. $\qquad\square$

5. Matrix-J-unitary formal power series: A multivariable non-commutative analogue of the circle case

In this section we study a multivariable non-commutative analogue of rational $\mathbb{C}^{q \times q}$-valued functions which are J-unitary on the unit circle $\mathbb{T}$.

5.1. Minimal Givone–Roesser realizations and the Stein equation

Let $n \in \mathbb{N}$. We denote by $\mathbb{T}^{n \times n}$ the *matrix unit circle*

$$\mathbb{T}^{n \times n} = \left\{ W \in \mathbb{C}^{n \times n} \ : \ WW^* = I_n \right\},$$

i.e., the family of unitary $n \times n$ complex matrices. We will call the set $(\mathbb{T}^{n \times n})^N$ the *matrix unit torus*. The set

$$\mathcal{T}_N = \coprod_{n \in \mathbb{N}} (\mathbb{T}^{n \times n})^N$$

serves as a multivariable non-commutative counterpart of the unit circle. Let $J = J^{-1} = J^* \in \mathbb{C}^{q \times q}$. We will say that a rational FPS f is *matrix-J-unitary on* $\mathcal{T}_N$ if for every $n \in \mathbb{N}$,

$$f(W)(J \otimes I_n)f(W)^* = J \otimes I_n$$

at all points $W = (W_1, \ldots, W_N) \in (\mathbb{T}^{n \times n})^N$ where it is defined. In the following theorem we establish the relationship between matrix-J-unitary rational FPSs on $\mathcal{J}_N$ and on $\mathcal{T}_N$, their minimal GR-realizations, and the structured Hermitian solutions of the corresponding Lyapunov and Stein equations.

Theorem 5.1. *Let f be a matrix-J-unitary on $\mathcal{T}_N$ rational FPS, with a minimal GR-realization α of the form (3.11), and let $a \in \mathbb{T}$ be such that $-\bar{a} \notin \sigma(A)$. Then*

$$F(z) = f(a(z_1 - 1)(z_1 + 1)^{-1}, \ldots, a(z_N - 1)(z_N + 1)^{-1}) \tag{5.1}$$

is well defined as a rational FPS which is matrix-J-unitary on $\mathcal{J}_N$, and $F = T_\beta^{\mathrm{nc}}$, where $\beta = (N; A_a, B_a, C_a, D_a; \mathbb{C}^\gamma = \bigoplus_{k=1}^N \mathbb{C}^{\gamma_k}, \mathbb{C}^q)$, with

$$\begin{aligned}
A_a &= (aA - I_\gamma)(aA + I_\gamma)^{-1}, & B_a &= \sqrt{2}(aA + I_\gamma)^{-1}aB, \\
C_a &= \sqrt{2}C(aA + I_\gamma)^{-1}, & D_a &= D - C(aA + I_\gamma)^{-1}aB.
\end{aligned} \tag{5.2}$$

A GR-node β is minimal, and its associated structured Hermitian matrix $H = \mathrm{diag}(H_1, \ldots, H_N)$ is the unique invertible structured Hermitian solution of

$$\begin{pmatrix} A & B \\ C & D \end{pmatrix}^* \begin{pmatrix} H & 0 \\ 0 & J \end{pmatrix} \begin{pmatrix} A & B \\ C & D \end{pmatrix} = \begin{pmatrix} H & 0 \\ 0 & J \end{pmatrix}. \tag{5.3}$$

Proof. For any $a \in \mathbb{T}$ and $n \in \mathbb{N}$ the Cayley transform

$$Z_0 \longmapsto W_0 = a(Z_0 - I_n)(Z_0 + I_n)^{-1}$$

maps $i\mathbb{H}^{n \times n}$ onto $\mathbb{T}^{n \times n}$, thus its simultaneous application to each matrix variable maps $(i\mathbb{H}^{n \times n})^N$ onto $(\mathbb{T}^{n \times n})^N$. Since the simultaneous application of the Cayley transform to each formal variable in a rational FPS gives a rational FPS, (5.1) defines a rational FPS F. Since f is matrix-J-unitary on $\mathcal{T}_N$, F is matrix-J-unitary on $\mathcal{J}_N$. Moreover,

$$\begin{aligned}
F(z) &= D + C\left(I_\gamma - a(\Delta(z) - I_\gamma)(\Delta(z) + I_\gamma)^{-1}A\right)^{-1} \\
&\quad \times a(\Delta(z) - I_\gamma)(\Delta(z) + I_\gamma)^{-1}B \\
&= D + C\left(\Delta(z) + I_\gamma - a(\Delta(z) - I_\gamma)A\right)^{-1} a(\Delta(z) - I_\gamma)B \\
&= D + C\left(aA + I_\gamma - \Delta(z)(aA - I_\gamma)\right)^{-1} a(\Delta(z) - I_\gamma)B \\
&= D + C(aA + I_\gamma)^{-1}\left(I_\gamma - \Delta(z)(aA - I_\gamma)(aA + I_\gamma)^{-1}\right)^{-1} \Delta(z)aB \\
&\quad - C(aA + I_\gamma)^{-1}\left(I_\gamma - \Delta(z)(aA - I_\gamma)(aA + I_\gamma)^{-1}\right)^{-1} aB \\
&= D - C(aA + I_\gamma)^{-1}aB + C(aA + I_\gamma)^{-1} \\
&\quad \times \left(I_\gamma - \Delta(z)(aA - I_\gamma)(aA + I_\gamma)^{-1}\right)^{-1} \\
&\quad \times \Delta(z)\left(I_\gamma - (aA - I_\gamma)(aA + I_\gamma)^{-1}\right)aB \\
&= D_a + C_a(I_\gamma - \Delta(z)A_a)^{-1}\Delta(z)B_a.
\end{aligned}$$

Thus, $F = T_\beta^{\mathrm{nc}}$. Let us remark that the FPS

$$\varphi_k^a(z) = C_a(I_\gamma - \Delta(z)A_a)^{-1}\big|_{\mathbb{C}^{\gamma_k}}$$

(c.f. (3.5)) has the coefficients

$$(\varphi_k^a)_w = (C_a\flat A_a)^{wg_k}, \quad w \in \mathcal{F}_N.$$

Remark also that

$$\tilde\varphi_k(z) := \varphi_k\left(a(z_1 - 1)(z_1 + 1)^{-1}, \ldots, a(z_N - 1)(z_N + 1)^{-1}\right)$$
$$= C\left(I_\gamma - a(\Delta(z) - I_\gamma)(\Delta(z) + I_\gamma)^{-1}A\right)^{-1}\big|_{\mathbb{C}^{\gamma_k}}$$
$$= C\left((\Delta(z) + I_\gamma) - a(\Delta(z) - I_\gamma)A\right)^{-1}(\Delta(z) + I_\gamma)\big|_{\mathbb{C}^{\gamma_k}}$$
$$= C\left((aA + I_\gamma) - \Delta(z)(aA - I_\gamma)\right)^{-1}(\Delta(z) + I_\gamma)\big|_{\mathbb{C}^{\gamma_k}}$$
$$= C(aA + I_\gamma)^{-1}\left(I_\gamma - \Delta(z)(aA - I_\gamma)(aA + I_\gamma)^{-1}\right)^{-1}(\Delta(z) + I_\gamma)\big|_{\mathbb{C}^{\gamma_k}}$$
$$= \frac{1}{\sqrt{2}}\left(C_a(I_\gamma - \Delta(z)A_a)^{-1}\big|_{\mathbb{C}^{\gamma_k}}\right)(z_k + 1)$$
$$= \frac{1}{\sqrt{2}}\left(\varphi_k^a(z) \cdot z_k + \varphi_k^a(z)\right).$$

Let $k \in \{1, \ldots, N\}$ be fixed. Suppose that $n \in \mathbb{N}$, $n \geq (q\gamma - 1)^{q\gamma - 1}$ (for $q\gamma - 1 = 0$ choose arbitrary $n \in \mathbb{N}$), and $x \in \bigcap_{Z \in \Gamma_n(\varepsilon)} \ker \varphi_k^a(Z)$, where $\Gamma_n(\varepsilon)$ is a neighborhood of the origin of $\mathbb{C}^{n \times n}$ where $\varphi_k^a(Z)$ is well defined, e.g., of the form (2.9) with $\varepsilon = \|A_a\|^{-1}$. Then, by Theorem 3.1 and Theorem 2.1, one has

$$\bigcap_{Z \in \Gamma_n(\varepsilon)} \ker \varphi_k^a(Z) = \left(\bigcap_{w \in \mathcal{F}_N : |w| \leq q\gamma - 1} \ker (\varphi_k^a)_w\right) \otimes \mathbb{C}^n$$

$$= \left(\bigcap_{w \in \mathcal{F}_N : |w| \leq q\gamma - 1} \ker (C_a\flat A_a)^{wg_k}\right) \otimes \mathbb{C}^n = \left(\ker \tilde{O}_k(\beta)\right) \otimes \mathbb{C}^n.$$

Thus, there exist $l \in \mathbb{N}$, $\{u^{(\mu)}\}_{\mu=1}^l \subset \ker \tilde{O}_k(\beta)$, $\{y^{(\mu)}\}_{\mu=1}^l \subset \mathbb{C}^n$ such that

$$x = \sum_{\mu=1}^l u^{(\mu)} \otimes y^{(\mu)}. \tag{5.4}$$

Since $\left(\varphi_k^a(z) \cdot z_k\right)_{wg_k} = (C_a\flat A_a)^{wg_k}$ for $w \in \mathcal{F}_N$, and $\left(\varphi_k^a(z) \cdot z_k\right)_{w'} = 0$ for $w' \neq wg_k$ with any $w \in \mathcal{F}_N$, (5.4) implies that $\varphi_k^a(Z)(I_{\gamma_k} \otimes Z_k)x \equiv 0$. Thus,

$$\tilde\varphi_k(Z)x = \frac{1}{\sqrt{2}}\left(\varphi_k^a(Z)(I_{\gamma_k} \otimes Z_k) + \varphi_k^a(Z)\right)x \equiv 0.$$

Since the Cayley transform $a(\Delta(z) - I_\gamma)(\Delta(z) + I_\gamma)^{-1}$ maps an open and dense subset of the set of matrices of the form $\Delta(Z) = \mathrm{diag}\,(Z_1, \ldots, Z_N)$, $Z_j \in \mathbb{C}^{\gamma_j \times \gamma_j}$, $j = 1, \ldots, N$, onto an open and dense subset of the same set,

$$\varphi_k(Z)x = (C \otimes I_n)(I_\gamma - \Delta(Z)(A \otimes I_n))^{-1}x \equiv 0.$$

Since the GR-node α is observable, by Theorem 3.7 we get $x = 0$. Therefore,

$$\bigcap_{Z \in \Gamma_n(\varepsilon)} \ker \varphi_k^a(Z) = 0, \quad k = 1, \ldots, N.$$

Applying Theorem 3.7 once again, we obtain the observability of the GR-node β. In the same way one can prove the controllability of β. Thus, β is minimal.

Note that

$$\begin{pmatrix} A & B \\ C & D \end{pmatrix}^* \begin{pmatrix} H & 0 \\ 0 & J \end{pmatrix} \begin{pmatrix} A & B \\ C & D \end{pmatrix} - \begin{pmatrix} H & 0 \\ 0 & J \end{pmatrix} =$$
$$= \begin{pmatrix} A^*HA + C^*JC - H & A^*HB + C^*JD \\ B^*HA + D^*JC & B^*HB + D^*JD - J \end{pmatrix}. \tag{5.5}$$

Since $-\bar{a} \notin \sigma(A)$, the matrix $(aA + I_\gamma)^{-1}$ is well defined, as well as $A_a = (aA - I_\gamma)(aA + I_\gamma)^{-1}$, and $I_\gamma - A_a = 2(aA + I_\gamma)^{-1}$ is invertible. Having this in mind, one can deduce from (5.2) the following relations:

$$A^*HA + C^*JC - H = 2(I_\gamma - A_a^*)^{-1}(A_a^*H + HA_a + C_a^*JC_a)(I_\gamma - A_a)^{-1}$$

$$\begin{aligned} B^*HA + D^*JC &= \sqrt{2}(B_a^*H + D_a^*JC_a)(I_\gamma - A_a)^{-1} \\ &+ \sqrt{2}B_a^*(I_\gamma - A_a^*)^{-1}(A_a^*H + HA_a + C_a^*JC_a)(I_\gamma - A_a)^{-1} \end{aligned}$$

$$\begin{aligned} B^*HB &+ D^*JD - J \\ &= B_a^*(I_\gamma - A_a^*)^{-1}(A_a^*H + HA_a + C_a^*JC_a)(I_\gamma - A_a)^{-1}B_a \\ &+ (B_a^*H + D_a^*JC_a)(I_\gamma - A_a)^{-1}B_a + B_a^*(I_\gamma - A_a^*)^{-1}(C_a^*JD_a + HB_a). \end{aligned}$$

Thus, A, B, C, D, H satisfy (5.3) if and only if A_a, B_a, C_a, D_a, H satisfy (4.3) and (4.4) (in the place of A, B, C, D, H therein), which completes the proof. $\qquad \square$

We will call the invertible Hermitian solution $H = \operatorname{diag}(H_1, \ldots, H_N)$ of (5.3), which is determined uniquely by a minimal GR-realization α of a matrix-J-unitary on $\mathcal{T}_N$ rational FPS f, the *associated structured Hermitian matrix* (associated with a minimal GR-realization α of f). Let us note also that since for the GR-node β from Theorem 5.1 a pair of the equalities (4.3) and (4.4) is equivalent to a pair of the equalities (4.5) and (4.6), the equality (5.3) is equivalent to

$$\begin{pmatrix} A & B \\ C & D \end{pmatrix} \begin{pmatrix} H^{-1} & 0 \\ 0 & J \end{pmatrix} \begin{pmatrix} A & B \\ C & D \end{pmatrix}^* = \begin{pmatrix} H^{-1} & 0 \\ 0 & J \end{pmatrix}. \tag{5.6}$$

Remark 5.2. Equality (5.3) can be replaced by the following three equalities:

$$\begin{aligned} H - A^*HA &= C^*JC, & (5.7) \\ D^*JC &= -B^*HA, & (5.8) \\ J - D^*JD &= B^*HB, & (5.9) \end{aligned}$$

and equality (5.6) can be replaced by

$$H^{-1} - AH^{-1}A^* \;=\; BJB^*, \tag{5.10}$$

$$DJB^* \;=\; -CH^{-1}A^*, \tag{5.11}$$

$$J - DJD^* \;=\; CH^{-1}C^*. \tag{5.12}$$

Theorem 5.1 allows to obtain a counterpart of the results from Section 4 in the setting of rational FPSs which are matrix-J-unitary on $\mathcal{T}_N$. We will skip the proofs when it is clear how to get them.

Theorem 5.3. *Let f be a rational FPS and α be its minimal GR-realization of the form (3.11). Then f is matrix-J-unitary on $\mathcal{T}_N$ if and only if there exists an invertible Hermitian matrix $H = \mathrm{diag}(H_1, \ldots, H_N)$, with $H_k \in \mathbb{C}^{\gamma_k \times \gamma_k}$, $k = 1, \ldots, N$, which satisfies (5.3), or equivalently, (5.6).*

Remark 5.4. In the same way as in [7, Theorem 3.1] one can show that if a rational FPS f has a (not necessarily minimal) GR-realization (3.8) which satisfies (5.3) (resp., (5.6)), with an Hermitian invertible matrix $H = \mathrm{diag}(H_1, \ldots, H_N)$, then for any $n \in \mathbb{N}$,

$$
\begin{aligned}
f(Z')^*(J \otimes I_n)f(Z) \;=\;\; & J \otimes I_n - (B^* \otimes I_n)\left(I_\gamma \otimes I_n - \Delta(Z'^*)(A^* \otimes I_n)\right)^{-1} \\
\times\;\; & (H \otimes I_n)(I_\gamma \otimes I_n - \Delta(Z')^*\Delta(Z)) \\
\times\;\; & (I_\gamma \otimes I_n - (A \otimes I_n)\Delta(Z))^{-1}(B \otimes I_n)
\end{aligned}
\tag{5.13}
$$

and respectively,

$$
\begin{aligned}
f(Z)(J \otimes I_n)f(Z')^* \;=\;\; & J \otimes I_n - (C \otimes I_n)\left(I_\gamma \otimes I_n - \Delta(Z)(A \otimes I_n)\right)^{-1} \\
\times\;\; & (I_\gamma \otimes I_n - \Delta(Z)\Delta(Z')^*)(H^{-1} \otimes I_n) \\
\times\;\; & (I_\gamma \otimes I_n - (A^* \otimes I_n)\Delta(Z')^*)^{-1}(C^* \otimes I_n),
\end{aligned}
\tag{5.14}
$$

at all the points $Z, Z' \in (\mathbb{C}^{n \times n})^N$ where it is defined, which implies that f is matrix-J-unitary on $\mathcal{T}_N$. Moreover, the same statement holds true if $H = \mathrm{diag}(H_1, \ldots, H_N)$ in (5.3) and (5.13) is not supposed to be invertible, and if $H^{-1} = \mathrm{diag}(H_1^{-1}, \ldots, H_N^{-1})$ in (5.6) and (5.14) is replaced by any Hermitian, not necessarily invertible matrix $Y = \mathrm{diag}(Y_1, \ldots, Y_N)$.

Theorem 5.5. *Let f be a matrix-J-unitary on $\mathcal{T}_N$ rational FPS, and α be its GR-realization. Let $H = \mathrm{diag}(H_1, \ldots, H_N)$ with $H_k \in \mathbb{C}^{r_k \times r_k}$, $k = 1, \ldots, N$, be an Hermitian invertible matrix satisfying (5.3) or, equivalently, (5.6). Then α is observable if and only if α is controllable.*

Proof. Let $a \in \mathbb{T}$, $-\bar{a} \notin \sigma(A)$. Then F defined by (5.1) is a matrix-J-unitary on $\mathcal{J}_N$ rational FPS, and (5.2) is its GR-realization. As shown in the proof of Theorem 5.1, α is observable (resp., controllable) if and only if so is β. Since by Theorem 5.1 the GR-node β satisfies (4.3) and (4.4) (equivalently, (4.5) and (4.6)), Theorem 4.4 implies the statement of the present theorem. $\qquad\square$

Theorem 5.6. *Let f be a matrix-J-unitary on $\mathcal{T}_N$ rational FPS and α be its minimal GR-realization of the form (3.11), with the associated structured Hermitian matrix H. If $D = f_\emptyset$ is invertible then so is A, and*

$$A^{-1} = H^{-1}(A^\times)^* H. \tag{5.15}$$

Proof. It follows from (5.8) that $C = -JD^{-*}B^* HA$. Then (5.7) turns into

$$H - A^* HA = C^* J(-JD^{-*}B^* HA) = -C^* D^{-*}B^* HA,$$

which implies that $H = (A^\times)^* HA$, and (5.15) follows. $\qquad\square$

The following two lemmas, which are used in the sequel, can be found in [7].

Lemma 5.7. *Let $A \in \mathbb{C}^{r\times r}$, $C \in \mathbb{C}^{q\times r}$, where A is invertible. Let H be an invertible Hermitian matrix and J be a signature matrix such that*

$$H - A^* HA = C^* JC.$$

Let $a \in \mathbb{T}$, $a \notin \sigma(A)$. Define

$$\begin{aligned}
D_a &= I_q - CH^{-1}(I_r - aA^*)^{-1}C^* J, \tag{5.16}\\
B_a &= -H^{-1}A^{-*}C^* JD_a. \tag{5.17}
\end{aligned}$$

Then

$$\begin{pmatrix} A & B_a \\ C & D_a \end{pmatrix}^* \begin{pmatrix} H & 0 \\ 0 & J \end{pmatrix} \begin{pmatrix} A & B_a \\ C & D_a \end{pmatrix} = \begin{pmatrix} H & 0 \\ 0 & J \end{pmatrix}.$$

Lemma 5.8. *Let $A \in \mathbb{C}^{r\times r}$, $B \in \mathbb{C}^{r\times q}$, where A is invertible. Let H be an invertible Hermitian matrix and J be a signature matrix such that*

$$H^{-1} - AH^{-1}A^* = BJB^*.$$

Let $a \in \mathbb{T}$, $a \notin \sigma(A)$. Define

$$\begin{aligned}
D'_a &= I_q - JB^*(I_r - aA^*)^{-1}HB, \tag{5.18}\\
C'_a &= -D'_a JB^* A^{-*}H. \tag{5.19}
\end{aligned}$$

Then

$$\begin{pmatrix} A & B \\ C'_a & D'_a \end{pmatrix} \begin{pmatrix} H^{-1} & 0 \\ 0 & J \end{pmatrix} \begin{pmatrix} A & B \\ C'_a & D'_a \end{pmatrix}^* = \begin{pmatrix} H^{-1} & 0 \\ 0 & J \end{pmatrix}.$$

Theorem 5.9. *Let (C, A) be an observable pair of matrices $C \in \mathbb{C}^{q\times r}, A \in \mathbb{C}^{r\times r}$ in the sense that $\mathbb{C}^r = \bigoplus_{k=1}^{N} \mathbb{C}^{r_k}$ and $\mathcal{O}_k$ has full column rank for each $k \in \{1, \ldots, N\}$. Let A be invertible and $J \in \mathbb{C}^{q\times q}$ be a signature matrix. Then there exists a matrix-J-unitary on $\mathcal{T}_N$ rational FPS f with a minimal GR-realization $\alpha = (N; A, B, C, D; \mathbb{C}^r = \bigoplus_{k=1}^{N} \mathbb{C}^{r_k}, \mathbb{C}^q)$ if and only if the Stein equation (5.7) has a structured solution $H = \mathrm{diag}(H_1, \ldots, H_N)$ which is both Hermitian and invertible. If such a solution H exists, possible choices of D and B are D_a and B_a defined in (5.16) and (5.17), respectively. For a given such H, all other choices of D and B differ from D_a and B_a by a right multiplicative J-unitary constant matrix.*

Proof. Let $H = \operatorname{diag}(H_1, \ldots, H_N)$ be a structured solution of the Stein equation (5.7) which is both Hermitian and invertible, D_a and B_a are defined as in (5.16) and (5.17), respectively, where $a \in \mathbb{T}$, $a \notin \sigma(A)$. Set $\alpha_a = (N; A, B_a, C, D_a; \mathbb{C}^r = \bigoplus_{k=1}^N \mathbb{C}^{r_k}, \mathbb{C}^q)$. By Lemma 5.7 and due to Remark 5.4, the transfer function $T_{\alpha_a}^{\mathrm{nc}}$ of α_a is a matrix-J-unitary on $\mathcal{T}_N$ rational FPS. Since α_a is observable, by Theorem 5.5 α_a is controllable, and thus, minimal.

Conversely, if $\alpha = (N; A, B, C, D; \mathbb{C}^r = \bigoplus_{k=1}^N \mathbb{C}^{r_k}, \mathbb{C}^q)$ is a minimal GR-node whose transfer function is matrix-J-unitary on $\mathcal{T}_N$ then by Theorem 5.3 there exists a solution $H = \operatorname{diag}(H_1, \ldots, H_N)$ of the Stein equation (5.7) which is both Hermitian and invertible. The rest of the proof is analogous to the one of Theorem 4.2. $\qquad\square$

Analogously, one can obtain the following.

Theorem 5.10. *Let (A, B) be a controllable pair of matrices $A \in \mathbb{C}^{r \times r}$, $B \in \mathbb{C}^{r \times q}$ in the sense that $\mathbb{C}^r = \bigoplus_{k=1}^N \mathbb{C}^{r_k}$ and C_k has full row rank for each $k \in \{1, \ldots, N\}$. Let A be invertible and $J \in \mathbb{C}^{q \times q}$ be a signature matrix. Then there exists a matrix-J-unitary on $\mathcal{T}_N$ rational FPS f with a minimal GR-realization $\alpha = (N; A, B, C, D; \mathbb{C}^r = \bigoplus_{k=1}^N \mathbb{C}^{r_k}, \mathbb{C}^q)$ if and only if the Stein equation*

$$G - AGA^* = BJB^* \tag{5.20}$$

has a structured solution $G = \operatorname{diag}(G_1, \ldots, G_N)$ which is both Hermitian and invertible. If such a solution G exists, possible choices of D and C are D'_a and C'_a defined in (5.16) and (5.17), respectively, where $H = G^{-1}$. For a given such G, all other choices of D and C differ from D'_a and C'_a by a left multiplicative J-unitary constant matrix.

5.2. The associated structured Hermitian matrix

In this subsection we give the analogue of the results of Section 4.2. The proofs are similar and will be omitted.

Lemma 5.11. *Let f be a matrix-J-unitary on $\mathcal{T}_N$ rational FPS and $\alpha^{(i)} = (N; A^{(i)}, B^{(i)}, C^{(i)}, D; \mathbb{C}^\gamma = \bigoplus_{k=1}^N \mathbb{C}^{\gamma_k}, \mathbb{C}^q)$ be its minimal GR-realizations, with the associated structured Hermitian matrices $H^{(i)} = \operatorname{diag}(H_1^{(i)}, \ldots, H_N^{(i)})$, $i = 1, 2$. Then $\alpha^{(1)}$ and $\alpha^{(2)}$ are similar, that is*

$$C^{(1)} = C^{(2)}T, \quad TA^{(1)} = A^{(2)}T, \quad \text{and} \quad TB^{(1)} = B^{(2)},$$

for a uniquely defined invertible matrix $T = \operatorname{diag}(T_1, \ldots, T_N) \in \mathbb{C}^{\gamma \times \gamma}$ and

$$H_k^{(1)} = T_k^* H_k^{(2)} T_k, \qquad k = 1, \ldots, N.$$

In particular, the matrices $H_k^{(1)}$ and $H_k^{(2)}$ have the same signature.

Theorem 5.12. *Let f be a matrix-J-unitary on $\mathcal{T}_N$ rational FPS, and let α be its minimal GR-realization of the form (3.11), with the associated structured Hermitian matrix $H = \operatorname{diag}(H_1, \ldots, H_N)$. Then for each $k \in \{1, \ldots, N\}$ the number of* *negative eigenvalues of the matrix H_k is equal to the number of negative squares of each of the kernels (on $\mathcal{F}_N$):*

$$
\begin{aligned}
K^{f,k}_{w,w'} &= (C\flat A)^{wg_k} H_k^{-1} (A^*\sharp C^*)^{g_k w'^T}, \\
K^{f^*,k}_{w,w'} &= (B^*\flat A^*)^{wg_k} H_k (A\sharp B)^{g_k w'^T}.
\end{aligned}
\tag{5.21}
$$

Finally, for $k \in \{1,\ldots,N\}$ let $\mathcal{K}_k(f)$ (resp., $\mathcal{K}_k(f^*)$) be the span of the functions $w \mapsto K^{f,k}_{w,w'}c$ (resp., $w \mapsto K^{f^*,k}_{w,w'}c$) where $w' \in \mathcal{F}_N$ and $c \in \mathbb{C}^q$. Then

$$
\dim \mathcal{K}_k(f) = \dim \mathcal{K}_k(f^*) = \gamma_k.
$$

We will denote by $\nu_k(f)$ the number of negative squares of either of the functions defined in (5.21).

Theorem 5.13. *Let f_i, $i = 1,2$, be two matrix-J-unitary on $\mathcal{T}_N$ rational FPSs, with minimal GR-realizations*

$$
\alpha^{(i)} = \left(N; A^{(i)}, B^{(i)}, C^{(i)}, D; \mathbb{C}^{\gamma^{(i)}} = \bigoplus_{k=1}^{N} \mathbb{C}^{\gamma_k^{(i)}}, \mathbb{C}^q \right)
$$

and the associated structured Hermitian matrices $H^{(i)} = \mathrm{diag}(H_1^{(i)},\ldots,H_N^{(i)})$. Assume that the product $\alpha = \alpha^{(1)}\alpha^{(2)}$ is a minimal GR-node. Then, for each $k \in \{1,\ldots,N\}$ the matrix

$$
H_k = \begin{pmatrix} H_k^{(1)} & 0 \\ 0 & H_k^{(2)} \end{pmatrix} \in \mathbb{C}^{(\gamma_k^{(1)}+\gamma_k^{(2)})\times(\gamma_k^{(1)}+\gamma_k^{(2)})}
\tag{5.22}
$$

is the associated kth Hermitian matrix for $\alpha = \alpha^{(1)}\alpha^{(2)}$.

Corollary 5.14. *Let f_1 and f_2 be two matrix-J-unitary on $\mathcal{T}_N$ rational FPSs, and assume that the factorization $f = f_1 f_2$ is minimal. Then,*

$$
\nu(f_1 f_2) = \nu(f_1) + \nu(f_2).
$$

5.3. Minimal matrix-J-unitary factorizations

In this subsection we consider minimal factorizations of matrix-J-unitary on $\mathcal{T}_N$ rational FPSs into two factors, both of which are also matrix-J-unitary on $\mathcal{T}_N$ rational FPSs. Such factorizations will be called *minimal matrix-J-unitary factorizations*.

The following theorem is analogous to its one-variable counterpart [7, Theorem 3.7] and proved in the same way.

Theorem 5.15. *Let f be a matrix-J-unitary on $\mathcal{T}_N$ rational FPS and α be its minimal GR-realization of the form (3.11), with the associated structured Hermitian matrix $H = \mathrm{diag}(H_1,\ldots,H_N)$, and assume that D is invertible. Let $\mathcal{M} = \bigoplus_{k=1}^{N} \mathcal{M}_k$ be an A-invariant subspace of $\mathbb{C}^\gamma$, which is non-degenerate in the associated inner product $[\,\cdot\,,\,\cdot\,]_H$ and such that $M_k \subset \mathbb{C}^{\gamma_k}$, $k = 1,\ldots,N$. Let $\Pi = \mathrm{diag}(\Pi_1,\ldots,\Pi_N)$ be a projection defined by*

$$
\ker \Pi = M, \quad and \quad \mathrm{ran}\,\Pi = M^{[\perp]},
$$

that is

$$\ker \Pi_k = M_k, \quad \text{and} \quad \operatorname{ran}\Pi_k = M_k^{[\perp]} \quad for \quad k = 1,\ldots,N.$$

Then $f(z) = f_1(z)f_2(z)$, *where*

$$f_1(z) = \left[I_q + C(I_\gamma - \Delta(z)A)^{-1}\Delta(z)(I_\gamma - \Pi)BD^{-1}\right]D_1, \qquad (5.23)$$

$$f_2(z) = D_2\left[I_q + D^{-1}C\Pi(I_\gamma - \Delta(z)A)^{-1}\Delta(z)B\right], \qquad (5.24)$$

with

$$D_1 = I_q - CH^{-1}(I_\gamma - aA^*)^{-1}C^*J, \qquad D = D_1 D_2,$$

where $a \in \mathbb{T}$ *belongs to the resolvent set of* A_1, *and where*

$$C_1 = C\big|_{\mathcal{M}}, \quad A_1 = A\big|_{\mathcal{M}}, \quad H_1 = P_{\mathcal{M}}H\big|_{\mathcal{M}}$$

(with $P_{\mathcal{M}}$ *being the orthogonal projection onto* $\mathcal{M}$ *in the standard metric of* $\mathbb{C}^\gamma$*), is a minimal matrix-J-unitary factorization of* f.

Conversely, any minimal matrix-J-unitary factorization of f *can be obtained in such a way, and the correspondence between minimal matrix-J-unitary factorizations of* f *with* $f_1(\bar{a},\ldots,\bar{a}) = I_q$ *and non-degenerate subspaces of* A *of the form* $\mathcal{M} = \bigoplus_{k=1}^N \mathcal{M}_k$, *with* $\mathcal{M}_k \subset \mathbb{C}^{\gamma_k}$, $k = 1,\ldots,N$, *is one-to-one.*

Remark 5.16. In the proof of Theorem 5.15, as well as of Theorem 4.10, we make use of Theorem 3.9 and Corollary 3.10.

Remark 5.17. Minimal matrix-J-unitary factorizations do not always exist, even in the case $N = 1$. See [7] for examples in that case.

5.4. Matrix-unitary rational formal power series

In this subsection we specialize some of the results in the present section to the case $J = I_q$. We shall call corresponding rational FPSs *matrix-unitary on* $\mathcal{T}_N$.

Theorem 5.18. *Let* f *be a rational FPS and* α *be its minimal GR-realization of the form* (3.11). *Then* f *is matrix-unitary on* $\mathcal{T}_N$ *if and only if:*

(a) There exists an Hermitian matrix $H = \operatorname{diag}(H_1,\ldots,H_N)$ *(with* $H_k \in \mathbb{C}^{\gamma_k \times \gamma_k}$, $k = 1,\ldots,N$*) such that*

$$\begin{pmatrix} A & B \\ C & D \end{pmatrix}^* \begin{pmatrix} H & 0 \\ 0 & I_q \end{pmatrix} \begin{pmatrix} A & B \\ C & D \end{pmatrix} = \begin{pmatrix} H & 0 \\ 0 & I_q \end{pmatrix}. \qquad (5.25)$$

Condition (a) *is equivalent to:*

(a') There exists an Hermitian matrix $G = \operatorname{diag}(G_1,\ldots,G_N)$ *(with* $G_k \in \mathbb{C}^{\gamma_k \times \gamma_k}$, $k = 1,\ldots,N$*) such that*

$$\begin{pmatrix} A & B \\ C & D \end{pmatrix} \begin{pmatrix} G & 0 \\ 0 & I_q \end{pmatrix} \begin{pmatrix} A & B \\ C & D \end{pmatrix}^* = \begin{pmatrix} G & 0 \\ 0 & I_q \end{pmatrix}. \qquad (5.26)$$

Proof. The necessity follows from Theorem 5.1. To prove the sufficiency, suppose that the Hermitian matrix $H = \operatorname{diag}(H_1,\ldots,H_N)$ satisfies (5.25) and let $a \in \mathbb{T}$ be such that $-\bar{a} \notin \sigma(A)$. Then, H satisfies conditions (4.18) and (4.19) for the GR-node $\beta = (N; A_a, B_a, C_a, D_a; \mathbb{C}^\gamma = \bigoplus_{k=1}^N \mathbb{C}^{\gamma_k}, \mathbb{C}^q)$ defined by (5.2) (this follows from the proof of Theorem 5.1). Thus, from Theorem 4.13 and Theorem 5.1 we obtain that f is matrix-unitary on $\mathcal{T}_N$. Analogously, condition (a') implies that the FPS f is matrix-unitary on $\mathcal{T}_N$. $\square$

A counterpart of Theorem 4.14 in the present case is the following theorem:

Theorem 5.19. *Let (C, A) be an observable pair of matrices in the sense that $\mathcal{O}_k$ has full column rank for each $k = 1, \ldots, N$. Assume that $A \in \mathbb{C}^{r \times r}$ is invertible. Then there exists a matrix-unitary on $\mathcal{T}_N$ rational FPS f with a minimal GR-realization $\alpha = (N; A, B, C, D; \mathbb{C}^r = \bigoplus_{k=1}^{N} \mathbb{C}^{r_k}, \mathbb{C}^q)$ if and only if the Stein equation*

$$H - A^* H A = C^* C \tag{5.27}$$

has an Hermitian solution $H = \mathrm{diag}(H_1, \ldots, H_N)$, with $H_k \in \mathbb{C}^{r_k \times r_k}$, $k = 1, \ldots, N$. If such a matrix H exists, it is invertible, and possible choices of D and B are D_a and B_a given by (5.16) and (5.17) with $J = I_q$. Finally, for a given $H = \mathrm{diag}(H_1, \ldots, H_N)$, all other choices of D and B differ from D_a and B_a by a right multiplicative unitary constant.

A counterpart of Theorem 4.15 is the following theorem:

Theorem 5.20. *Let (A, B) be a controllable pair of matrices, in the sense that $\mathcal{C}_k$ has full row rank for each $k = 1, \ldots, N$. Assume that $A \in \mathbb{C}^{r \times r}$ is invertible. Then there exists a matrix-unitary on $\mathcal{T}_N$ rational FPS f with a minimal GR-realization $\alpha = (N; A, B, C, D; \mathbb{C}^r = \bigoplus_{k=1}^{N} \mathbb{C}^{r_k}, \mathbb{C}^q)$ if and only if the Stein equation*

$$G - A G A^* = B B^* \tag{5.28}$$

has an Hermitian solution $G = \mathrm{diag}(G_1, \ldots, G_N)$ with $G_k \in \mathbb{G}^{r_k \times r_k}$, $k = 1, \ldots, N$. If such a matrix G exists, it is invertible, and possible choices of D and C are D'_a and C'_a given by (5.18) and (5.19) with $H = G^{-1}$ and $J = I_q$. Finally, for a given $G = \mathrm{diag}(G_1, \ldots, G_N)$, all other choices of D and C differ from D'_a and C'_a by a left multiplicative unitary constant.

A counterpart of Theorem 4.16 in the present case is the following:

Theorem 5.21. *Let f be a matrix-unitary on $\mathcal{T}_N$ rational FPS and α be its minimal GR-realization of the form (3.11), with the associated structured Hermitian matrix $H = \mathrm{diag}(H_1, \ldots, H_N)$ and the associated kth inner products $[\cdot, \cdot]_{H_k}$, $k = 1, \ldots, N$. Let P_k denote the orthogonal projection in $\mathbb{C}^\gamma$ onto the subspace $\{0\} \oplus \cdots \oplus \{0\} \oplus \mathbb{C}^{\gamma_k} \oplus \{0\} \oplus \cdots \oplus \{0\}$, and set $\overline{A_k} = A P_k$ for $k = 1, \ldots, N$. If $x \in \mathbb{C}^\gamma$ is a common eigenvector for $\overline{A} = \{\overline{A_1}, \ldots, \overline{A_N}\}$ corresponding to a common eigenvalue $\lambda = (\lambda_1, \ldots, \lambda_N) \in \mathbb{C}^N$, then there exists $j \in \{1, \ldots, N\}$ such that $|\lambda_j| \neq 1$ and $[P_j x, P_j x]_{H_j} \neq 0$. In particular $\overline{A}$ has no common eigenvalues on $\mathbb{T}^N$.*

The proof of this theorem relies on the equality

$$(1 - |\lambda_k|^2)[P_k x, P_k x]_{H_k} = \langle C P_k x, C P_k x \rangle, \quad k = 1, \ldots, N,$$

and follows the same argument as the proof of Theorem 4.16.

6. Matrix-J-inner rational formal power series

6.1. A multivariable non-commutative analogue of the half-plane case

Let $n \in \mathbb{N}$. We define the *matrix open right poly-half-plane* as the set

$$\left(\Pi^{n \times n}\right)^N = \left\{ Z = (Z_1, \ldots, Z_N) \in \left(\mathbb{C}^{n \times n}\right)^N : Z_k + Z_k^* > 0, \ k = 1, \ldots, N \right\},$$

and the *matrix closed right poly-half-plane* as the set

$$\operatorname{clos}\left(\Pi^{n \times n}\right)^N = \left(\operatorname{clos} \Pi^{n \times n}\right)^N$$
$$= \left\{ Z = (Z_1, \ldots, Z_N) \in \left(\mathbb{C}^{n \times n}\right)^N : Z_k + Z_k^* \geq 0, \ k = 1, \ldots, N \right\}.$$

We also introduce

$$\mathcal{P}_N = \coprod_{n \in \mathbb{N}} \left(\Pi^{n \times n}\right)^N \quad \text{and} \quad \operatorname{clos} \mathcal{P}_N = \coprod_{n \in \mathbb{N}} \operatorname{clos}\left(\Pi^{n \times n}\right)^N.$$

It is clear that

$$\left(i\mathbb{H}^{n \times n}\right)^N \subset \operatorname{clos}\left(\Pi^{n \times n}\right)^N$$

is the *essential* (or *Shilov*) *boundary* of the matrix poly-half-plane $\left(\Pi^{n \times n}\right)^N$ (see [45]) and that $\mathcal{J}_N \subset \operatorname{clos} \mathcal{P}_N$ (recall that $\mathcal{J}_N = \coprod_{n \in \mathbb{N}} \left(i\mathbb{H}^{n \times n}\right)^N$).

Let $J = J^{-1} = J^* \in \mathbb{C}^{q \times q}$. A matrix-$J$-unitary on $\mathcal{J}_N$ rational FPS F is called *matrix-J-inner (in $\mathcal{P}_N$)* if for each $n \in \mathbb{N}$:

$$F(Z)(J \otimes I_n)F(Z)^* \leq J \otimes I_n \tag{6.1}$$

at those points $Z \in \operatorname{clos}\left(\Pi^{n \times n}\right)^N$ where it is defined (the set of such points is open and dense, in the relative topology, in $\operatorname{clos}\left(\Pi^{n \times n}\right)^N$ since $F(Z)$ is a rational matrix-valued function of the complex variables $(Z_k)_{ij}$, $k = 1, \ldots, N$, $i, j = 1, \ldots, n$).

The following theorem is a counterpart of part $a)$ of Theorem 2.16 of [7].

Theorem 6.1. *Let F be a matrix-J-unitary on $\mathcal{J}_N$ rational FPS and α be its minimal GR-realization of the form (3.11). Then F is matrix-J-inner in $\mathcal{P}_N$ if and only if the associated structured Hermitian matrix $H = \operatorname{diag}(H_1, \ldots, H_N)$ is strictly positive.*

Proof. Let $n \in \mathbb{N}$. Equality (4.9) can be rewritten as

$$J \otimes I_n - F(Z)(J \otimes I_n)F(Z')^* = \varphi(Z)\Delta(Z + Z'^*)(H^{-1} \otimes I_n)\varphi(Z')^* \tag{6.2}$$

where φ is a FPS defined by

$$\varphi(z) := C(I_\gamma - \Delta(z)A)^{-1} \in \mathbb{C}^{q \times \gamma} \langle\langle z_1, \ldots, z_N \rangle\rangle_{\text{rat}},$$

and (6.2) is well defined at all points $Z, Z' \in (\mathbb{C}^{n \times n})^N$ for which

$$1 \notin \sigma\left(\Delta(Z)(A \otimes I_n)\right), \quad 1 \notin \sigma\left(\Delta(Z')(A \otimes I_n)\right).$$

Set $\varphi_k(z) := C(I_\gamma - \Delta(z)A)^{-1}\big|_{\mathbb{C}^{\gamma_k}} \in \mathbb{C}^{q \times \gamma_k} \langle\langle z_1, \ldots, z_N\rangle\rangle_{\mathrm{rat}}$, $k = 1, \ldots, N$. Then (6.2) becomes:

$$J \otimes I_n - F(Z)(J \otimes I_n)F(Z')^* = \sum_{k=1}^{N} \varphi_k(Z)(H_k^{-1} \otimes (Z_k + Z_k'^{\,*}))\varphi_k(Z')^*. \qquad (6.3)$$

Let $X \in \mathbb{H}^{n \times n}$ be some positive semidefinite matrix, let $Y \in (\mathbb{H}^{n \times n})^N$ be such that $1 \notin \sigma(\Delta(iY)(A \otimes I_n))$, and set for $k = 1, \ldots, N$:

$$e_k := (0, 0, \ldots, 0, 1, 0, \ldots, 0) \in \mathbb{C}^N$$

with 1 at the kth place. Then for $\lambda \in \mathbb{C}$ set

$$Z_{X,Y}^{(k)}(\lambda) := \lambda X \otimes e_k + iY = (iY_1, \ldots, iY_{k-1}, \lambda X + iY_k, iY_{k+1}, \ldots, iY_N).$$

Now, (6.3) implies that

$$J \otimes I_n - F(Z_{X,Y}^{(k)}(\lambda))(J \otimes I_n)F(Z_{X,Y}^{(k)}(\lambda'))^*$$
$$= (\lambda + \overline{\lambda'})\varphi_k(Z_{X,Y}^{(k)}(\lambda))(H_k^{-1} \otimes X)\varphi_k(Z_{X,Y}^{(k)}(\lambda'))^*. \qquad (6.4)$$

The function $h(\lambda) = F(Z_{X,Y}^{(k)}(\lambda))$ is a rational function of $\lambda \in \mathbb{C}$. It is easily seen from (6.4) that h is $(J \otimes I_n)$-inner in the open right half-plane. In particular, it is $(J \otimes I_n)$-contractive in the closed right half-plane (this also follows directly from (6.1)). Therefore (see, e.g., [22]) the function

$$\Psi(\lambda, \lambda') = \frac{J \otimes I_n - F(Z_{X,Y}^{(k)}(\lambda))(J \otimes I_n)F(Z_{X,Y}^{(k)})(\lambda')^*}{\lambda + \overline{\lambda'}} \qquad (6.5)$$

is a positive semidefinite kernel on $\mathbb{C}$: for every choice of $r \in \mathbb{N}$, of points $\lambda_1, \ldots, \lambda_r \in \mathbb{C}$ for which the matrices $\Psi(\lambda_j, \lambda_i)$ are well defined, and vectors $c_1, \ldots, c_r \in \mathbb{C}^q \otimes \mathbb{C}^n$ one has

$$\sum_{i,j=1}^{r} c_j^* \Psi(\lambda_j, \lambda_i)c_i \geq 0,$$

i.e., the matrix $(\Psi(\lambda_j, \lambda_i))_{i,j=1,\ldots,r}$ is positive semidefinite. Since $\varphi_k(Z_{X,Y}^{(k)}(0)) = \varphi_k(iY)$ is well defined, we obtain from (6.4) that $\Psi(0,0)$ is also well defined and

$$\Psi(0,0) = \varphi_k(iY)(H_k^{-1} \otimes X)\varphi_k(iY)^* \geq 0.$$

This inequality holds for every $n \in \mathbb{N}$, every positive semidefinite $X \in \mathbb{H}^{n \times n}$ and every $Y \in (\mathbb{H}^{n \times n})^N$. Thus, for an arbitrary $r \in \mathbb{N}$ we can define $\tilde{n} = nr$, $\tilde{Y} = (\tilde{Y}_1, \ldots, \tilde{Y}_N) \in (\mathbb{H}^{\tilde{n} \times \tilde{n}})^N$, where $\tilde{Y}_k = \mathrm{diag}(Y_k^{(1)}, \ldots, Y_k^{(r)})$ and $Y_k^{(j)} \in \mathbb{H}^{n \times n}$, $k = 1, \ldots, N$, $j = 1, \ldots, r$, such that $\varphi_k(i\tilde{Y})$ is well defined,

$$\tilde{X} = \begin{pmatrix} I_n & \cdots & I_n \\ \vdots & & \vdots \\ I_n & \cdots & I_n \end{pmatrix} \in \mathbb{C}^{n \times n} \otimes \mathbb{C}^{r \times r} \cong \mathbb{C}^{\tilde{n} \times \tilde{n}}$$

and get

$$0 \le \varphi_k(i\widetilde{Y})(H_k^{-1} \otimes \widetilde{X})\varphi_k(i\widetilde{Y})^*$$

$$= \mathrm{diag}(\varphi_k(iY^{(1)}), \ldots, \varphi_k(iY^{(r)}))\times$$

$$\times \left(H_k^{-1} \otimes \begin{pmatrix} I_n \\ \vdots \\ I_n \end{pmatrix} (I_n \quad \cdots \quad I_n) \right) \mathrm{diag}(\varphi_k(iY^{(1)})^*, \ldots, \varphi_k(iY^{(r)})^*)$$

$$= \begin{pmatrix} \varphi_k(iY^{(1)}) \\ \vdots \\ \varphi_k(iY^{(r)}) \end{pmatrix} (H_k^{-1} \otimes I_n) \left(\varphi_k(iY^{(1)})^* \quad \cdots \quad \varphi_k(iY^{(r)})^* \right)$$

$$= \left(\varphi_k(iY^{(\mu)})(H_k^{-1} \otimes I_n)\varphi_k(iY^{(\nu)})^* \right)_{\mu,\nu=1,\ldots,r}.$$

Therefore, the function

$$K_k(iY, iY') = \varphi_k(iY)(H_k^{-1} \otimes I_n)\varphi_k(iY')^*$$

is a positive semidefinite kernel on any subset of $(i\mathbb{H}^{n\times n})^N$ where it is defined, and in particular in some neighborhood of the origin. One can extend this function to

$$K_k(Z, Z') = \varphi_k(Z)(H_k^{-1} \otimes I_n)\varphi_k(Z')^* \tag{6.6}$$

at those points $Z, Z' \in (\mathbb{C}^{n\times n})^N \times (\mathbb{C}^{n\times n})^N$ where φ_k is defined. Thus, on some neighborhood Γ of the origin in $(\mathbb{C}^{n\times n})^N \times (\mathbb{C}^{n\times n})^N$, the function $K_k(Z, Z')$ is holomorphic in Z and anti-holomorphic in Z'. On the other hand, it is well known (see, e.g., [9]) that one can construct a reproducing kernel Hilbert space (which we will denote by $\mathcal{H}(K_k)$) with reproducing kernel $K_k(iY, iY')$, which is obtained as the completion of

$$\mathcal{H}_0 = \mathrm{span}\left\{ K_k(\cdot, iY)x \,;\, iY \in (i\mathbb{H}^{n\times n})^N \cap \Gamma, \ x \in \mathbb{C}^q \otimes \mathbb{C}^n \right\}$$

with respect to the inner product

$$\left\langle \sum_{\mu=1}^{r} K_k(\cdot, iY^{(\mu)})x_\mu, \sum_{\nu=0}^{\ell} K_k(\cdot, iY^{(\nu)})x_\nu \right\rangle_{\mathcal{H}_0}$$

$$= \sum_{\mu=1}^{r}\sum_{\nu=1}^{\ell} \left\langle K_k(iY^{(\nu)}, iY^{(\mu)})x_\mu, x_\nu \right\rangle_{\mathbb{C}^q \otimes \mathbb{C}^n}.$$

The reproducing kernel property reads:

$$\langle f(\cdot), K_k(\cdot, iY)x \rangle_{\mathcal{H}(K_k)} = \langle f(iY), x \rangle_{\mathbb{C}^q \otimes \mathbb{C}^n},$$

and thus $K_k(iY, iY') = \Phi(iY)\Phi(iY')^*$ where

$$\Phi(iY) : f(\cdot) \mapsto f(iY)$$

is the evaluation map. In view of (6.6), the kernel $K_k(\cdot, \cdot)$ is extendable on $\Gamma \times \Gamma$ to the function $K(Z, Z')$ which is holomorphic in Z and antiholomorphic in Z', all the elements of $\mathcal{H}(K_k)$ have holomorphic continuations to Γ, and so has the function $\Phi(\cdot)$. Thus,

$$K_k(Z, Z') = \Phi(Z)\Phi(Z')^*$$

and so $K_k(Z, Z')$ is a positive semidefinite kernel on Γ. (We could also use [3, Theorem 1.1.4, p.10] to obtain this conclusion.) Therefore, for any choice of $\ell \in \mathbb{N}$ and $Z^{(1)}, \ldots, Z^{(\ell)} \in \Gamma$ the matrix

$$\left(\varphi_k(Z^{(\mu)})(H_k^{-1} \otimes I_n)\varphi_k(Z^{(\nu)})^* \right)_{\mu,\nu=1,\ldots,\ell}$$

$$= \begin{pmatrix} \varphi_k(Z^{(1)}) \\ \vdots \\ \varphi_k(Z^{(\ell)}) \end{pmatrix} \cdot (H_k^{-1} \otimes I_n) \cdot \left(\varphi_k(Z^{(1)})^* \quad \cdots \quad \varphi_k(Z^{(\ell)})^* \right) \tag{6.7}$$

is positive semidefinite. Since the coefficients of the FPS φ_k are $(\varphi_k)_w = (C\flat A)^{wg_k}$, $w \in \mathcal{F}_N$, and since α is an observable GR-node, we have

$$\bigcap_{w \in \mathcal{F}_N} \ker(C\flat A)^{wg_k} = \{0\}.$$

Hence, by Theorem 2.1 we can chose $n, \ell \in \mathbb{N}$ and $Z^{(1)}, \ldots, Z^{(\ell)} \in \Gamma$ such that

$$\bigcap_{j=1}^{\ell} \ker \varphi_k(Z^{(j)}) = \{0\}.$$

Thus the matrix $\mathrm{col}_{j=1,\ldots,\ell} \left(\varphi_k(Z^{(j)}) \right)$ has full column rank. (We could also use Theorem 3.7.) From (6.7) it then follows that $H_k^{-1} > 0$. Since this holds for all $k \in \{1, \ldots, N\}$, we get $H > 0$.

Conversely, if $H > 0$ then it follows from (6.2) that for every $n \in \mathbb{N}$ and $Z \in (\Pi^{n \times n})^N$ for which $1 \notin \sigma(\Delta(Z)(A \otimes I_n))$, one has

$$J \otimes I_n - F(Z)(J \otimes I_n)F(Z)^* \geq 0.$$

Therefore F is matrix-J-inner in $\mathcal{P}_N$, and the proof is complete. $\square$

Theorem 6.2. *Let $F \in \mathbb{C}^{q \times q} \langle\langle z_1, \ldots, z_N \rangle\rangle_{\mathrm{rat}}$ be matrix-J-inner in $\mathcal{P}_N$. Then F has a minimal GR-realization of the form* (3.11) *with the associated structured Hermitian matrix $H = I_\gamma$. This realization is unique up to a unitary similarity.*

Proof. Let

$$\alpha^\circ = (N; A^\circ, B^\circ, C^\circ, D; \mathbb{C}^\gamma = \bigoplus_{k=1}^{N} \mathbb{C}^{\gamma_k}, \mathbb{C}^q)$$

be a minimal GR-realization of F, with the associated structured Hermitian matrix $H^\circ = \mathrm{diag}(H_1^\circ, \ldots, H_N^\circ)$. By Theorem 6.1 the matrix H° is strictly positive. Therefore, $(H^\circ)^{1/2} = \mathrm{diag}((H_1^\circ)^{1/2}, \ldots, (H_N^\circ)^{1/2})$ is well defined and strictly positive, and

$$\alpha = (N; A, B, C, D; \mathbb{C}^\gamma = \bigoplus_{k=1}^{N} \mathbb{C}^{\gamma_k}, \mathbb{C}^q),$$

where

$$A = (H^\circ)^{1/2} A^\circ (H^\circ)^{-1/2}, \quad B = (H^\circ)^{1/2} B^\circ, \quad C = C^\circ (H^\circ)^{-1/2}, \qquad (6.8)$$

is a minimal GR-realization of F satisfying

$$A^* + A \;=\; -C^* J C, \qquad (6.9)$$
$$B \;=\; -C^* J D, \qquad (6.10)$$

or equivalently,

$$A^* + A \;=\; -BJB^*, \qquad (6.11)$$
$$C \;=\; -DJB^*, \qquad (6.12)$$

and thus having the associated structured Hermitian matrix $H = I_\gamma$. Since in this case the inner product $[\,\cdot\,,\,\cdot\,]_H$ coincides with the standard inner product $\langle\,\cdot\,,\,\cdot\,\rangle$ of $\mathbb{C}^\gamma$, by Remark 4.6 this minimal GR-realization with the property $H = I_\gamma$ is unique up to unitary similarity. $\qquad\square$

We remark that a one-variable counterpart of the latter result is essentially contained in [20], [38] (see also [10, Section 4.2]).

6.2. A multivariable non-commutative analogue of the disk case

Let $n \in \mathbb{N}$. We define the *matrix open unit polydisk* as

$$\left(\mathbb{D}^{n\times n}\right)^N = \left\{ W = (W_1,\ldots,W_N) \in \left(\mathbb{C}^{n\times n}\right)^N \; : \; W_k W_k^* < I_n, \; k = 1,\ldots,N \right\},$$

and the *matrix closed unit polydisk* as

$$\operatorname{clos}\left(\mathbb{D}^{n\times n}\right)^N = \left(\operatorname{clos}\mathbb{D}^{n\times n}\right)^N$$
$$= \left\{ W = (W_1,\ldots,W_N) \in \left(\mathbb{C}^{n\times n}\right)^N \; : \; W_k W_k^* \le I_n, \; k = 1,\ldots,N \right\}.$$

The matrix unit torus $\left(\mathbb{T}^{n\times n}\right)^N$ is the essential (or Shilov) boundary of $\left(\mathbb{D}^{n\times n}\right)^N$ (see [45]). In our setting, the set

$$\mathcal{D}_N = \coprod_{n\in\mathbb{N}} \left(\mathbb{D}^{n\times n}\right)^N \quad \left(\text{resp.,} \quad \operatorname{clos}\mathcal{D}_N = \coprod_{n\in\mathbb{N}} \operatorname{clos}\left(\mathbb{D}^{n\times n}\right)^N\right)$$

is a multivariable non-commutative counterpart of the open (resp., closed) unit disk.

Let $J = J^{-1} = J^* \in \mathbb{C}^{q\times q}$. A rational FPS f which is matrix-J-unitary on $\mathcal{T}_N$ is called *matrix-J-inner in $\mathcal{D}_N$* if for every $n \in \mathbb{N}$:

$$f(W)(J \otimes I_n)f(W)^* \le J \otimes I_n \qquad (6.13)$$

at those points $W \in \operatorname{clos}\left(\mathbb{D}^{n\times n}\right)^N$ where it is defined. We note that the set of such points is open and dense (in the relative topology) in $\operatorname{clos}\left(\mathbb{D}^{n\times n}\right)^N$ since $f(W)$ is a rational matrix-valued function of the complex variables $(W_k)_{ij}$, $k = 1,\ldots,N$, $i,j = 1,\ldots,n$.

Theorem 6.3. *Let f be a rational FPS which is matrix-J-unitary on $\mathcal{T}_N$, and let α be its minimal GR-realization of the form (3.11). Then f is matrix-J-inner in $\mathcal{D}_N$ if and only if the associated structured Hermitian matrix $H = \mathrm{diag}(H_1, \ldots, H_N)$ is strictly positive.*

Proof. The statement of this theorem follows from Theorem 6.1 and Theorem 5.1, since the Cayley transform defined in Theorem 5.1 maps each open matrix unit polydisk $(\mathbb{D}^{n \times n})^N$ onto the open right matrix poly-half-plane $(\Pi^{n \times n})^N$, and the inequality (6.13) turns into (6.1) for the function F defined in (5.1). $\qquad\square$

The following theorem is an analogue of Theorem 6.2.

Theorem 6.4. *Let f be a rational FPS which is matrix-J-inner in $\mathcal{D}_N$. Then there exists its minimal GR-realization α of the form (3.11), with the associated structured Hermitian matrix $H = I_\gamma$. Such a realization is unique up to a unitary similarity.*

In the special case of Theorem 6.4 where $J = I_q$ the FPS f is called *matrix-inner*, and the GR-node α satisfies

$$\begin{pmatrix} A & B \\ C & D \end{pmatrix}^* \begin{pmatrix} A & B \\ C & D \end{pmatrix} = I_{\gamma+q},$$

i.e., α is a *unitary GR-node*, which has been considered first by J. Agler in [1]. In what follows we will show that Theorem 6.4 for $J = I_q$ is a special case of the theorem of J. A. Ball, G. Groenewald and T. Malakorn on unitary GR-realizations of FPSs from the non-commutative Schur–Agler class [12], which becomes in several aspects stronger in this special case.

Let $\mathcal{U}$ and $\mathcal{Y}$ be Hilbert spaces. Denote by $L(\mathcal{U}, \mathcal{Y})$ the Banach space of bounded linear operators from $\mathcal{U}$ into $\mathcal{Y}$. A GR-node in the general setting of Hilbert spaces is

$$\alpha = \left(N; A, B, C, D; \mathcal{X} = \bigoplus_{k=1}^{N} \mathcal{X}_k, \mathcal{U}, \mathcal{Y} \right),$$

i.e., a collection of Hilbert spaces $\mathcal{X}, \mathcal{X}_1, \ldots, \mathcal{X}_N, \mathcal{U}, \mathcal{Y}$ and operators $A \in L(\mathcal{X}) = L(\mathcal{X}, \mathcal{X})$, $B \in L(\mathcal{U}, \mathcal{X})$, $C \in L(\mathcal{X}, \mathcal{Y})$, and $D \in L(\mathcal{U}, \mathcal{Y})$. Such a GR-node α is called *unitary* if

$$\begin{pmatrix} A & B \\ C & D \end{pmatrix}^* \begin{pmatrix} A & B \\ C & D \end{pmatrix} = I_{\mathcal{X} \oplus \mathcal{U}}, \qquad \begin{pmatrix} A & B \\ C & D \end{pmatrix} \begin{pmatrix} A & B \\ C & D \end{pmatrix}^* = I_{\mathcal{X} \oplus \mathcal{Y}},$$

i.e., $\begin{pmatrix} A & B \\ C & D \end{pmatrix}$ is a unitary operator from $\mathcal{X} \oplus \mathcal{U}$ onto $\mathcal{X} \oplus \mathcal{Y}$. The *non-commutative transfer function of* α is

$$T_\alpha^{\mathrm{nc}}(z) = D + C(I - \Delta(z)A)^{-1}\Delta(z)B, \tag{6.14}$$

where the expression (6.14) is understood as a FPS from $L(\mathcal{U}, \mathcal{Y}) \langle\langle z_1, \ldots, z_N \rangle\rangle$ given by

$$T_\alpha^{\mathrm{nc}}(z) = D + \sum_{w \in \mathcal{F}_N \setminus \{\emptyset\}} (C\flat A\sharp B)^w \, z^w = D + \sum_{k=0}^{\infty} C \left(\Delta(z) A\right)^k \Delta(z) B. \qquad (6.15)$$

The non-commutative Schur–Agler class $\mathcal{SA}_N^{\mathrm{nc}}(\mathcal{U}, \mathcal{Y})$ *consists of all FPSs* $f \in L(\mathcal{U}, \mathcal{Y}) \langle\langle z_1, \ldots, z_N \rangle\rangle$ *such that for any separable Hilbert space* $\mathcal{K}$ *and any N-tuple* $\delta = (\delta_1, \ldots, \delta_N)$ *of strict contractions in* $\mathcal{K}$ *the limit in the operator norm topology*

$$f(\delta) = \lim_{m \to \infty} \sum_{w \in \mathcal{F}_N : \, |w| \leq m} f_w \otimes \delta^w$$

exists and defines a contractive operator $f(\delta) \in L(\mathcal{U} \otimes \mathcal{K}, \mathcal{Y} \otimes \mathcal{K})$. We note that the non-commutative Schur–Agler class was defined in [12] also for a more general class of operator N-tuples δ.

Consider another set of non-commuting indeterminates $z' = (z_1', \ldots, z_N')$. For $f(z) \in L(\mathcal{V}, \mathcal{Y}) \langle\langle z_1, \ldots, z_N \rangle\rangle$ and $f'(z') \in L(\mathcal{V}, \mathcal{U}) \langle\langle z_1', \ldots, z_N' \rangle\rangle$ we define a FPS

$$f(z)f'(z')^* \in L(\mathcal{U}, \mathcal{Y}) \langle\langle z_1, \ldots, z_N, z_1', \ldots, z_N' \rangle\rangle$$

by

$$f(z)f'(z')^* = \sum_{w, w' \in \mathcal{F}_N} f_w (f_{w'}')^* z^w z'^{w'^T}. \qquad (6.16)$$

In [12] the class $\mathcal{SA}_N^{\mathrm{nc}}(\mathcal{U}, \mathcal{Y})$ was characterized as follows:

Theorem 6.5. *Let* $f \in L(\mathcal{U}, \mathcal{Y}) \langle\langle z_1, \ldots, z_N \rangle\rangle$. *The following statements are equivalent:*

(1) $f \in \mathcal{SA}_N^{\mathrm{nc}}(\mathcal{U}, \mathcal{Y})$;

(2) *there exist auxiliary Hilbert spaces* $\mathcal{H}, \mathcal{H}_1, \ldots, \mathcal{H}_N$ *which are related by* $\mathcal{H} = \bigoplus_{k=1}^{N} \mathcal{H}_k$, *and a FPS* $\varphi \in L(\mathcal{H}, \mathcal{Y}) \langle\langle z_1, \ldots, z_N \rangle\rangle$ *such that*

$$I_{\mathcal{Y}} - f(z)f(z')^* = \varphi(z)(I_{\mathcal{H}} - \Delta(z)\Delta(z')^*)\varphi(z')^*; \qquad (6.17)$$

(3) *there exists a unitary GR-node* $\alpha = (N; A, B, C, D; \mathcal{X} = \bigoplus_{k=1}^{N} \mathcal{X}_k, \mathcal{U}, \mathcal{Y})$ *such that* $f = T_\alpha^{\mathrm{nc}}$.

We now give another characterization of the Schur–Agler class $\mathcal{SA}_N^{\mathrm{nc}}(\mathcal{U}, \mathcal{Y})$.

Theorem 6.6. *A FPS* f *belongs to* $\mathcal{SA}_N^{\mathrm{nc}}(\mathcal{U}, \mathcal{Y})$ *if and only if for every* $n \in \mathbb{N}$ *and* $W \in (\mathbb{D}^{n \times n})^N$ *the limit in the operator norm topology*

$$f(W) = \lim_{m \to \infty} \sum_{w \in \mathcal{F}_N : \, |w| \leq m} f_w \otimes W^w \qquad (6.18)$$

exists and $\|f(W)\| \leq 1$.

Proof. The necessity is clear. We prove the sufficiency. We set

$$f_k(z) = \sum_{w \in \mathcal{F}_N : \, |w| = k} f_w z^w, \qquad k = 0, 1, \ldots.$$

Then for every $n \in \mathbb{N}$ and $W \in (\mathbb{D}^{n \times n})^N$, (6.18) becomes

$$f(W) = \lim_{m \to \infty} \sum_{k=0}^{m} f_k(W), \tag{6.19}$$

where the limit is taken in the operator norm topology. Let $r \in (0,1)$ and choose $\tau > 0$ such that $r + \tau < 1$. Let $W \in (\mathbb{D}^{n \times n})^N$ be such that $\|W_j\| \leq r$, $j = 1, \ldots, N$. Then, for every $x \in \mathcal{U} \otimes \mathbb{C}^n$ the series

$$f\left(\frac{r+\tau}{r}\lambda W\right) x = \sum_{k=0}^{\infty} \lambda^k f_k\left(\frac{r+\tau}{r}W\right) x$$

converges uniformly in $\lambda \in \operatorname{clos}\mathbb{D}$ to a $\mathcal{Y} \otimes \mathbb{C}^n$-valued function holomorphic on $\operatorname{clos}\mathbb{D}$. Furthermore,

$$\left\| f_k\left(\frac{r+\tau}{r}W\right) x \right\| = \left\| \frac{1}{2\pi i} \int_{\mathbb{T}} f\left(\frac{r+\tau}{r}\lambda W\right) x \lambda^{-k-1} d\lambda \right\| \leq \|x\|,$$

and therefore

$$\|f_k(W)\| = \left\| f_k\left(\frac{r+\tau}{r}W\right)\left(\frac{r}{r+\tau}\right)^k \right\| \leq \left(\frac{r}{r+\tau}\right)^k. \tag{6.20}$$

Thus we have

$$\left\| f(W) - \sum_{k=0}^{m} f_k(W) \right\| \leq \sum_{k=m+1}^{\infty} \|f_k(W)\| \leq \sum_{k=m+1}^{\infty} \left(\frac{r}{r+\tau}\right)^k < \infty.$$

We observe that the limit in (6.19) is uniform in $n \in \mathbb{N}$ and $W \in (\mathbb{D}^{n \times n})^N$ such that $\|W_j\| \leq r$, $j = 1, \ldots, N$. Without loss of generality we may assume that in the definition of the Schur–Agler class the space $\mathcal{K}$ is taken to be the space ℓ_2 of square summable sequences $s = (s_j)_{j=1}^{\infty}$ of complex numbers indexed by $\mathbb{N}$: $\sum_{j=1}^{\infty} |s_j|^2 < \infty$. We denote by P_n the orthogonal projection from ℓ_2 onto the subspace of sequences for which $s_j = 0$ for $j > n$. This subspace is isomorphic to $\mathbb{C}^n$, and thus for every $\delta = (\delta_1, \ldots, \delta_N) \in L(\ell_2)^N$ such that $\|\delta_j\| \leq r$, $j = 1, \ldots, N$, we may use (6.20) and write

$$\|f_k(P_n \delta_1 P_n, \ldots, P_n \delta_N P_n)\| \leq \left(\frac{r}{r+\tau}\right)^k. \tag{6.21}$$

Since the sequence P_n converges to I_{ℓ_2} in the strong operator topology (see, e.g., [2]), and since strong limits of finite sums and products of operator sequences are equal to the corresponding sums and products of strong limits of these sequences, we obtain that

$$s - \lim_{n \to \infty} f_k(P_n \delta_1 P_n, \ldots, P_n \delta_N P_n) = f_k(\delta).$$

Thus from (6.21) we obtain

$$\|f_k(\delta)\| \leq \left(\frac{r}{r+\tau}\right)^k.$$

Therefore, the limit in the operator norm topology

$$f(\delta) = \lim_{m \to \infty} \sum_{k=0}^{m} f_k(\delta)$$

does exist, and

$$\|f(\delta)\| \leq \sum_{k=0}^{\infty} \|f_k(\delta)\| \leq \sum_{k=0}^{\infty} \left(\frac{r}{r+\tau}\right)^k < \infty.$$

Moreover, since the limit in (6.19) is uniform in $n \in \mathbb{N}$ and $W \in (\mathbb{D}^{n \times n})^N$ such that $\|W_j\| \leq r < 1$, $j = 1, \ldots, N$, the rearrangement of limits in the following chain of equalities is justified:

$$\lim_{n \to \infty} f(P_n \delta_1 P_n, \ldots, P_n \delta_N P_n)h = \lim_{n \to \infty} \lim_{m \to \infty} \sum_{k=0}^{m} f_k(P_n \delta_1 P_n, \ldots, P_n \delta_N P_n)h$$

$$= \lim_{m \to \infty} \lim_{n \to \infty} \sum_{k=0}^{m} f_k(P_n \delta_1 P_n, \ldots, P_n \delta_N P_n)h = \lim_{m \to \infty} \sum_{k=0}^{m} f_k(\delta)h = f(\delta)h$$

(here h is an arbitrary vector in $\mathcal{U} \otimes \ell_2$ and $\delta \in L(\ell_2)^N$ such that $\|\delta_j\| \leq r$, $j = 1, \ldots, N$). Thus for every $\delta \in L(\ell_2)^N$ such that $\|\delta_j\| < 1$, $j = 1, \ldots, N$, we obtain $\|f(\delta)\| \leq 1$, i.e., $f \in \mathcal{SA}_N^{\mathrm{nc}}(\mathcal{U}, \mathcal{Y})$. $\qquad \square$

Remark 6.7. One can see from the proof of Theorem 6.6 that for arbitrary $f \in \mathcal{SA}_N^{\mathrm{nc}}(\mathcal{U}, \mathcal{Y})$ and $r : 0 < r < 1$, the series

$$f(\delta) = \sum_{k=0}^{\infty} f_k(\delta)$$

converges uniformly and absolutely in $\delta \in L(\mathcal{K})^N$ such that $\|\delta_j\| \leq r$, $j = 1, \ldots, N$, where $\mathcal{K}$ is any separable Hilbert space.

Corollary 6.8. *A matrix-inner in $\mathcal{D}_N$ rational FPS f belongs to the class* $\mathcal{SA}_N^{\mathrm{nc}}(\mathbb{C}^q) = \mathcal{SA}_N^{\mathrm{nc}}(\mathbb{C}^q, \mathbb{C}^q)$.

Thus, for the case $J = I_q$, Theorem 6.4 establishes the existence of a unitary GR-realization for an arbitrary matrix-inner rational FPS, i.e., recovers Theorem 6.5 for the case of a *matrix-inner rational* FPS. However, it says even more than Theorem 6.5 in this case, namely that such a unitary realization can be found minimal, thus finite-dimensional, and that this minimal unitary realization is unique up to a unitary similarity. The representation (6.17) with the rational FPS $\varphi \in \mathbb{C}^{q \times \gamma} \langle\langle z_1, \ldots, z_N \rangle\rangle_{\mathrm{rat}}$ given by

$$\varphi(z) = C(I_\gamma - \Delta(z)A)^{-1}$$

is obtained from (5.14) by making use of Corollary 2.2.

7. Matrix-selfadjoint rational formal power series

7.1. A multivariable non-commutative analogue of the line case

A rational FPS $\Phi \in \mathbb{C}^{q \times q} \langle\langle z_1, \ldots, z_N \rangle\rangle_{\mathrm{rat}}$ will be called *matrix-selfadjoint on* $\mathcal{J}_N$ if for every $n \in \mathbb{N}$:

$$\Phi(Z) = \Phi(Z)^*$$

at all points $Z \in \left(i\mathbb{H}^{n \times n}\right)^N$ where it is defined.

The following theorem is a multivariable non-commutative counterpart of Theorem 4.1 from [7] which was originally proved in [28].

Theorem 7.1. *Let* $\Phi \in \mathbb{C}^{q \times q} \langle\langle z_1, \ldots, z_N \rangle\rangle_{\mathrm{rat}}$, *and let* α *be a minimal GR-realization of* Φ *of the form* (3.11). *Then* Φ *is matrix-selfadjoint on* $\mathcal{J}_N$ *if and only if the following conditions hold:*

(a) *the matrix* D *is Hermitian, that is,* $D = D^*$;

(b) *there exists an invertible Hermitian matrix* $H = \mathrm{diag}(H_1, \ldots, H_N)$ *with* $H_k \in \mathbb{C}^{\gamma_k \times \gamma_k}$, $k = 1, \ldots, N$, *and such that*

$$A^*H + HA \ = \ 0, \tag{7.1}$$
$$C \ = \ iB^*H. \tag{7.2}$$

Proof. We first observe that Φ is matrix-selfadjoint on $\mathcal{J}_N$ if and only if the FPS $F \in \mathbb{C}^{2q \times 2q} \langle\langle z_1, \ldots, z_N \rangle\rangle_{\mathrm{rat}}$ given by

$$F(z) = \begin{pmatrix} I_q & i\Phi(z) \\ 0 & I_q \end{pmatrix} \tag{7.3}$$

is matrix-J_1-unitary on $\mathcal{J}_N$, where

$$J_1 = \begin{pmatrix} 0 & I_q \\ I_q & 0 \end{pmatrix}. \tag{7.4}$$

Moreover, F admits the GR-realization

$$\beta = \left(N; A, \begin{pmatrix} 0 & B \end{pmatrix}, \begin{pmatrix} iC \\ 0 \end{pmatrix}, \begin{pmatrix} I_q & iD \\ 0 & I_q \end{pmatrix}; \mathbb{C}^\gamma = \bigoplus_{k=1}^{N} \mathbb{C}^{\gamma_k}, \mathbb{C}^{2q}\right).$$

This realization is minimal. Indeed, the kth truncated observability (resp., controllability) matrix of β is equal to

$$\widetilde{\mathscr{O}_k}(\beta) = \begin{pmatrix} i\widetilde{\mathscr{O}_k}(\alpha) \\ 0 \end{pmatrix} \tag{7.5}$$

and, resp.,

$$\widetilde{\mathscr{C}_k}(\beta) = \begin{pmatrix} 0 & \widetilde{\mathscr{C}_k}(\alpha) \end{pmatrix}, \tag{7.6}$$

and therefore has full column (resp., row) rank. Using Theorem 4.1 of the present paper we see that Φ is matrix-selfadjoint on $\mathcal{J}_N$ if and only if:

(1) the matrix $\begin{pmatrix} I_q & iD \\ 0 & I_q \end{pmatrix}$ is J_1-unitary;

(2) there exists an invertible Hermitian matrix $H = \mathrm{diag}(H_1,\ldots,H_N)$, with $H_k \in \mathbb{C}^{\gamma_k \times \gamma_k}$, $k = 1,\ldots,N$, such that

$$A^*H + HA = -\begin{pmatrix} iC \\ 0 \end{pmatrix}^* J_1 \begin{pmatrix} iC \\ 0 \end{pmatrix},$$

$$(0 \quad B) = -H^{-1}\begin{pmatrix} iC \\ 0 \end{pmatrix}^* J_1 \begin{pmatrix} I_q & iD \\ 0 & I_q \end{pmatrix}.$$

These conditions are in turn readily seen to be equivalent to conditions (a) and (b) in the statement of the theorem. $\qquad\square$

From Theorem 4.1 it follows that the matrix $H = \mathrm{diag}(H_1,\ldots,H_N)$ is uniquely determined by the given minimal GR-realization of Φ. In a similar way as in Section 4, it can be shown that H_k, $k = 1,\ldots,N$, are given by the formulas

$$H_k = -\left(\mathrm{col}_{w\in\mathbb{F}_N:\,|w|\leq q\gamma-1}(B^*\flat(-A^*))^{wg_k}\right)^+ \left(\mathrm{col}_{w\in\mathbb{F}_N:\,|w|\leq q\gamma-1}(C\flat A)^{wg_k}\right)$$

$$= \left(\mathrm{row}_{w\in\mathcal{F}_N:\,|w|\leq q\gamma-1}((-A^*)\sharp C^*)^{g_k w^T}\right)\left(\mathrm{row}_{w\in\mathcal{F}_N:\,|w|\leq q\gamma-1}(A\sharp B)^{g_k w^T}\right)^\dagger.$$

The matrix $H = \mathrm{diag}(H_1,\ldots,H_N)$ is called in this case *the associated structured Hermitian matrix* (associated with a minimal GR-realization of the FPS Φ).

It follows from (7.1) and (7.2) that for $n \in \mathbb{N}$ and $Z, Z' \in (i\mathbb{H}^{n\times n})^N$ we have:

$$\Phi(Z) - \Phi(Z')^* = i(C \otimes I_n)\left(I_\gamma \otimes I_n - \Delta(Z)(A \otimes I_n)\right)^{-1} \tag{7.7}$$
$$\times \Delta(Z + Z'^*)\left(H^{-1} \otimes I_n\right)\left(I_\gamma \otimes I_n - (A^* \otimes I_n)\Delta(Z'^*)\right)^{-1}(C^* \otimes I_n),$$

$$\Phi(Z) - \Phi(Z')^* = i(B^* \otimes I_n)\left(I_\gamma \otimes I_n - \Delta(Z'^*)(A^* \otimes I_n)\right)^{-1} \tag{7.8}$$
$$\times \Delta(Z + Z'^*)\left(H \otimes I_n\right)\left(I_\gamma \otimes I_n - (A \otimes I_n)\Delta(Z)\right)^{-1}(B \otimes I_n).$$

Note that if A, B and C are matrices which satisfy (7.1) and (7.2) for some (not necessarily invertible) Hermitian matrix H, and if D is Hermitian, then

$$\Phi(z) = D + C(I - \Delta(z)A)^{-1}\Delta(z)B$$

is a rational FPS which is matrix-selfadjoint on $\mathcal{J}_N$. This follows from the fact that (7.8) is still valid in this case (the corresponding GR-realization of Φ is, in general, not minimal).

If A, B and C satisfy the equalities

$$GA^* + AG = 0, \tag{7.9}$$

$$B = iGC^* \tag{7.10}$$

for some (not necessarily invertible) Hermitian matrix $G = \mathrm{diag}(G_1,\ldots,G_N)$ then (7.7) is valid with H^{-1} replaced by G (the diagonal structures of G, $\Delta(Z)$ and $\Delta(Z')$ are compatible), and hence Φ is matrix-selfadjoint on $\mathcal{J}_N$.

As in Section 4, we can solve inverse problems using Theorem 7.1. The proofs are easy and omitted.

Theorem 7.2. *Let (C, A) be an observable pair of matrices, in the sense that $\mathcal{O}_k$ has a full column rank for all $k \in \{1, \dots, N\}$. Then there exists a rational FPS which is matrix-selfadjoint on $\mathcal{J}_N$ with a minimal GR-realization α of the form (3.11) if and only if the equation*

$$A^* H + H A = 0$$

has a solution $H = \mathrm{diag}(H_1, \dots, H_N)$ (with $H_k \in \mathbb{C}^{\gamma_k \times \gamma_k}$, $k = 1, \dots, N$) which is both Hermitian and invertible. When such a solution exists, D can be any Hermitian matrix and $B = iH^{-1}C^*$.

Theorem 7.3. *Let (A, B) be a controllable pair of matrices, in the sense that $\mathcal{C}_k$ has a full row rank for all $k \in \{1, \dots, N\}$. Then there exists a rational FPS which is matrix-selfadjoint on $\mathcal{J}_N$ with a minimal GR-realization α of the form (3.11) if and only if the equation*

$$G A^* + A G = 0$$

has a solution $G = \mathrm{diag}(G_1, \dots, G_N)$ (with $G_k \in \mathbb{C}^{\gamma_k \times \gamma_k}$, $k = 1, \dots, N$) which is both Hermitian and invertible. When such a solution exists, D can be any Hermitian matrix and $C = iB^*G^{-1}$.

From (7.5) and (7.6) obtained in Theorem 7.1, and from Theorem 4.4 we obtain the following result:

Theorem 7.4. *Let Φ be a matrix-selfadjoint on $\mathcal{J}_N$ rational FPS with a GR-realization α of the form (3.8). Let $H = \mathrm{diag}(H_1, \dots, H_N)$ (with $H_k \in \mathbb{C}^{r_k \times r_k}$, $k = 1, \dots, N$) be both Hermitian and invertible and satisfy (7.1) and (7.2). Then the GR-node α is observable if and only if it is controllable.*

The following Lemma is an analogue of Lemma 4.5. It is easily proved by applying Lemma 4.5 to the matrix-J_1-unitary on $\mathcal{J}_N$ function F defined in (7.3).

Lemma 7.5. *Let $\Phi \in \mathbb{C}^{q \times q} \langle\langle z_1, \dots, z_N \rangle\rangle_{\mathrm{rat}}$ be matrix-selfadjoint on $\mathcal{J}_N$, and let $\alpha^{(i)} = (N; A^{(i)}, B^{(i)}, C^{(i)}, D; \mathbb{C}^\gamma = \bigoplus_{k=1}^{N} \mathbb{C}^{\gamma_k}, \mathbb{C}^q)$ be two minimal GR-realizations of Φ, with the associated structured Hermitian matrices $H^{(i)} = \mathrm{diag}(H_1^{(i)}, \dots, H_N^{(i)})$, $i = 1, 2$. Then these two realizations and associated matrices $H^{(i)}$ are linked by (2.8) and (4.14). In particular, for each $k \in \{1, \dots, N\}$ the matrices $H_k^{(1)}$ and $H_k^{(2)}$ have the same signature.*

For $n \in \mathbb{N}$, points $Z, Z' \in (\mathbb{C}^{n \times n})^N$ where $\Phi(Z)$ and $\Phi(Z')$ are well defined, F given by (7.3), and J_1 defined by (7.4) we have:

$$J_1 \otimes I_n - F(Z)(J_1 \otimes I_n)F(Z')^* = \begin{pmatrix} \frac{\Phi(Z) - \Phi(Z')^*}{i} & 0 \\ 0 & 0 \end{pmatrix} \tag{7.11}$$

and

$$J_1 \otimes I_n - F(Z')^*(J_1 \otimes I_n)F(Z) = \begin{pmatrix} 0 & 0 \\ 0 & \frac{\Phi(Z) - \Phi(Z')^*}{i} \end{pmatrix}. \tag{7.12}$$

Combining these equalities with (7.7) and (7.8) and using Corollary 2.2 we obtain the following analogue of Theorem 4.7.

Theorem 7.6. *Let Φ be a matrix-selfadjoint on $\mathcal{J}_N$ rational FPS, and let α be its minimal GR-realization of the form (3.11), with the associated structured Hermitian matrix $H = \mathrm{diag}(H_1, \ldots, H_N)$. Then for each $k \in \{1, \ldots, N\}$ the number of negative eigenvalues of the matrix H_k is equal to the number of negative squares of the kernels*

$$
\begin{aligned}
K^{\Phi,k}_{w,w'} &= (C\flat A)^{wg_k} H_k^{-1} (A^* \sharp C^*)^{g_k w'^T} \\
K^{\Phi^*,k}_{w,w'} &= (B^* \flat A^*)^{wg_k} H_k (A \sharp B)^{g_k w'^T},
\end{aligned}
\qquad w, w' \in \mathcal{F}_N.
\tag{7.13}
$$

Finally, for $k \in \{1, \ldots, N\}$, let $\mathcal{K}_k(\Phi)$ (resp., $\mathcal{K}_k(\Phi^*)$) denote the span of the functions $w \mapsto K^{\Phi,k}_{w,w'}$ (resp., $w \mapsto K^{\Phi^*,k}_{w,w'}$) where $w' \in \mathcal{F}_N$ and $c \in \mathbb{C}^q$. Then,

$$
\dim \mathcal{K}_k(\Phi) = \dim \mathcal{K}_k(\Phi^*) = \gamma_k.
$$

Let Φ_1 and Φ_2 be two FPSs from $\mathbb{C}^{q \times q} \langle\langle z_1, \ldots, z_N \rangle\rangle_{\mathrm{rat}}$. The *additive decomposition*

$$
\Phi = \Phi_1 + \Phi_2
$$

is called *minimal* if

$$
\gamma_k(\Phi) = \gamma_k(\Phi_1) + \gamma_k(\Phi_2), \quad k = 1, \ldots, N,
$$

where $\gamma_k(\Phi), \gamma_k(\Phi_1)$ and $\gamma_k(\Phi_2)$ denote the dimensions of the kth component of the state space of a minimal GR-realization of Φ, Φ_1 and Φ_2, respectively. The following theorem is an analogue of Theorem 4.8.

Theorem 7.7. *Let Φ_i, $i = 1, 2$, be matrix-selfadjoint on $\mathcal{J}_N$ rational FPSs, with minimal GR-realizations $\alpha^{(i)} = (N; A^{(i)}, B^{(i)}, C^{(i)}, D^{(i)}; \mathbb{C}^{\gamma^{(i)}} = \bigoplus_{k=1}^N \mathbb{C}^{\gamma_k^{(i)}}, \mathbb{C}^q)$ and the associated structured Hermitian matrices $H^{(i)} = \mathrm{diag}(H_1^{(i)}, \ldots, H_N^{(i)})$. Assume that the additive decomposition $\Phi = \Phi_1 + \Phi_2$ is minimal. Then the GR-node $\alpha = (N; A, B, C, D; \mathbb{C}^{\gamma} = \bigoplus_{k=1}^N \mathbb{C}^{\gamma_k}, \mathbb{C}^q)$ defined by*

$$
D = D^{(1)} + D^{(2)}, \qquad \gamma_k = \gamma_k^{(1)} + \gamma_k^{(2)}, \quad k = 1, \ldots, N,
$$

and with respect to the decomposition $\mathbb{C}^{\gamma} = \mathbb{C}^{\gamma^{(1)}} \oplus \mathbb{C}^{\gamma^{(2)}}$,

$$
A = \begin{pmatrix} A^{(1)} & 0 \\ 0 & A^{(2)} \end{pmatrix}, \quad
B = \begin{pmatrix} B^{(1)} \\ B^{(2)} \end{pmatrix}, \quad
C = \begin{pmatrix} C^{(1)} & C^{(2)} \end{pmatrix},
\tag{7.14}
$$

is a minimal GR-realization of Φ, with the associated structured Hermitian matrix $H = \mathrm{diag}(H_1, \ldots, H_N)$ such that for each $k \in \{1, \ldots, N\}$:

$$
H_k = \begin{pmatrix} H_k^{(1)} & 0 \\ 0 & H_k^{(2)} \end{pmatrix}.
$$

Let $\nu_k(\Phi)$ denote the number of negative squares of either of the functions defined in (7.13). In view of Theorem 7.6 and Theorem 7.1 these numbers are uniquely determined by Φ.

Corollary 7.8. *Let Φ_1 and Φ_2 be matrix-selfadjoint on $\mathcal{J}_N$ rational FPSs, and assume that the additive decomposition $\Phi = \Phi_1 + \Phi_2$ is minimal. Then*

$$\nu_k(\Phi) = \nu_k(\Phi_1) + \nu_k(\Phi_2), \quad k = 1, 2, \ldots, N.$$

An additive decomposition of a matrix-selfadjoint on $\mathcal{J}_N$ rational FPS Φ is called a *minimal matrix-selfadjoint decomposition* if it is minimal and both Φ_1 and Φ_2 are matrix-selfadjoint on $\mathcal{J}_N$ rational FPSs. The set of all minimal matrix-selfadjoint decompositions of a matrix-selfadjoint on $\mathcal{J}_N$ rational FPS is given by the following theorem, which is a multivariable non-commutative counterpart of [7, Theorem 4.6]. The proof uses Theorem 4.10 applied to the FPS F defined by (7.3), and follows the same argument as one in the proof of Theorem 4.6 in [7].

Theorem 7.9. *Let Φ be a matrix-selfadjoint on $\mathcal{J}_N$ rational FPS, with a minimal GR-realization α of the form (3.11) and the associated structured Hermitian matrix $H = \operatorname{diag}(H_1, \ldots, H_N)$. Let $\mathcal{M} = \bigoplus_{k=1}^{N} \mathcal{M}_k$ be an A-invariant subspace, with $\mathcal{M}_k \subset \mathbb{C}^{\gamma_k}$, $k = 1, \ldots, N$, and assume that $\mathcal{M}$ is non-degenerate in the associated inner product $[\,\cdot\,,\,\cdot\,]_H$. Let $\Pi = \operatorname{diag}(\Pi_1, \ldots, \Pi_N)$ be the projection defined by*

$$\ker \Pi = \mathcal{M}, \qquad \operatorname{ran} \Pi = \mathcal{M}^{[\perp]},$$

that is,

$$\ker \Pi_k = \mathcal{M}_k, \qquad \operatorname{ran} \Pi_k = \mathcal{M}_k^{[\perp]}, \quad k = 1, \ldots, N.$$

Let $D = D_1 + D_2$ be a decomposition of D into two Hermitian matrices. Then the decomposition $\Phi = \Phi_1 + \Phi_2$, where

$$\Phi_1(z) = D_1 + C(I_\gamma - \Delta(z)A)^{-1}\Delta(z)(I_\gamma - \Pi)B,$$

$$\Phi_2(z) = D_2 + C\Pi(I_\gamma - \Delta(z)A)^{-1}\Delta(z)B,$$

is a minimal matrix-selfadjoint decomposition of Φ.

Conversely, any minimal matrix-selfadjoint decomposition of Φ can be obtained in such a way, and with a fixed decomposition $D = D_1 + D_2$, the correspondence between minimal matrix-selfadjoint decompositions of Φ and non-degenerate A-invariant subspaces of the form $\mathcal{M} = \bigoplus_{k=1}^{N} \mathcal{M}_k$, where $\mathcal{M}_k \subset \mathbb{C}^{\gamma_k}$, $k = 1, \ldots, N$, is one-to-one.

Remark 7.10. Minimal matrix-selfadjoint decompositions do not always exist, even in the case $N = 1$. For counterexamples see [7].

7.2. A multivariable non-commutative analogue of the circle case

In this subsection we briefly review some analogues of the theorems presented in Section 7.1.

Theorem 7.11. *Let Ψ be a rational FPS and α be its minimal GR-realization of the form (3.11). Then Ψ is matrix-selfadjoint on $\mathcal{T}_N$ (that is, for all $n \in \mathbb{N}$ one has $\Psi(Z) = \Psi(Z)^*$ at all points $Z \in (\mathbb{T}^{n \times n})^N$ where Ψ is defined) if and only if there exists an invertible Hermitian matrix $H = \operatorname{diag}(H_1, \ldots, H_N)$, with $H_k \in \mathbb{C}^{\gamma_k \times \gamma_k}$, $k = 1, \ldots, N$, such that*

$$A^* H A = H, \quad D - D^* = iB^* H B, \quad C = iB^* H A. \tag{7.15}$$

Proof. Consider the FPS $f \in \mathbb{C}^{2q \times 2q} \langle\langle z_1, \ldots, z_N \rangle\rangle_{\mathrm{rat}}$ defined by

$$f(z) = \begin{pmatrix} I_q & i\Psi(z) \\ 0 & I_q \end{pmatrix}. \tag{7.16}$$

Using Theorem 5.3, we see that f is matrix-J_1-unitary on $\mathcal{T}_N$, with

$$J_1 = \begin{pmatrix} 0 & I_q \\ I_q & 0 \end{pmatrix}, \tag{7.17}$$

if and only if its GR-realization

$$\beta = \left(N; A, \begin{pmatrix} 0 & B \end{pmatrix}, \begin{pmatrix} iC \\ 0 \end{pmatrix}, \begin{pmatrix} I_q & iD \\ 0 & I_q \end{pmatrix}; \mathbb{C}^\gamma = \oplus_{j=1}^N \mathbb{C}^{\gamma_j}, \mathbb{C}^{2q} \right)$$

(which turns out to be minimal, as can be shown in the same way as in Theorem 7.1) satisfies the following condition: there exists an Hermitian invertible matrix $H = \mathrm{diag}(H_1, \ldots, H_N)$, with $H_k \in \mathbb{C}^{\gamma_k \times \gamma_k}$, $k = 1, \ldots, N$, such that

$$\begin{pmatrix} A & 0 & B \\ iC & I_q & iD \\ 0 & 0 & I_q \end{pmatrix}^* \begin{pmatrix} H & 0 & 0 \\ 0 & 0 & I_q \\ 0 & I_q & 0 \end{pmatrix} \begin{pmatrix} A & 0 & B \\ iC & I_q & iD \\ 0 & 0 & I_q \end{pmatrix} = \begin{pmatrix} H & 0 & 0 \\ 0 & 0 & I_q \\ 0 & I_q & 0 \end{pmatrix},$$

which is equivalent to the condition stated in the theorem. $\qquad\square$

For a given minimal GR-realization of Ψ the matrix H is unique, as follows from Theorem 5.1. It is called *the associated structured Hermitian matrix* of Ψ.

The set of all minimal matrix-selfadjoint additive decompositions of a given matrix-selfadjoint on $\mathcal{T}_N$ rational FPS is described by the following theorem, which is a multivariable non-commutative counterpart of [7, Theorem 5.2], and is proved by applying Theorem 5.15 to the matrix-J_1-unitary on $\mathcal{T}_N$ FPS f defined by (7.16), where J_1 is defined by (7.17). (We omit the proof.)

Theorem 7.12. *Let Ψ be a matrix-selfadjoint on $\mathcal{T}_N$ rational FPS and α be its minimal GR-realization of the form (3.11), with the associated structured Hermitian matrix $H = \mathrm{diag}(H_1, \ldots, H_N)$. Let $\mathcal{M} = \bigoplus_{k=1}^N \mathcal{M}_k$ be an A-invariant subspace, with $\mathcal{M}_k \subset \mathbb{C}^{\gamma_k}$, $k = 1, \ldots, N$, and assume that $\mathcal{M}$ is non-degenerate in the associated inner product $[\cdot, \cdot]_H$. Let $\Pi = \mathrm{diag}(\Pi_1, \ldots, \Pi_N)$ be the projection defined by*

$$\ker \Pi = \mathcal{M}, \qquad \mathrm{ran}\, \Pi = \mathcal{M}^{[\perp]},$$

that is,

$$\ker \Pi_k = \mathcal{M}_k, \qquad \mathrm{ran}\, \Pi_k = \mathcal{M}_k^{[\perp]}, \qquad k = 1, \ldots, N.$$

Then the decomposition $\Psi = \Psi_1 + \Psi_2$, where

$$\Psi_1(z) = D_1 + C(I_\gamma - \Delta(z)A)^{-1}\Delta(z)(I_\gamma - \Pi)B,$$

$$\Psi_2(z) = D_2 + C\Pi(I_\gamma - \Delta(z)A)^{-1}\Delta(z)B,$$

with $D_1 = \frac{i}{2}B_1^* H^{(1)} B_1 + S$, the matrix S being an arbitrary Hermitian matrix, and

$$B_1 = P_\mathcal{M} B, \qquad H^{(1)} = P_\mathcal{M} H \big|_\mathcal{M},$$

is a minimal matrix-selfadjoint additive decomposition of Ψ *(here* $P_{\mathcal{M}}$ *denotes the orthogonal projection onto* $\mathcal{M}$ *in the standard metric of* $\mathbb{C}^\gamma$ *).*

Conversely, any minimal matrix-selfadjoint additive decomposition of Ψ *is obtained in such a way, and for a fixed* S*, the correspondence between minimal matrix-selfadjoint additive decompositions of* Ψ *and non-degenerate* A*-invariant subspaces of the form* $\mathcal{M} = \bigoplus_{k=1}^{N} \mathcal{M}_k$*, where* $\mathcal{M}_k \subset \mathbb{C}^{\gamma_k}$*,* $k = 1, \ldots, N$*, is one-to-one.*

8. Finite-dimensional de Branges–Rovnyak spaces and backward shift realizations: The multivariable non-commutative setting

In this section we describe certain model realizations of matrix-J-unitary rational FPSs. We restrict ourselves to the case of FPSs which are matrix-J-unitary on $\mathcal{J}_N$. Analogous realizations can be constructed for rational FPSs which are matrix-J-unitary on $\mathcal{T}_N$ or matrix-selfadjoint either on $\mathcal{J}_N$ or $\mathcal{T}_N$.

8.1. Non-commutative formal reproducing kernel Pontryagin spaces

Let F be a matrix-J-unitary on $\mathcal{J}_N$ rational FPS and α be its minimal GR-realization of the form (3.11), with the associated structured Hermitian matrix $H = \mathrm{diag}(H_1, \ldots, H_N)$. Then by Theorem 4.7, for each $k \in \{1, \ldots, N\}$ the kernel (4.15) has the number $\nu_k(F)$ of negative eigenvalues equal to the number of negative squares of H_k. Lemma 4.5 implies that the kernel $K^{F,k}_{w,w'}$ from (4.15) does not depend on the choice of a minimal realization of F. Theorem 4.7 also asserts that the span of the functions

$$w \mapsto K^{F,k}_{w,w'}c, \quad \text{where} \quad w' \in \mathcal{F}_N \quad \text{and} \quad c \in \mathbb{C}^q,$$

is the space $\mathcal{K}_k(F)$ with $\dim \mathcal{K}_k(F) = \gamma_k$, $k = 1, \ldots, N$. One can introduce a new metric on each of the spaces $\mathcal{K}_k(F)$ as follows. First, define an Hermitian form $[\,\cdot\,,\,\cdot\,]_{F,k}$ by:

$$[K^{F,k}_{\cdot,w'}c', K^{F,k}_{\cdot,w}c]_{F,k} = c^* K^{F,k}_{w,w'}c'.$$

This form is easily seen to be well defined on the whole space $\mathcal{K}_k(F)$, that is, if f and h belong to $\mathcal{K}_k(F)$ and

$$f_w = \sum_j K^{F,k}_{w,w_j}c_j = \sum_\ell K^{F,k}_{w,w'_\ell}c'_\ell$$

and

$$h_w = \sum_s K^{F,k}_{w,v_s}d_s = \sum_t K^{F,k}_{w,v'_t}d'_t,$$

where all the sums are finite, then

$$[f,h]_{F,k} = \left[\sum_j K^{F,k}_{\cdot,w_j}c_j, \sum_s K^{F,k}_{\cdot,v_s}d_s\right]_{F,k} = \left[\sum_\ell K^{F,k}_{\cdot,w'_\ell}c'_\ell, \sum_t K^{F,k}_{\cdot,v'_t}d'_t\right]_{F,k}.$$

Thus, the space $\mathcal{K}_k(F)$ endowed with this new (indefinite) metric is a finite-dimensional reproducing kernel Pontryagin space (RKPS) of functions on $\mathcal{F}_N$ with the reproducing kernel $K^{F,k}_{w,w'}$. We refer to [46, 4, 3] for more information on the theory of reproducing kernel Pontryagin spaces. In a similar way, the space $\mathcal{K}(F) = \bigoplus_{k=1}^{N} \mathcal{K}_k(F)$ endowed with the indefinite inner product

$$[f, h]_F = \sum_{k=1}^{N} [f_k, h_k]_{F,k}.$$

where $f = \operatorname{col}(f_1, \ldots, f_N)$ and $h = \operatorname{col}(h_1, \ldots, h_N)$, becomes a reproducing kernel Pontryagin space with the reproducing kernel

$$K^F_{w,w'} = \operatorname{diag}(K^{F,1}_{w,w'}, \ldots, K^{F,N}_{w,w'}), \quad w, w' \in \mathcal{F}^N.$$

Rather than the kernels $K^{F,k}_{w,w'}$, $k = 1, \ldots N$, and $K^F_{w,w'}$ we prefer to use the FPS kernels

$$K^{F,k}(z, z') = \sum_{w,w' \in \mathcal{F}^N} K^{F,k}_{w,w'} z^w z'^{w'^T}, \quad k = 1, \ldots, N, \tag{8.1}$$

$$K^F(z, z') = \sum_{w,w' \in \mathcal{F}^N} K^F_{w,w'} z^w z'^{w'^T}, \tag{8.2}$$

and instead of the reproducing kernel Pontryagin spaces $\mathcal{K}_k(F)$ and $\mathcal{K}(F)$ we will use the notion of *non-commutative formal reproducing kernel Pontryagin spaces* (*NFRKPS* for short; we will use the same notations for these spaces) which we introduce below in a way analogous to the way J.A. Ball and V. Vinnikov introduce non-commutative formal reproducing kernel Hilbert spaces (NFRKHS for short) in [14].

Consider a FPS

$$K(z, z') = \sum_{w,w' \in \mathcal{F}_N} K_{w,w'} z^w z'^{w'^T} \in L(\mathcal{C}) \langle\langle z_1, \ldots, z_N, z'_1, \ldots, z'_N \rangle\rangle_{\mathrm{rat}},$$

where $\mathcal{C}$ is a Hilbert space. Suppose that

$$K(z', z) = K(z, z')^* = \sum_{w,w' \in \mathcal{F}_N} K^*_{w,w'} z'^{w'} z^{w^T}.$$

Then $K^*_{w,w'} = K_{w',w}$ for all $w, w' \in \mathcal{F}_N$. Let $\kappa \in \mathbb{N}$. We will say that the *FPS* $K(z, z')$ is a *kernel with κ negative squares* if $K_{w,w'}$ is a kernel on $\mathcal{F}_N$ with κ negative squares, i.e., for every integer ℓ and every choice of $w_1, \ldots, w_\ell \in \mathcal{F}_N$ and $c_1, \ldots, c_\ell \in \mathcal{C}$ the $\ell \times \ell$ Hermitian matrix with (i, j)th entry equal to $c_i^* K_{w_i, w_j} c_j$ has at most κ strictly negative eigenvalues, and exactly κ such eigenvalues for some choice of $\ell, w_1, \ldots, w_\ell, c_1, \ldots, c_\ell$.

Define on the space $\mathcal{G}$ of finite sums of FPSs of the form

$$K_{w'}(z)c = \sum_{w \in \mathcal{F}_N} K_{w,w'} z^w c,$$

where $w' \in \mathcal{F}_N$ and $c \in \mathcal{C}$, the inner product as follows:

$$\left[\sum_i K_{w_i}(z) c_i, \sum_j K_{w'_j}(z) c'_j \right]_{\mathcal{G}} = \sum_{i,j} \langle K_{w'_j, w_i} c_i, c'_j \rangle_{\mathcal{C}}.$$

It is easily seen to be well defined. The space $\mathcal{G}$ endowed with this inner product can be completed in a unique way to a Pontryagin space $\mathcal{P}(K)$ of FPSs, and in $\mathcal{P}(K)$ the reproducing kernel property is

$$[f, K_w(\cdot)c]_{\mathcal{P}(K)} = \langle f_w, c \rangle_{\mathcal{C}}. \tag{8.3}$$

See [4, Theorem 6.4] for more details on such completions.

Define the pairings $[\cdot, \cdot]_{\mathcal{P}(K) \times \mathcal{P}(K) \langle\langle z_1, \ldots, z_N \rangle\rangle}$ and $\langle \cdot, \cdot \rangle_{\mathcal{C}\langle\langle z_1, \ldots, z_N \rangle\rangle \times \mathcal{C}}$ as mappings $\mathcal{P}(K) \times \mathcal{P}(K) \langle\langle z_1, \ldots, z_N \rangle\rangle \to \mathbb{C} \langle\langle z_1, \ldots, z_N \rangle\rangle$ and $\mathcal{C} \langle\langle z_1, \ldots, z_N \rangle\rangle \times \mathcal{C} \to \mathbb{C} \langle\langle z_1, \ldots, z_N \rangle\rangle$ by

$$\left[f, \sum_{w \in \mathcal{F}_N} g_w z^w \right]_{\mathcal{P}(K) \times \mathcal{P}(K) \langle\langle z_1, \ldots, z_N \rangle\rangle} = \sum_{w \in \mathcal{F}_N} [f, g_w]_{\mathcal{P}(K)} \, z^{w^T},$$

$$\left\langle \sum_{w \in \mathcal{F}_N} f_w z^w, c \right\rangle_{\mathcal{C}\langle\langle z_1, \ldots, z_N \rangle\rangle \times \mathcal{C}} = \sum_{w \in \mathcal{F}_N} \langle f_w, c \rangle_{\mathcal{C}} \, z^w.$$

Then the reproducing kernel property (8.3) can be rewritten as

$$[f, K(\cdot, z)c]_{\mathcal{P}(K) \times \mathcal{P}(K) \langle\langle z_1, \ldots, z_N \rangle\rangle} = \langle f(z), c \rangle_{\mathcal{C}\langle\langle z_1, \ldots, z_N \rangle\rangle \times \mathcal{C}}. \tag{8.4}$$

The space $\mathcal{P}(K)$ endowed with the metric $[\cdot, \cdot]_{\mathcal{P}(K)}$ will be said to be a *NFRKPS* associated with the FPS kernel $K(z, z')$. It is clear that this space is isomorphic to the RKPS associated with the kernel $K_{w,w'}$ on $\mathcal{F}_N$, and this isomorphism is well defined by

$$K_{w'}(\cdot)c \mapsto K_{\cdot, w'}c, \quad w' \in \mathcal{F}_N, c \in \mathcal{C}.$$

Let us now come back to the kernels (8.1) and (8.2) (see also (4.15)). Clearly, they can be rewritten as

$$K^{F,k}(z, z') = \varphi_k(z) H_k^{-1} \varphi_k(z')^*, \quad k = 1, \ldots, N, \tag{8.5}$$

$$K^F(z, z') = \varphi(z) H^{-1} \varphi(z')^*, \tag{8.6}$$

where rational FPSs φ_k, $k = 1, \ldots, N$, and φ are determined by a given minimal GR-realization α of the FPS F as

$$\varphi(z) = C(I_\gamma - \Delta(z)A)^{-1},$$

$$\varphi_k(z) = \varphi(z)\big|_{\mathbb{C}^{\gamma_k}}, \quad k = 1, \ldots, N.$$

For a model minimal GR-realization of F, we will start, conversely, with establishing an explicit formula for the kernels (8.1) and (8.2) in terms of F and then define a minimal GR-realization via these kernels.

Suppose that for a fixed $k \in \{1, \ldots, N\}$, (8.5) holds with some rational FPS φ_k. Recall that

$$J - F(z)JF(z')^* = \sum_{k=1}^{N} \varphi_k(z)H_k^{-1}(z_k + (z_k')^*)\varphi_k(z')^* \qquad (8.7)$$

(note that $(z_k')^* = z_k'$). Then for any $n \in \mathbb{N}$ and $Z, Z' \in \mathbb{C}^{n \times n}$:

$$J \otimes I_n - F(Z)(J \otimes I_n)F(Z')^* = \sum_{k=1}^{N} \varphi_k(Z)(H_k^{-1} \otimes (Z_k + (Z_k')^*))\varphi_k(Z')^*. \qquad (8.8)$$

Therefore, for $\lambda \in \mathbb{C}$:

$$J \otimes I_{2n} - F(\Lambda_{Z,Z'}(\lambda))(J \otimes I_{2n})F(\operatorname{diag}(-Z^*, Z'))^*$$
$$= \quad \lambda \varphi_k(\Lambda_{Z,Z'}(\lambda)) \left\{ H_k^{-1} \otimes \begin{pmatrix} I_n & I_n \\ I_n & I_n \end{pmatrix} \right\} \varphi_k(\operatorname{diag}(-Z^*, Z'))^*, \qquad (8.9)$$

where

$$\Lambda_{Z,Z'}(\lambda) := \lambda \begin{pmatrix} I_n & I_n \\ I_n & I_n \end{pmatrix} \otimes e_k + \begin{pmatrix} Z & 0 \\ 0 & -Z'^* \end{pmatrix}$$
$$= \left(\begin{pmatrix} Z_1 & 0 \\ 0 & -(Z_1')^* \end{pmatrix}, \ldots, \begin{pmatrix} Z_{k-1} & 0 \\ 0 & -(Z_{k-1}')^* \end{pmatrix}, \begin{pmatrix} \lambda I_n + Z_k & \lambda I_n \\ \lambda I_n & \lambda I_n - (Z_k')^* \end{pmatrix}, \right.$$
$$\left. \begin{pmatrix} Z_{k+1} & 0 \\ 0 & -(Z_{k+1}')^* \end{pmatrix}, \ldots, \begin{pmatrix} Z_N & 0 \\ 0 & -(Z_N')^* \end{pmatrix} \right),$$

$$\operatorname{diag}(-Z^*, Z') := \left(\begin{pmatrix} -Z_1^* & 0 \\ 0 & Z_1' \end{pmatrix}, \ldots, \begin{pmatrix} -Z_N^* & 0 \\ 0 & Z_N' \end{pmatrix} \right),$$

and, in particular,

$$\Lambda_{Z,Z'}(0) = \operatorname{diag}(Z, -Z'^*).$$

For Z and Z' where both F and φ_k are holomorphic, $\varphi_k(\Lambda_{Z,Z'}(\lambda))$ is continuous in λ, and $F(\Lambda_{Z,Z'}(\lambda))$ is holomorphic in λ at $\lambda = 0$. Thus, dividing by λ the expressions in both sides of (8.9) and passing to the limit as $\lambda \to 0$, we get

$$-\frac{d}{d\lambda}\{F(\Lambda_{Z,Z'}(\lambda))\}\big|_{\lambda=0}(J \otimes I_{2n})F(\operatorname{diag}(-Z^*, Z'))^*$$
$$= \quad \varphi_k\left(\operatorname{diag}(Z, -Z'^*)\right) \left\{ H_k^{-1} \otimes \begin{pmatrix} I_n & I_n \\ I_n & I_n \end{pmatrix} \right\} \varphi_k(\operatorname{diag}(-Z^*, Z'))^*$$
$$= \quad \begin{pmatrix} \varphi_k(Z) \\ \varphi_k(-Z'^*) \end{pmatrix} (H_k^{-1} \otimes I_n) \begin{pmatrix} \varphi_k(-Z^*)^* & \varphi_k(Z')^* \end{pmatrix}.$$

Taking the $(1,2)$th entry of the 2×2 block matrices in this equality, we get:

$$K^{F,k}(Z, Z') = -\frac{d}{d\lambda}\{F(\Lambda_{Z,Z'}(\lambda))_{12}\}\big|_{\lambda=0}(J \otimes I_n)F(Z')^*. \qquad (8.10)$$

Using the FPS representation for F we obtain from (8.10) the representation

$$K^{F,k}(Z,Z') = \sum_{w,w' \in \mathcal{F}_N} \left(\sum_{v,v' \in \mathcal{F}_N:\, vv'=w'} (-1)^{|v'|+1} F_{wg_k v'^T} J F_v \right) \otimes Z^w \left(Z'^*\right)^{w'^T}.$$

From Corollary 2.2 we get the expression for a FPS $K^{F,k}(z,z')$, namely:

$$K^{F,k}(z,z') = \sum_{w,w' \in \mathcal{F}_N} \left(\sum_{v,v' \in \mathcal{F}_N:\, vv'=w'} (-1)^{|v'|+1} F_{wg_k v'^T} J F_v \right) z^w z'^{w'^T}. \qquad (8.11)$$

Using formal differentiation with respect to λ we can also represent this kernel as

$$K^{F,k}(z,z') = -\frac{d}{d\lambda}\left\{ F\left(\Lambda_{z,z'}(\lambda)\right)_{12} \right\}\Big|_{\lambda=0} J F(z')^*. \qquad (8.12)$$

We note that one gets (8.11) and (8.12) from (8.7) using the same argument applied to FPSs.

Let us now consider the NFRKPSs $\mathcal{K}_k(F)$, $k = 1, \ldots, N$, and $\mathcal{K}(F) = \bigoplus_{k=1}^{N} \mathcal{K}_k(F)$. They are finite-dimensional and isomorphic to the reproducing kernel Pontryagin spaces on $\mathcal{F}_N$ which were denoted above with the same notation. Thus

$$\begin{aligned} \dim \mathcal{K}_k(F) &= \gamma_k, \qquad k = 1, \ldots, N, \\ \dim \mathcal{K}(F) &= \gamma. \end{aligned} \qquad (8.13)$$

The space $\mathcal{K}(F)$ is a multivariable non-commutative analogue of a certain de Branges–Rovnyak space (see [19, p. 24], [4, Section 6.3], and [7, p. 217]).

8.2. Minimal realizations in non-commutative de Branges–Rovnyak spaces

Let us define for every $k \in \{1, \ldots, N\}$ the backward shift operator

$$R_k : \; \mathbb{C}^q \left\langle\!\left\langle z_1, \ldots, z_N \right\rangle\!\right\rangle_{\mathrm{rat}} \longrightarrow \mathbb{C}^q \left\langle\!\left\langle z_1, \ldots, z_N \right\rangle\!\right\rangle_{\mathrm{rat}}$$

by

$$R_k : \; \sum_{w \in \mathcal{F}_N} f_w z^w \longmapsto \sum_{w \in \mathcal{F}_N} f_{wg_k} z^w.$$

(Compare with the one-variable backward shift operator R_0 considered in Section 1.)

Lemma 8.1. *Let F be a matrix-J-unitary on $\mathcal{J}_N$ rational FPS. Then for every $k \in \{1, \ldots, N\}$ the following is true:*

1. *$R_k F(z)c \in \mathcal{K}_k(F)$ for every $c \in \mathbb{C}^q$;*
2. *$R_k \mathcal{K}_j(F) \subset \mathcal{K}_k(F)$ for every $j \in \{1, \ldots, N\}$.*

Proof. From (8.7) and the J-unitarity of $F_\emptyset$ we get

$$
J - F(z)JF_\emptyset^* = (F_\emptyset - F(z))JF_\emptyset^* = -\sum_{k=1}^{N} R_k F(z) z_k J F_\emptyset^*
$$

$$
= \sum_{k=1}^{N} \varphi_k(z) H_k^{-1} z_k \, (\varphi_k)_\emptyset^* ,
$$

and therefore for every $k \in \{1, \ldots, N\}$ and every $c \in \mathbb{C}^q$ we get

$$
R_k F(z)c = -\varphi_k(z) H_k^{-1} (\varphi_k)_\emptyset^* J F_\emptyset c = K_\emptyset^{F,k}(z)\,(-JF_\emptyset c) \in \mathcal{K}_k(F).
$$

Thus, the first statement of this Lemma is true. To prove the second statement we start again from (8.7) and get for a fixed $j \in \{1, \ldots, N\}$ and $w \in \mathcal{F}_N$:

$$
-F(z)JF_{wg_j}^* = \varphi_j(z) H_j^{-1} (\varphi_j)_w^* + \sum_{k=1}^{N} \varphi_k(z) H_k^{-1} z_k (\varphi_k)_{wg_j}^* ,
$$

and therefore for any $c \in \mathbb{C}^q$:

$$
-\sum_{k=1}^{N} \left(R_k F(z) J F_{wg_j}^* c \right) z_k = \sum_{k=1}^{N} \left(R_k K_w^{F,j}(z)c \right) z_k + \sum_{k=1}^{N} \left(K_{wg_j}^{F,k}(z)c \right) z_k.
$$

Hence, one has for every $k \in \{1, \ldots, N\}$:

$$
R_k K_w^{F,j}(z)c = -R_k F(z) J F_{wg_j}^* c - K_{wg_j}^{F,k}(z)c, \tag{8.14}
$$

and from the first statement of this Lemma we obtain that the right-hand side of this equality belongs to $\mathcal{K}_k(F)$. Thus, the second statement is true, too. $\qquad\square$

We now define operators $A_{kj} : \mathcal{K}_j(F) \to \mathcal{K}_k(F)$, $A : \mathcal{K}(F) \to \mathcal{K}(F)$, $B : \mathbb{C}^q \to \mathcal{K}(F)$, $C : \mathcal{K}(F) \to \mathbb{C}^q$, $D : \mathbb{C}^q \to \mathbb{C}^q$ by

$$
A_{kj} = R_k\big|_{\mathcal{K}_j(F)}, \quad k, j = 1, \ldots, N, \tag{8.15}
$$

$$
A = (A_{kj})_{k,j=1,\ldots,N}, \tag{8.16}
$$

$$
B : c \longmapsto \begin{pmatrix} R_1 F(z)c \\ \vdots \\ R_N F(z)c \end{pmatrix}, \tag{8.17}
$$

$$
C : \begin{pmatrix} f_1(z) \\ \vdots \\ f_N(z) \end{pmatrix} \longmapsto \sum_{k=1}^{N} (f_k)_\emptyset, \tag{8.18}
$$

$$
D = F_\emptyset. \tag{8.19}
$$

These definitions make sense in view of Lemma 8.1.

Theorem 8.2. *Let F be a matrix-J-unitary on $\mathcal{J}_N$ rational FPS. Then the GR-node $\alpha = (N; A, B, C, D; \mathcal{K}(F) = \bigoplus_{k=1}^{N} \mathcal{K}_k(F), \mathbb{C}^q)$, with operators defined by (8.15)–(8.19), is a minimal GR-realization of F.*

Proof. We first check that for every $w \in \mathcal{F}_N : w \neq \emptyset$ we have

$$F_w = (C\flat A\sharp B)^w. \tag{8.20}$$

Let $w = g_k$ for some $k \in \{1, \ldots, N\}$. Then for $c \in \mathbb{C}^q$:

$$(C\flat A\sharp B)^w c = C_k B_k c = (R_k F(z)c)_\emptyset = \left(\sum_{w \in \mathcal{F}_N} F_{wg_k} z^w c \right)_\emptyset = F_{g_k} c.$$

Assume now that $|w| > 1$, $w = g_{j_1} \ldots g_{j_{|w|}}$. Then for $c \in \mathbb{C}^q$:

$$\begin{aligned}
(C\flat A\sharp B)^w c &= C_{j_1} A_{j_1, j_2} \cdots A_{j_{|w|-1}, j_{|w|}} B_{j_{|w|}} c \\
&= \left(R_{j_1} \cdots R_{j_{|w|}} F(z)c \right)_\emptyset \\
&= \left(\sum_{w' \in \mathcal{F}_N} F_{w' g_{j_1} \cdots g_{j_{|w|}}} z^{w'} c \right)_\emptyset \\
&= F_{g_{j_1} \cdots g_{j_{|w|}}} c \\
&= F_w c.
\end{aligned}$$

Since $F_\emptyset = D$, we obtain that

$$F(z) = D + C(I - \Delta(z)A)^{-1}\Delta(z)B,$$

that is, α is a GR-realization of F. The minimality of α follows from (8.13). $\square$

Let us now show how the associated structured Hermitian matrix $H = \mathrm{diag}(H_1, \ldots, H_N)$ arises from this special realization. Let

$$h = \mathrm{col}_{1 \leq j \leq N}(K_{w_j}^{F,j}(\cdot)c_j) \quad \text{and} \quad h' = \mathrm{col}_{1 \leq j \leq N}(K_{w_j'}^{F,j}(\cdot)c_j').$$

Using (8.14), we obtain

$$\begin{aligned}
[A_{kj}h_j, h_k']_{F,k} &+ [h_j, A_{jk}h_k']_{F,j} \\
&= [R_k K_{w_j}^{F,j}(\cdot)c_j, K_{w_k'}^{F,k}(\cdot)c_k']_{F,k} + [K_{w_j}^{F,j}(\cdot)c_j, R_j K_{w_k'}^{F,k}(\cdot)c_k']_{F,j} \\
&= (c_k')^* \left(K_{w_k' g_k, w_j}^{F,j} + K_{w_k', w_j g_j}^{F,k} \right) c_j. \tag{8.21}
\end{aligned}$$

Let $\overset{\circ}{\alpha} = (N; \overset{\circ}{A}, \overset{\circ}{B}, \overset{\circ}{C}, \overset{\circ}{D}; \mathbb{C}^\gamma = \bigoplus_{k=1}^N \mathbb{C}^{\gamma_k}, \mathbb{C}^q)$ be any minimal GR-realization of F, with the associated structured Hermitian matrix $\overset{\circ}{H} = \mathrm{diag}(\overset{\circ}{H}_1, \ldots, \overset{\circ}{H}_N)$. Then the right-hand side of (8.21) can be rewritten as

$$(c_k')^* \left(K^{F,j}_{w_k' g_k, w_j} + K^{F,k}_{w_k', w_j g_j} \right) c_j$$

$$= (c_k')^* \left(\left(\overset{\circ}{C} \flat \overset{\circ}{A} \right)^{w_k' g_k g_j} \left(\overset{\circ}{H_j} \right)^{-1} \left(\overset{\circ}{A^*} \sharp \overset{\circ}{C^*} \right)^{g_j w_j^T} \right.$$

$$\left. + \left(\overset{\circ}{C} \flat \overset{\circ}{A} \right)^{w_k' g_k} \left(\overset{\circ}{H_k} \right)^{-1} \left(\overset{\circ}{A^*} \sharp \overset{\circ}{C^*} \right)^{g_k g_j w_j^T} \right) c_j$$

$$= (c_k')^* \left(\overset{\circ}{C} \flat \overset{\circ}{A} \right)^{w_k' g_k} \left(\overset{\circ}{A_{kj}} \left(\overset{\circ}{H_j} \right)^{-1} + \left(\overset{\circ}{H_k} \right)^{-1} \left(\overset{\circ}{A_{kj}} \right)^* \right) \left(\overset{\circ}{A^*} \sharp \overset{\circ}{C^*} \right)^{g_j w_j^T} c_j$$

$$= -(c_k')^* \left(\overset{\circ}{C} \flat \overset{\circ}{A} \right)^{w_k' g_k} \overset{\circ}{B_k} J \left(\overset{\circ}{B_j} \right)^* \left(\overset{\circ}{A^*} \sharp \overset{\circ}{C^*} \right)^{g_j w_j^T} c_j$$

$$= -(c_k')^* \left(\overset{\circ}{C} \flat \overset{\circ}{A} \right)^{w_k' g_k} \left(\overset{\circ}{H_k} \right)^{-1} \left(\overset{\circ}{C_k} \right)^* J \overset{\circ}{C_j} \left(\overset{\circ}{H_j} \right)^{-1} \left(\overset{\circ}{A^*} \sharp \overset{\circ}{C^*} \right)^{g_j w_j^T} c_j$$

$$= -(c_k')^* K^{F,k}_{w_k', \emptyset} J K^{F,j}_{\emptyset, w_j} c_j$$

$$= -(c_k')^* \left(K^{F,k}_{\emptyset, w_k'} \right)^* J K^{F,j}_{\emptyset, w_j} c_j$$

$$= -(h_k')^*_\emptyset J (h_j)_\emptyset.$$

In this chain of equalities we have exploited the relationship between $\overset{\circ}{A}, \overset{\circ}{B}, \overset{\circ}{C}, \overset{\circ}{D}, J$ and $\overset{\circ}{H}$ from Theorem 4.1 applied to a GR-node $\overset{\circ}{\alpha}$. Thus we have for all $k, j \in \{1, \ldots, N\}$:

$$[A_{kj} h_j, h_k']_{F,k} + [h_j, A_{jk} h_k']_{F,j} = -(h_k')^* C_k^* J C_j h_j. \tag{8.22}$$

Since this equality holds for generating elements of the spaces $\mathcal{K}_k(F)$, $k = 1, \ldots, N$) it extends by linearity to arbitrary elements $h = \mathrm{col}(h_1, \ldots, h_N)$ and $h' = \mathrm{col}(h_1', \ldots, h_N')$ in $\mathcal{K}(F)$. For $k = 1, \ldots, N$, let $\langle \cdot, \cdot \rangle_{F,k}$ be any inner product for which $\mathcal{K}_k(F)$ is a Hilbert space. Thus, $\mathcal{K}(F)$ is a Hilbert space with respect to the inner product

$$\langle h, h' \rangle_F := \sum_{k=1}^{N} \langle h_k, h_k' \rangle_{F,k} .$$

Then there exist uniquely defined linear operators $H_k : \mathcal{K}_k(F) \to \mathcal{K}_k(F)$ such that:

$$[h_k, h_k']_{F,k} = \langle H_k h_k, h_k' \rangle_{F,k}, \quad k = 1, \ldots N,$$

and so with $H := \mathrm{diag}(H_1, \ldots, H_N) : \mathcal{K}(F) \to \mathcal{K}(F)$ we have:

$$[h, h']_F = \langle Hh, h' \rangle_F.$$

Since the spaces $\mathcal{K}_k(F)$ are non-degenerate (see [4]), the operators H_k are invertible and (8.22) can be rewritten as:

$$(A^*)_{kj}H_j + H_k A_{kj} = -C_k^* J C_j, \quad k,j = 1,\ldots N,$$

which is equivalent to (4.3).

Now, for arbitrary $c, c' \in \mathbb{C}^q$ and $w \in \mathcal{F}_N$ we have:

$$\langle H_k B_k c, K_{w'}^{F,k}(\cdot)c' \rangle_{F,k} = [R_k F(\cdot)c, K_{w'}^{F,k}(\cdot)c']_{F,k} = c'^* F_{w'g_k} c.$$

On the other hand,

$$- \langle C_k^* J D c, K_{w'}^{F,k}(\cdot)c' \rangle_{F,k} = -\langle J F_\emptyset c, C_k K_{w'}^{F,k}(\cdot)c' \rangle_{F,k} = -\langle J F_\emptyset c, K_{\emptyset,w'}^{F,k} c' \rangle_{\mathbb{C}^q}$$

$$= -c'^* K_{w',\emptyset}^{F,k} J F_\emptyset c = -c'^* (\overset{\circ}{C} \flat \overset{\circ}{A})^{w'g_k} \left(\overset{\circ}{H}_k\right)^{-1} \left(\overset{\circ}{C}_k\right)^* J \overset{\circ}{D} c$$

$$= c'^* \left(\overset{\circ}{C} \flat \overset{\circ}{A}\right)^{w'g_k} \overset{\circ}{B}_k c = c'^* \left(\overset{\circ}{C} \flat \overset{\circ}{A} \sharp \overset{\circ}{B}\right)^{w'g_k} c = c'^* F_{w'g_k} c.$$

Here we have used the relation (4.4) for an arbitrary minimal GR-realization $\overset{\circ}{\alpha} = (N; \overset{\circ}{A}, \overset{\circ}{B}, \overset{\circ}{C}, \overset{\circ}{D}; \mathbb{C}^\gamma = \bigoplus_{k=1}^N \mathbb{C}^{\gamma_k}, \mathbb{C}^q)$ of F, with the associated structured Hermitian matrix $\overset{\circ}{H} = \mathrm{diag}(\overset{\circ}{H}_1, \ldots, \overset{\circ}{H}_N)$. Thus, $H_k B_k = -C_k^* J D$, $k = 1, \ldots, N$, that is, $B = -H^{-1} C^* J D$, and (4.4) holds for the GR-node α. Finally, by Theorem 4.1, we may conclude that $H = \mathrm{diag}(H_1, \ldots, H_N)$ is the associated structured Hermitian matrix of the special GR-realization α.

8.3. Examples

In this subsection we give certain examples of matrix-inner rational FPSs on $\mathcal{J}_2$ with scalar coefficients (i.e., $N = 2$, $q = 1$, and $J = 1$). We also present the corresponding non-commutative positive kernels $K^{F,1}(z, z')$ and $K^{F,2}(z, z')$ computed using formula (8.12).

Example 1. $F(z) = (z_1 + 1)^{-1}(z_1 - 1)(z_2 + 1)^{-1}(z_2 - 1).$

$$K^{F,1}(z, z') = 2(z_1 + 1)^{-1}(z_1' + 1)^{-1},$$

$$K^{F,2}(z, z') = 2(z_1 + 1)^{-1}(z_1 - 1)(z_2 + 1)^{-1}(z_2' + 1)^{-1}(z_1' - 1)(z_1' + 1)^{-1}.$$

Example 2. $F(z) = (z_1 + z_2 + 1)^{-1}(z_1 + z_2 - 1).$

$$K^{F,1}(z, z') = K^{F,2}(z, z') = 2(z_1 + z_2 + 1)^{-1}(z_1' + z_2' + 1)^{-1}.$$

Example 3.

$$F(z) = \left(z_1 + (z_2 + i)^{-1} + 1\right)^{-1} \left(z_1 + (z_2 + i)^{-1} - 1\right)$$

$$= ((z_2 + i)(z_1 + 1) + 1)^{-1} ((z_2 + i)(z_1 - 1) + 1).$$

$$K^{F,1}(z, z') = 2 ((z_2 + i)(z_1 + 1) + 1)^{-1} (z_2 + i)(z_2' - i) ((z_1' + 1)(z_2' - i) + 1)^{-1},$$

$$K^{F,2}(z, z') = 2 ((z_2 + i)(z_1 + 1) + 1)^{-1} ((z_1' + 1)(z_2' - i) + 1)^{-1}.$$

References

[1] J. Agler, *On the representation of certain holomorphic functions defined on a polydisk*, Oper. Theory Adv. Appl., vol. 48, pp. 47–66, Birkhäuser Verlag, Basel, 1990.

[2] N.I. Akhiezer and I.M. Glazman, *Theory of linear operators in Hilbert space*, Dover Publications Inc., New York, 1993, Translated from the Russian and with a preface by Merlynd Nestell, Reprint of the 1961 and 1963 translations.

[3] D. Alpay, A. Dijksma, J. Rovnyak, and H. de Snoo, *Schur functions, operator colligations, and reproducing kernel Pontryagin spaces*, Oper. Theory Adv. Appl., vol. 96, Birkhäuser Verlag, Basel, 1997.

[4] D. Alpay and H. Dym, *On applications of reproducing kernel spaces to the Schur algorithm and rational J-unitary factorization*, I. Schur methods in operator theory and signal processing, Oper. Theory Adv. Appl., vol. 18, Birkhäuser, Basel, 1986, pp. 89–159.

[5] D. Alpay and H. Dym, *On a new class of realization formulas and their application*, Proceedings of the Fourth Conference of the International Linear Algebra Society (Rotterdam, 1994), vol. 241/243, 1996, pp. 3–84.

[6] D. Alpay and I. Gohberg, *On orthogonal matrix polynomials*, Orthogonal matrix-valued polynomials and applications (Tel Aviv, 1987–88), Oper. Theory Adv. Appl., vol. 34, Birkhäuser, Basel, 1988, pp. 25–46.

[7] D. Alpay and I. Gohberg, *Unitary rational matrix functions*, Topics in interpolation theory of rational matrix-valued functions, Oper. Theory Adv. Appl., vol. 33, Birkhäuser, Basel, 1988, pp. 175–222.

[8] D. Alpay and D.S. Kalyuzhnyĭ-Verbovetzkiĭ, *On the intersection of null spaces for matrix substitutions in a non-commutative rational formal power series*, C. R. Math. Acad. Sci. Paris **339** (2004), no. 8, 533–538.

[9] N. Aronszajn, *Theory of reproducing kernels*, Trans. Amer. Math. Soc. **68** (1950), 337–404.

[10] D.Z. Arov, *Passive linear steady-state dynamical systems*, Sibirsk. Mat. Zh. **20** (1979), no. 2, 211–228, 457, (Russian).

[11] J.A. Ball, G. Groenewald, and T. Malakorn, *Structured noncommutative multidimensional linear systems*, Preprint.

[12] J.A. Ball, G. Groenewald, and T. Malakorn, *Conservative structured noncommutative multidimensional linear systems*, In this volume.

[13] J.A. Ball, G. Groenewald, and T. Malakorn, *Bounded Real Lemma for structured noncommutative multidimensional linear systems and robust control*, Preprint.

[14] J.A. Ball and V. Vinnikov, *Formal reproducing kernel Hilbert spaces: The commutative and noncommutative settings*, Reproducing kernel spaces and applications, Oper. Theory Adv. Appl., vol. 143, Birkhäuser, Basel, 2003, pp. 77–134.

[15] H. Bart, I. Gohberg, and M.A. Kaashoek, *Minimal factorization of matrix and operator functions*, Oper. Theory Adv. Appl., vol. 1, Birkhäuser Verlag, Basel, 1979.

[16] C. Beck, *On formal power series representations for uncertain systems*, IEEE Trans. Automat. Control **46** (2001), no. 2, 314–319.

[17] C.L. Beck and J. Doyle, *A necessary and sufficient minimality condition for uncertain systems*, IEEE Trans. Automat. Control **44** (1999), no. 10, 1802–1813.

[18] J. Berstel and C. Reutenauer, *Rational series and their languages*, EATCS Monographs on Theoretical Computer Science, vol. 12, Springer-Verlag, Berlin, 1988.

[19] L. de Branges and J. Rovnyak, *Square summable power series*, Holt, Rinehart and Winston, New York, 1966.

[20] M.S. Brodskiĭ, *Triangular and Jordan representations of linear operators*, American Mathematical Society, Providence, R.I., 1971, Translated from the Russian by J.M. Danskin, Translations of Mathematical Monographs, Vol. 32.

[21] J.F. Camino, J.W. Helton, R.E. Skelton, and J. Ye, *Matrix inequalities: a symbolic procedure to determine convexity automatically*, Integral Equations Operator Theory **46** (2003), no. 4, 399–454.

[22] H. Dym, *J contractive matrix functions, reproducing kernel Hilbert spaces and interpolation*, CBMS Regional Conference Series in Mathematics, vol. 71, Published for the Conference Board of the Mathematical Sciences, Washington, DC, 1989.

[23] A.V. Efimov and V.P. Potapov, *J-expanding matrix-valued functions, and their role in the analytic theory of electrical circuits*, Uspehi Mat. Nauk **28** (1973), no. 1(169), 65–130, (Russian).

[24] M. Fliess, *Matrices de Hankel*, J. Math. Pures Appl. (9) **53** (1974), 197–222.

[25] E. Fornasini and G. Marchesini, *On the problems of constructing minimal realizations for two-dimensional filters*, IEEE Trans. Pattern Analysis and Machine Intelligence **PAMI-2** (1980), no. 2, 172–176.

[26] D.D. Givone and R.P. Roesser, *Multidimensional linear iterative circuits-general properties*, IEEE Trans. Computers **C-21** (1972), 1067–1073.

[27] D.D. Givone and R.P. Roesser, *Minimization of multidimensional linear iterative circuits*, IEEE Trans. Computers **C-22** (1973), 673–678.

[28] I. Gohberg, P. Lancaster, and L. Rodman, *Matrices and indefinite scalar products*, Oper. Theory Adv. Appl., vol. 8, Birkhäuser Verlag, Basel, 1983.

[29] J.W. Helton, *"Positive" noncommutative polynomials are sums of squares*, Ann. of Math. (2) **156** (2002), no. 2, 675–694.

[30] J.W. Helton, *Manipulating matrix inequalities automatically*, Mathematical systems theory in biology, communications, computation, and finance (Notre Dame, IN, 2002), IMA Vol. Math. Appl., vol. 134, Springer, New York, 2003, pp. 237–256.

[31] J.W. Helton and S.A. McCullough, *A Positivstellensatz for non-commutative polynomials*, Trans. Amer. Math. Soc. **356** (2004), no. 9, 3721–3737 (electronic).

[32] J.W. Helton, S.A. McCullough, and M. Putinar, *A non-commutative Positivstellensatz on isometries*, J. Reine Angew. Math. **568** (2004), 71–80.

[33] D.S. Kalyuzhniy, *On the notions of dilation, controllability, observability, and minimality in the theory of dissipative scattering linear nD systems*, Proceedings of the International Symposium MTNS-2000 (A. El Jai and M. Fliess, Eds.), CD-ROM (Perpignan, France), 2000, http://www.univ-perp.fr/mtns2000/articles/I13_3.pdf.

[34] D.S. Kalyuzhnyĭ-Verbovetzkiĭ and V. Vinnikov, *Non-commutative positive kernels and their matrix evaluations*, Proc. Amer. Math. Soc., to appear.

[35] S.C. Kleene, *Representation of events in nerve nets and finite automata*, Automata studies, Annals of mathematics studies, no. 34, Princeton University Press, Princeton, N. J., 1956, pp. 3–41.

[36] I.V. Kovališina, and V.P. Potapov, *Multiplicative structure of analytic real J-dilative matrix-functions*, Izv. Akad. Nauk Armjan. SSR Ser. Fiz.-Mat. Nau **18** (1965), no. 6, 3–10, (Russian).

[37] M.G. Kreĭn and H. Langer, *Über die verallgemeinerten Resolventen und die charakteristische Funktion eines isometrischen Operators im Raume Π_κ*, Hilbert space operators and operator algebras (Proc. Internat. Conf., Tihany, 1970), North-Holland, Amsterdam, 1972, pp. 353–399. Colloq. Math. Soc. János Bolyai, 5.

[38] M.S. Livšic, *Operators, oscillations, waves (open systems)*, American Mathematical Society, Providence, R.I., 1973, Translated from the Russian by Scripta Technica, Ltd. English translation edited by R. Herden, Translations of Mathematical Monographs, Vol. 34.

[39] T. Malakorn, *Multidimensional linear systems and robust control*, Ph.D. thesis, Virginia Polytechnic Institute and State University, Blacksburg, Virginia, 2003.

[40] S. McCullough, *Factorization of operator-valued polynomials in several non-commuting variables*, Linear Algebra Appl. **326** (2001), no. 1-3, 193–203.

[41] A.C.M. Ran, *Minimal factorization of selfadjoint rational matrix functions*, Integral Equations Operator Theory **5** (1982), no. 6, 850–869.

[42] R.P. Roesser, *A discrete state-space model for linear image processing*, IEEE Trans. Automatic Control **AC–20** (1975), 1–10.

[43] L.A. Sakhnovich, *Factorization problems and operator identities*, Russian Mathematical Surveys **41** (1986), no. 1, 1–64.

[44] M.P. Schützenberger, *On the definition of a family of automata*, Information and Control **4** (1961), 245–270.

[45] B.V. Shabat, *Introduction to complex analysis. Part II*, Translations of Mathematical Monographs, vol. 110, American Mathematical Society, Providence, RI, 1992, Functions of several variables, Translated from the third (1985) Russian edition by J. S. Joel.

[46] P. Sorjonen, *Pontrjaginräume mit einem reproduzierenden Kern*, Ann. Acad. Sci. Fenn. Ser. A I Math. **594** (1975), 30.

[47] K. Zhou, J.C. Doyle, and K. Glover, *Robust and optimal control*, Prentice-Hall, Upper Saddle River, NJ, 1996.

D. Alpay
Department of Mathematics
Ben-Gurion University of the Negev
Beer-Sheva 84105, Israel
e-mail: `dany@math.bgu.ac.il`

D.S. Kalyuzhnyĭ-Verbovetzkiĭ
Department of Mathematics
Ben-Gurion University of the Negev
Beer-Sheva 84105, Israel
e-mail: `dmitryk@math.bgu.ac.il`

Operator Theory:
Advances and Applications, Vol. 161, 115–177

State/Signal Linear Time-Invariant Systems Theory, Part I: Discrete Time Systems

Damir Z. Arov and Olof J. Staffans

Abstract. This is the first paper in a series of several papers in which we develop a state/signal linear time-invariant systems theory. In this first part we shall present the general state/signal setting in discrete time. Our following papers will deal with conservative and passive state/signal systems in discrete time, the general state/signal setting in continuous time, and conservative and passive state/signal systems in continuous time, respectively. The state/signal theory that we develop differs from the standard input/state/output theory in the sense that we do not distinguish between input signals and output signals, only between the "internal" states x and the "external" signals w. In the development of the general state/signal systems theory we take both the state space $\mathcal{X}$ and the signal space $\mathcal{W}$ to be Hilbert spaces. In later papers where we discuss conservative and passive systems we assume that the signal space $\mathcal{W}$ has an additional Kreĭn space structure. The definition of a state/signal system has been designed in such a way that to any state/signal system there exists at least one decomposition of the signal space $\mathcal{W}$ as the direct sum $\mathcal{W} = \mathcal{Y} \dotplus \mathcal{U}$ such that the evolution of the system can be described by the standard input/state/output system of equations with input space $\mathcal{U}$ and output space $\mathcal{Y}$. (In a passive state/signal system we may take $\mathcal{U}$ and $\mathcal{Y}$ to be the positive and negative parts, respectively, of a fundamental decomposition of the Kreĭn space $\mathcal{W}$.) Thus, to each state/signal system corresponds infinitely many input/state/output systems constructed in the way described above. A state/signal system consists of a state/signal node and the set of trajectories generated by this node. A state/signal node is a triple $\Sigma = (V; \mathcal{X}, \mathcal{W})$, where V is a subspace with appropriate properties of the product space $\mathcal{X} \times \mathcal{X} \times \mathcal{W}$. In this first paper we extend standard input/state/output notions, such as existence and uniqueness of solutions, continuous dependence on initial data, observability, controllability, stabilizability, detectability, and minimality to the state/signal setting. Three classes of representations of state/signal systems are presented (one of which is the class of input/state/output representations), and the families of all the transfer functions of these representations are studied. We also discuss realizations of signal behaviors by state/signal systems, as well as dilations and compressions of these systems. (Duality will be discussed later in connection with passivity and conservativity.)

Mathematics Subject Classification (2000). Primary 47A48, 93A05; Secondary 94C05.

Keywords. State/signal node, driving variable, output nulling, input/state/output, linear fractional transformation, transfer function, behavior, external equivalence, realization, dilation, compression, outgoing invariant, strongly invariant, controllability, observability, minimality, stabilizability, detectability.

Contents

1. Introduction

The main motivation for this work comes from the notion of a multi-port network. Such a network consists of *internal branches*, where the evolution of the data is described by, e.g., systems of ordinary or partial differential equations involving *state variables* (lumped or distributed), and *external branches* (*ports*), where the evolution of the *port variables* is only partially restricted by the network equations. Typically one part of the port variables can be prescribed in an arbitrary way (this is the "input" part), after which the remaining "output" part of the port variables can be computed from the network equations. However, the splitting of the port variables into an input part and output part is not specified, and many different choices are possible.

To be a little more concrete, let us consider a two-port Kirchhoff network, i.e., a Kirchhoff network with two external branches. To each of these branches we associate at each time instant t a normalized voltage/current pair $(v_1(t), i_1(t))$, respectively, $(v_2(t), i_2(t))$ (normalization means that we divide each voltage by $\sqrt{R}$ and multiply each current by $\sqrt{R}$, where R is a fixed resistance). Thus, the complete set of port variables is the four-dimensional vector $w(t) = (v_1(t), i_1(t), v_2(t), i_2(t))$.

Sometimes we may use $u(t) = (v_1(t), i_1(t))$ as the input data, and regard $(v_2(t), i_2(t))$ as the output data (or the other way around). This case is called the *transmission* case, and it is used, e.g., in the cascade synthesis of two-ports. However, this choice of input and output data is not always possible or reasonable. Another possibility is to choose $u(t) = (i_1(t), i_2(t))$ as the input data and $y(t) = (v_1(t), v_2(t))$ as the output data (or the other way around). These cases are referred to as the *impedance* and *admittance* cases, and they are used, e.g., in series and parallel connections of networks. Neither is this choice of input and output data always possible or reasonable. In his development of the theory of passive Kirchhoff networks V. Belevitch [Bel68] proposed the use of the incoming wave data $u(t) = (\frac{1}{\sqrt{2}}(v_1(t) + i_1(t)), \frac{1}{\sqrt{2}}(v_2(t) + i_2(t)))$ as input data and the outgoing wave data $u(t) = (\frac{1}{\sqrt{2}}(v_1(t) - i_1(t)), \frac{1}{\sqrt{2}}(v_2(t) - i_2(t)))$ as output. This case is called the *scattering* case, and this particular decomposition is always possible and meaningful for passive Kirchhoff networks. In all these cases the physical network is the same, but depending on the decomposition of $w(t) = (v_1(t), i_1(t), v_2(t), i_2(t))$ into an input part and an output part we get very different input/state/output characteristics.

The idea of considering the evolution of external signals $w(t)$ without an explicit decomposition into an input part $u(t)$ and an output part $y(t)$ is the most fundamental ingredient in the behavioral theory initiated by J. Willems (see, e.g., [PW98] for a recent presentation of behavioral theory). Our approach differs from the standard behavioral approach in the sense that we always include a state variable in the equations describing the evolution of the system, and we more or less ignore polynomial descriptions as well as dynamics generated by ordinary differential equations. It is genuinely infinite-dimensional, and it appears to be applicable to a large class of infinite-dimensional problems. A first step in this direction was taken by J. Ball and O. Staffans [BS05], where the main notion of a state/signal node and its trajectories are found in an implicit way.

A state/signal system consists of a state/signal node and the set of trajectories generated by this node. A state/signal node is a triple $\Sigma = (V; \mathcal{X}, \mathcal{W})$, where $\mathcal{X}$ (the state space) and $\mathcal{W}$ (the signal space) are Hilbert spaces, and V is a subspace of the product space $\begin{bmatrix} \mathcal{X} \\ \mathcal{X} \\ \mathcal{W} \end{bmatrix}$ with appropriate properties. In this paper we shall only discuss systems with discrete time. The list of properties that the subspace V should satisfy in this case is given in Definition 2.1. By a *trajectory* $(x(\cdot), w(\cdot))$ of Σ on $\mathbb{Z}^+ = \{0, 1, 2, \ldots\}$ we mean a pair of sequences $\{x(n)\}_{n=0}^{\infty}$ and $\{w(n)\}_{n=0}^{\infty}$ satisfying

$$\begin{bmatrix} x(n+1) \\ x(n) \\ w(n) \end{bmatrix} \in V, \quad n \in \mathbb{Z}^+. \tag{1.1}$$

The properties of the subspace V have been chosen in such a way that there exists at least one *admissible* decomposition (actually infinitely many decompositions) of the signal space $\mathcal{W}$ as the direct sum $\mathcal{W} = \mathcal{Y} \dotplus \mathcal{U}$ of an input space $\mathcal{U}$ and an output space $\mathcal{Y}$ such that trajectories are defined by a usual input/state/output

118   D.Z. Arov and O.J. Staffans

system of equations

$$x(n+1) = Ax(n) + Bu(n),$$
$$y(n) = Cx(n) + Du(n), \qquad n \in \mathbb{Z}^+, \tag{1.2}$$
$$x(0) = x_0,$$

where the coefficients A, B, C, and D are bounded linear operators between the respective Hilbert spaces, i.e., $\left[\begin{smallmatrix} A & B \\ C & D \end{smallmatrix}\right] \in \mathcal{B}([\begin{smallmatrix} \mathcal{X} \\ \mathcal{U} \end{smallmatrix}]; [\begin{smallmatrix} \mathcal{X} \\ \mathcal{Y} \end{smallmatrix}])$. The set of all trajectories $(x(\cdot), w(\cdot))$ of the state/signal system (1.1) can be obtained from the set of trajectories of (1.2) by taking the state sequence $x(\cdot)$ to be the same and taking $w(\cdot) = y(\cdot) + u(\cdot)$. The latter equation we write alternatively in the form $w(\cdot) = \left[\begin{smallmatrix} y(\cdot) \\ u(\cdot) \end{smallmatrix}\right]$, and likewise, instead of $\mathcal{W} = \mathcal{Y} \dotplus \mathcal{U}$ we write alternatively $\mathcal{W} = [\begin{smallmatrix} \mathcal{Y} \\ \mathcal{U} \end{smallmatrix}]$.

In addition to these input/state/output representations, there are two other useful types of representations, namely *driving variable* and *output nulling* representations. In a *driving variable representation* we parameterize the trajectories by using an extra driving variable ℓ with values in an auxiliary *driving variable* Hilbert space $\mathcal{L}$. The trajectories of the system are described by a system of equations

$$x(n+1) = A'x(n) + B'\ell(n),$$
$$w(n) = C'x(n) + D'\ell(n), \qquad n \in \mathbb{Z}^+, \tag{1.3}$$
$$x(0) = x_0,$$

where the coefficients (A', B', C', D') are bounded linear operators between the respective Hilbert spaces, i.e., $\left[\begin{smallmatrix} A' & B' \\ C' & D' \end{smallmatrix}\right] \in \mathcal{B}([\begin{smallmatrix} \mathcal{X} \\ \mathcal{L} \end{smallmatrix}]; [\begin{smallmatrix} \mathcal{X} \\ \mathcal{W} \end{smallmatrix}])$, and D' is injective and has closed range. The set of all trajectories $(x(\cdot), w(\cdot))$ of the state/signal system Σ can be obtained from the set of trajectories $(x(\cdot), \ell(\cdot), w(\cdot))$ of (1.3) by simply dropping the driving variable $\ell(\cdot)$. In an *output nulling representation* we formally consider the signal component w as an input which is restricted by an additional equation posed in an auxiliary *error space* $\mathcal{K}$. The trajectories of this new input/state/output system are described by a system of equations

$$x(n+1) = A''x(n) + B''w(n),$$
$$e(n) = C''x(n) + D''w(n), \qquad n \in \mathbb{Z}^+, \tag{1.4}$$
$$x(0) = x_0,$$

where the coefficients (A'', B'', C'', D'') are bounded linear operators between the respective Hilbert spaces, i.e., $\left[\begin{smallmatrix} A'' & B'' \\ C'' & D'' \end{smallmatrix}\right] \in \mathcal{B}([\begin{smallmatrix} \mathcal{X} \\ \mathcal{W} \end{smallmatrix}]; [\begin{smallmatrix} \mathcal{X} \\ \mathcal{K} \end{smallmatrix}])$, and D'' is surjective. The reason for the name "output nulling" for this representation is that $(x(\cdot), w(\cdot))$ is a trajectory of Σ if and only if $(x(\cdot), w(\cdot), e(\cdot))$ with $e(n) = 0$ for all n is a trajectory of the input/state/output system described by (1.4).

To each state/signal system there corresponds infinitely many representations of each of the three types described above. We prove the existence of these three types of representations, discuss their properties, and also discuss the relationships between different representations of the same type or of different types.

Each input/state/output representation (1.2) of a given state/signal system has a $\mathcal{B}(\mathcal{U};\mathcal{Y})$-valued transfer function given by

$$\mathfrak{D}(z) = D + zC(1_{\mathcal{X}} - zA)^{-1}B, \quad z \in \Lambda_A, \tag{1.5}$$

where Λ_A is the set of points $z \in \mathbb{C}$ for which $(1_{\mathcal{X}} - zA)$ has a bounded inverse, plus the point at infinity if A is boundedly invertible. Thus, each state/signal system has infinitely many such transfer functions, one corresponding to each input/state/output representation. All of these transfer functions can be obtained from one fixed input/state/output representation through the use of a linear fractional transformation. More precisely, let $\mathcal{W} = \mathcal{Y}\dotplus\mathcal{U}$ and $\mathcal{W} = \mathcal{Y}_1\dotplus\mathcal{U}_1$ be two admissible input/output decompositions of the signal space $\mathcal{W}$ of a given state/signal system Σ, and denote the corresponding transfer functions by $\mathfrak{D}$ and $\mathfrak{D}_1$, respectively. Let

$$\Theta = \begin{bmatrix} \Theta_{11} & \Theta_{12} \\ \Theta_{21} & \Theta_{22} \end{bmatrix} = \begin{bmatrix} P_{\mathcal{Y}_1}^{\mathcal{U}_1}|_{\mathcal{Y}} & P_{\mathcal{Y}_1}^{\mathcal{U}_1}|_{\mathcal{U}} \\ P_{\mathcal{U}_1}^{\mathcal{Y}_1}|_{\mathcal{Y}} & P_{\mathcal{U}_1}^{\mathcal{Y}_1}|_{\mathcal{U}} \end{bmatrix}, \tag{1.6}$$

where $P_{\mathcal{Y}_1}^{\mathcal{U}_1}|_{\mathcal{Y}}$ is the restriction to $\mathcal{Y}$ of the projection of $\mathcal{W}$ onto $\mathcal{Y}_1$ along $\mathcal{U}_1$, etc. (Note that Θ can be interpreted as a decomposition of the identity in $\mathcal{W}$ with respect to the two sum decompositions $\mathcal{W} = \mathcal{Y}\dotplus\mathcal{U} = \mathcal{Y}_1\dotplus\mathcal{U}_1$.) Then $\mathfrak{D}_1$ is the value of the linear fractional transform of $\mathfrak{D}$ with coefficient matrix Θ, i.e.,

$$\mathfrak{D}_1(z) = [\Theta_{11}\mathfrak{D}(z) + \Theta_{12}][\Theta_{21}\mathfrak{D}(z) + \Theta_{22}]^{-1}, \quad z \in \Lambda_A \cap \Lambda_{A_1}. \tag{1.7}$$

We also introduce notions of controllability, observability, and minimality of state/signal systems. These notions are defined in terms of the properties of its trajectories, without any reference to the various representations described above, but it is possible to give equivalent conditions for controllability and observability in terms of the different types of representations described above. In particular, we prove that a state/signal system is controllable (or observable, or minimal) if and only if at least one corresponding input/state/output system (1.2) (hence all of them) has the same property.

In Section 2 we discuss the main notions of the theory: state/signal nodes, the corresponding trajectories, and their basic properties. In Sections 3 and 4 we study driving variable and output nulling representations, respectively. Here we also define the notions of controllability and observability and develop tests for controllability and observability in terms of driving variable and output nulling representations. Input/state/output representations are studied in Section 5. Here we also give criteria for the admissibility of a decomposition of the signal space $\mathcal{W}$ into an input space $\mathcal{U}$ and an output space $\mathcal{Y}$ and describe the connections between different representations. Different kinds of transfer functions related to different representations of state/signal systems and their connections are studied in Section 6. In Section 7 we introduce and study signal behaviors and their realizations by means of state/signal systems. Dilations of state/signal systems are studied in depth in Section 8. In particular, we show that a dilation of a state/signal system has the same signal behavior and also the same set of input/output transfer functions (restricted to a neighborhood of zero) as the original system.

The main result of this section characterizes dilations in terms of the existence of a decomposition of the state space into parts with certain invariance properties. All the proofs are given in the state/signal setting, and we obtain standard input/state/output results as corollaries of our main results. Finally, Section 9 is devoted to a study of different stabilizability properties of state/signal systems in terms of the existence of stable representations of driving variable, output nulling, or input/state/output type. Not only power stability, but also strong stability is studied.

Notation. The space of bounded linear operators from one normed space $\mathcal{X}$ to another normed space $\mathcal{Y}$ is denoted by $\mathcal{B}(\mathcal{X};\mathcal{Y})$, and we abbreviate $\mathcal{B}(\mathcal{X};\mathcal{X})$ to $\mathcal{B}(\mathcal{X})$. The domain of a linear operator A is denoted by $\mathcal{D}(A)$, its range by $\mathcal{R}(A)$, and its kernel by $\mathcal{N}(A)$. The restriction of A to some subspace $\mathcal{Z} \subset \mathcal{D}(A)$ is denoted by $A|_{\mathcal{Z}}$. The identity operator on $\mathcal{X}$ is denoted by $1_{\mathcal{X}}$. For each $A \in \mathcal{B}(\mathcal{X})$ we let Λ_A be the set of points $z \in \mathbb{C}$ for which $(1_{\mathcal{X}} - zA)$ has a bounded inverse, plus the point at infinity if A is boundedly invertible.

$\mathbb{C}$ is the complex plane, $\mathbb{D}$ is the open unit disk in $\mathbb{C}$, $\mathbb{Z} = \{0, \pm1, \pm2, \ldots\}$, $\mathbb{Z}^+ = \{0, 1, 2, \ldots\}$, and $\mathbb{Z}^- = \{-1, -2, \ldots\}$. The space $H^2(\mathbb{D};\mathcal{U})$, where $\mathcal{U}$ is a Hilbert space, consists of all analytic $\mathcal{U}$-valued functions ϕ on $\mathbb{D}$ which satisfy $\|\phi\|^2 := \sup_{0 \le r < 1} \frac{1}{2\pi} \oint_{|z|=r} \|\phi(z)\|^2 |dz| < \infty$. The space $H^\infty(\mathbb{D};\mathcal{U},\mathcal{Y})$, where $\mathcal{U}$ and $\mathcal{Y}$ are Hilbert spaces, consists of all bounded analytic $\mathcal{B}(\mathcal{U};\mathcal{Y})$-valued functions on $\mathbb{D}$. The sequence spaces $\ell^1(\mathbb{Z}^+;\mathcal{U})$ and $\ell^2(\mathbb{Z}^+;\mathcal{U})$ contain those $\mathcal{U}$-valued sequences $u(\cdot)$ on $\mathbb{Z}^+$ which satisfy $\sum_{n \in \mathbb{Z}^+} \|u(n)\| < \infty$, respectively, $\sum_{n \in \mathbb{Z}^+} \|u(n)\|^2 < \infty$, and $\ell^\infty(\mathbb{Z}^+;\mathcal{U})$ consists of all bounded $\mathcal{U}$-valued sequences on $\mathbb{Z}^+$.

We denote the projection onto a closed subspace $\mathcal{Y}$ of a space $\mathcal{X}$ along some complementary subspace $\mathcal{U}$ by $P_{\mathcal{Y}}^{\mathcal{U}}$. The closed linear span or linear span of a sequence of subsets $\mathfrak{R}_n \subset \mathcal{X}$ where n runs over some index set Λ is denoted by $\bigvee_{n \in \Lambda} \mathfrak{R}_n$ and $\operatorname{span}_{n \in \Lambda} \mathfrak{R}_n$, respectively.

We denote the product of the two locally convex topological vector spaces $\mathcal{X}$ and $\mathcal{Y}$ by $\left[\begin{smallmatrix}\mathcal{X}\\\mathcal{Y}\end{smallmatrix}\right]$. In particular, although $\mathcal{X}$ and $\mathcal{Y}$ may be Hilbert spaces (in which case the product topology in $\left[\begin{smallmatrix}\mathcal{X}\\\mathcal{Y}\end{smallmatrix}\right]$ is induced by an inner product), we shall not require that $\left[\begin{smallmatrix}\mathcal{X}\\0\end{smallmatrix}\right] \perp \left[\begin{smallmatrix}0\\\mathcal{Y}\end{smallmatrix}\right]$ in $\left[\begin{smallmatrix}\mathcal{X}\\\mathcal{Y}\end{smallmatrix}\right]$. Furthermore, in this case we identify a vector $\left[\begin{smallmatrix}x\\0\end{smallmatrix}\right] \in \left[\begin{smallmatrix}\mathcal{X}\\0\end{smallmatrix}\right]$ with $x \in \mathcal{X}$ and a vector $\left[\begin{smallmatrix}0\\y\end{smallmatrix}\right] \in \left[\begin{smallmatrix}0\\\mathcal{Y}\end{smallmatrix}\right]$ with $y \in \mathcal{Y}$. (Thus, we also denote the ordered direct sum $\mathcal{X} \dotplus \mathcal{Y}$ by $\left[\begin{smallmatrix}\mathcal{X}\\\mathcal{Y}\end{smallmatrix}\right]$.)

2. State/signal nodes and trajectories

In this section we shall study time-invariant linear systems in discrete time induced by something that we call a *state/signal node*.

Definition 2.1. A triple $\Sigma = (V; \mathcal{X}, \mathcal{W})$, where the (*internal*) *state space* $\mathcal{X}$ and the (*external*) *signal space* $\mathcal{W}$ are Hilbert spaces and V is a subspace of the product space $\mathfrak{K} := \begin{bmatrix} \mathcal{X} \\ \mathcal{X} \\ \mathcal{W} \end{bmatrix}$ is called a *state/signal node* if it has the following properties:[1]

  (i) V is closed in $\mathfrak{K}$;

  (ii) For every $x \in \mathcal{X}$ there is some $\begin{bmatrix} z \\ w \end{bmatrix} \in \begin{bmatrix} \mathcal{X} \\ \mathcal{W} \end{bmatrix}$ such that $\begin{bmatrix} z \\ x \\ w \end{bmatrix} \in V$;

  (iii) If $\begin{bmatrix} z \\ 0 \\ 0 \end{bmatrix} \in V$, then $z = 0$;

  (iv) The set $\left\{ \begin{bmatrix} x \\ w \end{bmatrix} \in \begin{bmatrix} \mathcal{X} \\ \mathcal{W} \end{bmatrix} \ \middle| \ \begin{bmatrix} z \\ x \\ w \end{bmatrix} \in V \text{ for some } z \in \mathcal{X} \right\}$ is closed in $\begin{bmatrix} \mathcal{X} \\ \mathcal{W} \end{bmatrix}$.

We call $\mathfrak{K}$ the *node space* and V the *generating subspace*.

As we shall see in a moment (in Proposition 2.2, Lemmas 2.3–2.4 and Theorem 2.5), all of these conditions have a clear meaning related to the fact that we shall use the generating subspace V as the main tool in our definition of a *trajectory*. To *define* such a trajectory it is not important that (i)–(iv) hold.

We define a *trajectory* $(x(\cdot), w(\cdot))$ *along an arbitrary subspace* V of $\mathfrak{K}$ on the time interval $[n_1, n_2]$, where $n_1, n_2 \in \mathbb{Z}$, $n_1 \le n_2$, to be a pair of sequences $\{x(k)\}_{k=n_1}^{n_2+1}$ and $\{w(k)\}_{k=n_1}^{n_2}$ satisfying

$$\begin{bmatrix} x(k+1) \\ x(k) \\ w(k) \end{bmatrix} \in V, \qquad n_1 \le k \le n_2. \tag{2.1}$$

We shall also allow $n_1 = -\infty$ or $n_2 = \infty$, in which case we replace $\le$ by $<$ in the formula above. Most of our trajectories will be considered on $\mathbb{Z}^+$. We shall refer to the sequence $x(\cdot)$ as the *state component* and to the sequence $w(\cdot)$ as the *signal component* of the trajectory $(x(\cdot), w(\cdot))$. In the case where n_1 is finite we shall call $x(n_1)$ the *initial state* of this trajectory.

It follows immediately from Definition 2.1 that the set of trajectories along a given subspace V of $\mathfrak{K}$ has the following two properties:

1) *if* $(x(\cdot), w(\cdot))$ *is a trajectory along* V *on* $[n_1, n_2]$, *then for each* $k \in \mathbb{Z}$, *the shifted pair of sequences* $(x(\cdot + k), w(\cdot + k))$ *is a trajectory along* V *on* $[n_1 - k, n_2 - k]$.

2) *if* $(x_1(\cdot), w_1(\cdot))$ *is a trajectory along* V *on* $[n_1, n_2]$, *if* $(x_2(\cdot), w_2(\cdot))$ *is a trajectory along* V *on* $[n_2 + 1, n_3]$, *and if* $x_1(n_2 + 1) = x_2(n_2 + 1)$, *then the concatenation* $(x(\cdot), w(\cdot))$ *defined by* $(x(k), w(k)) = (x_1(k), w_1(k))$ *for* $k \in [n_1, n_2]$, $(x(k), w(k)) = (x_2(k), w_2(k))$ *for* $k \in [n_2 + 1, n_3]$, *and* $x(n_3 + 1) = x_2(n_3 + 1)$, *is a trajectory along* V *on* $[n_1, n_3]$.

Property 1) means that the set of trajectories along V is *time-invariant*, and property 2) says that x has the *state property*; cf. [PW98, p. 119].

---

[1] Recall that we denote the direct product $\mathcal{X} \times \mathcal{X} \times \mathcal{W}$ by $\begin{bmatrix} \mathcal{X} \\ \mathcal{X} \\ \mathcal{W} \end{bmatrix}$. Later when we introduce *passive* nodes we shall require $\mathcal{X}$ to be a Hilbert space, $\mathcal{W}$ to be a Kreĭn space, and equip $\mathfrak{K}$ with a particular Kreĭn space structure rather than the Hilbert space structure that it inherits from $\mathcal{X}$ and $\mathcal{W}$. This is the reason why we throughout ignore the Hilbert space inner product in $\mathfrak{K}$ induced by the inner products in $\mathcal{X}$ and $\mathcal{W}$. The only way in which we use the fact that $\mathcal{X}$ and $\mathcal{W}$ are Hilbert spaces is in the assertion that every closed subspace of $\mathfrak{K}$ has a complementary subspace. The same comments applies to all other Hilbert spaces and their products that appear in this paper.

Properties (ii) and (iii) in Definition 2.1 are reflected in the properties of the set of all trajectories along V as follows:

Proposition 2.2. *Let V be a subspace of the product space $\mathfrak{K} := \begin{bmatrix} \mathcal{X} \\ \mathcal{X} \\ \mathcal{W} \end{bmatrix}$.*

1) *The following three statements are equivalent:*
   (a) *V has property (ii) in Definition 2.1;*
   (b) *for every $x_0 \in \mathcal{X}$ there is a trajectory $(x(\cdot), w(\cdot))$ along V on $\mathbb{Z}^+$ with $x(0) = x_0$;*
   (c) *every trajectory $(x(\cdot), w(\cdot))$ along V defined on some interval $[0, n_2]$ can be extended to a trajectory on $\mathbb{Z}^+$.*

2) *The following four statements are equivalent:*
   (a) *V has property (iii) in Definition 2.1;*
   (b) *if $(x(\cdot), w(\cdot))$ is a trajectory on $[n_1, n_2]$ along V, then for every $k \in [n_1, n_2]$, the value of $x(k+1)$ is determined uniquely by $\begin{bmatrix} x(k) \\ w(k) \end{bmatrix}$;*
   (c) *if $(x(\cdot), w(\cdot))$ is a trajectory on $[n_1, n_2]$ along V, then the value of $x(n_2 + 1)$ is determined uniquely by $x(n_1)$ and $w(k)$, $n_1 \le k \le n_2$.*
   (d) *if $(x(\cdot), w(\cdot))$ is a trajectory on $[n_1, n_2]$ along V with $x(n_1) = 0$, then the value of $x(n_2 + 1)$ is determined uniquely by $w(k)$, $n_1 \le k \le n_2$.*

Proof. Proof of 1): The implications (b) $\Rightarrow$ (a) and (c) $\Rightarrow$ (a) are obvious.

We next prove that (a) $\Rightarrow$ (b). Suppose that (a) holds. Let $x_0 \in \mathcal{X}$, and define $x(0) = x_0$. It follows from property (ii) in Definition 2.1 that there exist $x(1)$ and $w(0)$ such that $\begin{bmatrix} x(1) \\ x(0) \\ w(0) \end{bmatrix} \in V$. By the same argument with $x(0)$ replaced by $x(1)$, there exist $x(2)$ and $w(1)$ such that $\begin{bmatrix} x(2) \\ x(1) \\ w(1) \end{bmatrix} \in V$. By induction, we will obtain (b).

The proof of the fact that (a) $\Rightarrow$ (c) is the same as the proof of the implication (a) $\Rightarrow$ (b) given above, except that we start from time $n + 1$ and the initial value $x(n + 1)$ (instead of time zero and initial value x_0).

The proof of 2) is left to the reader. $\qquad\square$

By the *state/signal system generated by the state/signal node* $\Sigma = (V; \mathcal{X}, \mathcal{W})$ we mean this node itself together with the set of all trajectories along V. For simplicity we use the same notation Σ for the system as we used for the original node. We shall also refer to the trajectories along V as *the trajectories of Σ*.

We shall next develop certain representations of the subspace V in Definition 2.1, and begin with the following lemmas.

Lemma 2.3. *Let V be a subspace of the product space $\mathfrak{K} := \begin{bmatrix} \mathcal{X} \\ \mathcal{X} \\ \mathcal{W} \end{bmatrix}$. Let $G_{2,3} \colon V \to \begin{bmatrix} \mathcal{X} \\ \mathcal{W} \end{bmatrix}$ be the bounded linear operator that maps the vector $\begin{bmatrix} z \\ x \\ w \end{bmatrix} \in V$ into $\begin{bmatrix} x \\ w \end{bmatrix} \in \begin{bmatrix} \mathcal{X} \\ \mathcal{W} \end{bmatrix}$. Then the following conditions are equivalent:*

1) *V has property (iii);*
2) *$G_{2,3}$ is injective;*

3) V has a graph representation over the last two components $\left[\begin{smallmatrix} \mathcal{X} \\ \mathcal{W} \end{smallmatrix}\right]$ of $\mathfrak{K}$, i.e., there exists a linear operator F, mapping $\mathcal{D}(F) \subset \left[\begin{smallmatrix} \mathcal{X} \\ \mathcal{W} \end{smallmatrix}\right]$ into $\mathcal{X}$ such that $\left[\begin{smallmatrix} z \\ x \\ w \end{smallmatrix}\right] \in V$ if and only if $\left[\begin{smallmatrix} x \\ w \end{smallmatrix}\right] \in \mathcal{D}(F)$ and $z = F\left[\begin{smallmatrix} x \\ w \end{smallmatrix}\right]$.

Assuming 1), with $G_{2,3}$ and F defined as in 2) and 3), the operator F is uniquely determined by V (hence so is $\mathcal{D}(F)$), $\mathcal{R}\left(G_{2,3}\right) = \mathcal{D}(F)$, $G_{2,3}^{-1}\colon \mathcal{D}(F) \to V$ is given by $G_{2,3}^{-1} = \left[\begin{smallmatrix} F & \\ 1_{\mathcal{X}} & 0 \\ 0 & 1_{\mathcal{W}} \end{smallmatrix}\right]$, and

$$
V = G_{2,3}^{-1}\mathcal{D}(F) = \left\{ \begin{bmatrix} z \\ x \\ w \end{bmatrix} \;\middle|\; z = F\left(\begin{bmatrix} x \\ w \end{bmatrix}\right),\; \begin{bmatrix} x \\ w \end{bmatrix} \in \mathcal{D}(F) \right\} \tag{2.2}
$$

Lemma 2.4. *Let V be a subspace of the product space $\mathfrak{K} := \left[\begin{smallmatrix} \mathcal{X} \\ \mathcal{X} \\ \mathcal{W} \end{smallmatrix}\right]$. Assume that V has property (iii), and let F be the operator defined in Lemma 2.3. Then*

1) *V has property (i) if and only if F is closed,*
2) *V has property (ii) if and only if the linear operator $\mathcal{D}(F) \to \mathcal{X}$ that maps $\left[\begin{smallmatrix} x \\ w \end{smallmatrix}\right] \in \mathcal{D}(F)$ into $x \in \mathcal{X}$ is surjective,*
3) *V has property (iv) if and only if $\mathcal{D}(F)$ is closed,*
4) *V has properties (i) and (iv) if and only if F is bounded and $\mathcal{D}(F)$ is closed.*

We leave the straightforward proofs of Lemmas 2.3 and 2.4 to the reader. By combining Lemmas 2.3 and 2.4 we get the following theorem:

Theorem 2.5. *Let V be a subspace of the product space $\mathfrak{K} := \left[\begin{smallmatrix} \mathcal{X} \\ \mathcal{X} \\ \mathcal{W} \end{smallmatrix}\right]$. Then V has properties (i)–(iv) listed in Definition 2.1, i.e., $\Sigma = (V; \mathcal{X}, \mathcal{W})$ is a state/signal node, if and only if V has a graph representation over the last two components $\left[\begin{smallmatrix} \mathcal{X} \\ \mathcal{W} \end{smallmatrix}\right]$ of $\mathfrak{K}$ with a bounded linear operator $F\colon \mathcal{D}(F) \subset \left[\begin{smallmatrix} \mathcal{X} \\ \mathcal{W} \end{smallmatrix}\right] \to \mathcal{X}$ with closed domain, i.e.,*

$$
V = \left\{ \begin{bmatrix} z \\ x \\ w \end{bmatrix} \;\middle|\; z = F\left(\begin{bmatrix} x \\ w \end{bmatrix}\right),\; \begin{bmatrix} x \\ w \end{bmatrix} \in \mathcal{D}(F) \right\}, \tag{2.3}
$$

with the additional property that the linear operator $\mathcal{D}(F) \to \mathcal{X}$ that maps $\left[\begin{smallmatrix} x \\ w \end{smallmatrix}\right] \in \mathcal{D}(F)$ into $x \in \mathcal{X}$ is surjective.

In the next three sections we shall develop three different types of representations of a state/signal system Σ: *driving variable* representations, *output nulling* representations, and *input/state/output* representations. They complement each other, and all of them are important in slightly different connections.

3. The driving variable representation

In our first representation of the generating subspace V we write V as the image of a bounded linear injective operator of the following type.

Lemma 3.1. *Let V be a subspace of the product space $\mathfrak{K} := \begin{bmatrix} \mathcal{X} \\ \mathcal{X} \\ \mathcal{W} \end{bmatrix}$, where $\mathcal{X}$ and $\mathcal{W}$ are Hilbert spaces. If there exists a Hilbert space $\mathcal{L}$ and four operators*

$$A' \in \mathcal{B}(\mathcal{X}), \ B' \in \mathcal{B}(\mathcal{L};\mathcal{X}), \ C' \in \mathcal{B}(\mathcal{X},\mathcal{W}), \ and \ D' \in \mathcal{B}(\mathcal{L};\mathcal{W}), \tag{3.1}$$

where

$$D' \ is \ injective \ and \ has \ a \ closed \ range \tag{3.2}$$

such that

$$V = \mathcal{R}\left(\begin{bmatrix} A' & B' \\ 1_\mathcal{X} & 0 \\ C' & D' \end{bmatrix}\right) = \left\{ \begin{bmatrix} A'x + B'\ell \\ x \\ C'x + D'\ell \end{bmatrix} \ \middle| \ x \in \mathcal{X}, \ \ell \in \mathcal{L} \right\}, \tag{3.3}$$

then V has properties (i)–(iv) listed in Definition 2.1, i.e., $(V;\mathcal{X},\mathcal{W})$ is a state/ signal node. Conversely, if V has properties (i)–(iv) listed in Definition 2.1 then V is given by (3.3) for some Hilbert space $\mathcal{L}$ and some operators A', B', C', and D' satisfying (3.1) and (3.2).

Proof. We begin by proving that the representation (3.1)–(3.3) implies that V has properties (i)–(iv) in Definition 2.1. Trivially, (3.1) and (3.3) imply (ii). It is also clear that the injectivity of D' implies that the operator $\begin{bmatrix} 1_\mathcal{X} & 0 \\ C' & D' \end{bmatrix}$ is injective. Thus, by defining $\mathcal{D}(F) = \mathcal{R}\left(\begin{bmatrix} 1_\mathcal{X} & 0 \\ C' & D' \end{bmatrix}\right)$ and $F = \begin{bmatrix} A' & B' \end{bmatrix} \begin{bmatrix} 1_\mathcal{X} & 0 \\ C' & D' \end{bmatrix}^{-1}$ we get the graph representation (2.3) of V. According to Lemma 2.3, this implies that V has property (iii). The closedness of $\mathcal{R}(D')$ implies that also $\mathcal{R}\left(\begin{bmatrix} 1_\mathcal{X} & 0 \\ C' & D' \end{bmatrix}\right)$ is closed, because $\mathcal{R}\left(\begin{bmatrix} 1_\mathcal{X} & 0 \\ C' & D' \end{bmatrix}\right) = \begin{bmatrix} 1_\mathcal{X} & 0 \\ C' & 1_\mathcal{W} \end{bmatrix}\begin{bmatrix} \mathcal{X} \\ \mathcal{R}(D') \end{bmatrix}$, where $\begin{bmatrix} 1_\mathcal{X} & 0 \\ C' & 1_\mathcal{W} \end{bmatrix}$ is boundedly invertible. Finally, the closed graph theorem implies that $\begin{bmatrix} 1_\mathcal{X} & 0 \\ C' & D' \end{bmatrix}^{-1}$ is bounded on $\mathcal{D}(F)$, hence so is F, and by part 4) of Lemma 2.4, V has properties (i) and (iv). We have now showed that V has all the properties (i)–(iv).

Conversely, suppose that V has properties (i)–(iv) in Definition 2.1. Let $G_2 \in \mathcal{B}(V;\mathcal{X})$ be the bounded linear operator that maps $\begin{bmatrix} z \\ x \\ w \end{bmatrix} \in V$ into $x \in \mathcal{X}$. We take $\mathcal{L} = \mathcal{N}(G_2)$, and define $B' \in \mathcal{B}(\mathcal{L};\mathcal{X})$ and $D' \in \mathcal{B}(\mathcal{X};\mathcal{W})$ by $B' \begin{bmatrix} z \\ 0 \\ w \end{bmatrix} = z$ and $D'\begin{bmatrix} z \\ 0 \\ w \end{bmatrix} = w$ for each $\begin{bmatrix} z \\ 0 \\ w \end{bmatrix} \in \mathcal{L}$. Clearly $\ell = \begin{bmatrix} B' \\ 0 \\ D' \end{bmatrix}\ell$ for all $\ell \in \mathcal{L}$, $\begin{bmatrix} B' \\ D' \end{bmatrix}$ is injective on $\mathcal{L}$, and the range of $\begin{bmatrix} B' \\ D' \end{bmatrix}$ is closed in $\begin{bmatrix} \mathcal{X} \\ \mathcal{W} \end{bmatrix}$. By property (ii) in Definition 2.1, G_2 maps V *onto* $\mathcal{X}$. Let $G_{2,\mathrm{right}}^{-1} \in \mathcal{B}(\mathcal{X};V)$ be an arbitrary right-inverse of G_2 (such a bounded right-inverse exists since V is closed). This right-inverse must be of the form $G_{2,\mathrm{right}}^{-1} = \begin{bmatrix} A' \\ 1_\mathcal{X} \\ C' \end{bmatrix}$ (the middle component must be the identity operator since $G_2 \begin{bmatrix} z \\ x \\ w \end{bmatrix} = x$ for all $\begin{bmatrix} z \\ x \\ w \end{bmatrix} \in V$). By property (i), $V = \mathcal{R}\left(G_{2,\mathrm{right}}^{-1}\right) \dotplus \mathcal{L}$, hence

$$V = \begin{bmatrix} A' \\ 1_\mathcal{X} \\ C' \end{bmatrix}\mathcal{X} \dotplus \mathcal{L} = \begin{bmatrix} A' \\ 1_\mathcal{X} \\ C' \end{bmatrix}\mathcal{X} \dotplus \begin{bmatrix} B' \\ 0 \\ D' \end{bmatrix}\mathcal{L}.$$

This implies (3.3).

We still have to show that D' is injective and has closed range, and for this we need properties (iii) and (iv) (which we have not used up to now). By construction the operator $\begin{bmatrix} A' & B' \\ 1_{\mathcal{X}} & 0 \\ C' & D' \end{bmatrix}$ is injective. It then follows from Lemma 2.3 that the operator

$$G_{2,3} \begin{bmatrix} A' & B' \\ 1_{\mathcal{X}} & 0 \\ C' & D' \end{bmatrix} = \begin{bmatrix} 1_{\mathcal{X}} & 0 \\ C' & D' \end{bmatrix} : \begin{bmatrix} \mathcal{X} \\ \mathcal{L} \end{bmatrix} \to \begin{bmatrix} \mathcal{X} \\ \mathcal{W} \end{bmatrix}$$

also must be injective (since we now assume (iii)). This implies that D' is injective. That the range is closed follows from (iv), i.e., from the closedness of $\mathcal{D}(F) = \mathcal{R}\left(\begin{bmatrix} 1_{\mathcal{X}} & 0 \\ C' & D' \end{bmatrix}\right)$, since (as we observed above)

$$\mathcal{R}\left(\begin{bmatrix} 1_{\mathcal{X}} & 0 \\ C' & D' \end{bmatrix}\right) = \begin{bmatrix} 1_{\mathcal{X}} & 0 \\ C' & 1_{\mathcal{W}} \end{bmatrix}\begin{bmatrix} \mathcal{X} \\ \mathcal{R}(D') \end{bmatrix},$$

where $\begin{bmatrix} 1_{\mathcal{X}} & 0 \\ C' & 1_{\mathcal{W}} \end{bmatrix}$ is boundedly invertible. $\qquad\square$

We shall call a colligation $\Sigma_{dv/s/s} := \left(\begin{bmatrix} A' & B' \\ C' & D' \end{bmatrix}; \mathcal{X}, \mathcal{L}, \mathcal{W}\right)$, where $\mathcal{L}$ is a Hilbert space and A', B', C', and D' satisfy (3.1)–(3.3) a *driving variable representation of the state/signal node* $\Sigma = (V; \mathcal{X}, \mathcal{W})$. We shall also refer to $\Sigma_{dv/s/s}$ as a *driving-variable/state/signal node*. By the driving-variable/state/signal *system* $\Sigma_{dv/s/s}$ we mean the node $\Sigma_{dv/s/s}$ itself together with the set of all trajectories $(x(\cdot), \ell(\cdot), w(\cdot))$ generated by this node through the equations

$$\begin{aligned} x(k+1) &= A'x(k) + B'\ell(k), \\ w(k) &= C'x(k) + D'\ell(k), \qquad n_1 \le k \le n_2. \end{aligned} \tag{3.4}$$

The space $\mathcal{L}$ considered above is called a *driving variable space*, and the vector $\ell \in \mathcal{L}$ in (3.3) is called a *driving variable*. (The notion of a driving variable is known in the finite-dimensional setting from the theory of behaviors; see, e.g., [WT02].) From each trajectory $(x(\cdot), \ell(\cdot), w(\cdot))$ of the driving-variable/state/signal system $\Sigma_{dv/s/s}$ we get a trajectory $(x(\cdot), w(\cdot))$ of the state/signal system Σ by simply deleting the driving variable component ℓ. It follows from part 3) of Proposition 3.2 below that this correspondence between the trajectories of the two types of systems is one-to-one.

Let us next point out some important properties of driving variable representations.

Proposition 3.2. *Let* $\Sigma = (V; \mathcal{X}, \mathcal{W})$ *be a state/signal node with the driving variable representation* $\Sigma_{dv/s/s} = \left(\begin{bmatrix} A' & B' \\ C' & D' \end{bmatrix}; \mathcal{X}, \mathcal{L}, \mathcal{W}\right)$, *and let* $F \colon \mathcal{D}(F) \to \mathcal{X}$ *be the linear operator defined in Lemma 2.3. Then the following assertions are true.*

1) $\mathcal{R}\left(\begin{bmatrix} 1_{\mathcal{X}} & 0 \\ C' & D' \end{bmatrix}\right) = \mathcal{D}(F)$, $\mathcal{R}(B') = \mathfrak{R}_0$, $\mathcal{R}(D') = \mathcal{U}_0$, *and the preimage of* $\mathcal{R}(D')$ *under* C' *is given by* $\mathfrak{U}_0$, *where*

$$\begin{aligned} \mathfrak{R}_0 &= \left\{ F\begin{bmatrix} 0 \\ w \end{bmatrix} \,\middle|\, \begin{bmatrix} 0 \\ w \end{bmatrix} \in \mathcal{D}(F) \right\} \\ &= \left\{ z \in \mathcal{X} \,\middle|\, \begin{bmatrix} z \\ 0 \\ w \end{bmatrix} \in V \text{ for some } w \in \mathcal{W} \right\}, \end{aligned} \tag{3.5}$$

$$\begin{aligned} \mathcal{U}_0 &= \left\{ w \in \mathcal{W} \,\middle|\, \begin{bmatrix} 0 \\ w \end{bmatrix} \in \mathcal{D}(F) \right\} \\ &= \left\{ w \in \mathcal{W} \,\middle|\, \begin{bmatrix} z \\ 0 \\ w \end{bmatrix} \in V \text{ for some } z \in \mathcal{X} \right\}, \end{aligned} \tag{3.6}$$

$$\mathfrak{U}_0 = \left\{ x \in \mathcal{X} \mid \begin{bmatrix} x \\ 0 \end{bmatrix} \in \mathcal{D}(F) \right\}$$
$$= \left\{ x \in \mathcal{X} \mid \begin{bmatrix} z \\ x \\ 0 \end{bmatrix} \in V \text{ for some } z \in \mathcal{X} \right\}. \tag{3.7}$$

Consequently, the ranges of B', D', and $\begin{bmatrix} 1_{\mathcal{X}} & 0 \\ C' & D' \end{bmatrix}$ do not depend on the particular choice of $\Sigma_{dv/s/s}$.

2) *The space $\mathcal{L}$ is isomorphic to the space $\mathcal{U}_0$ defined in (3.6).*

3) *The operator $\begin{bmatrix} 1_{\mathcal{X}} & 0 \\ C' & D' \end{bmatrix}$ has a bounded inverse mapping $\mathcal{D}(F)$ one-to-one onto $\begin{bmatrix} \mathcal{X} \\ \mathcal{L} \end{bmatrix}$, and the vector ℓ in the representation (3.3) is uniquely determined by $\begin{bmatrix} x \\ w \end{bmatrix}$ via*

$$\begin{bmatrix} x \\ \ell \end{bmatrix} = \begin{bmatrix} 1_{\mathcal{X}} & 0 \\ C' & D' \end{bmatrix}^{-1} \begin{bmatrix} x \\ w \end{bmatrix}, \qquad \begin{bmatrix} x \\ w \end{bmatrix} \in \mathcal{D}(F). \tag{3.8}$$

4) *The operator $\begin{bmatrix} A' & B' \end{bmatrix}$ is given by*

$$\begin{bmatrix} A' & B' \end{bmatrix} = F \begin{bmatrix} 1_{\mathcal{X}} & 0 \\ C' & D' \end{bmatrix}. \tag{3.9}$$

Consequently, A' is determined uniquely by C' and B' is determined uniquely by D'.

Proof. Assertion 1) follows from (3.3) and the definition of F. To see that assertion 2) holds it suffices to note that the operator D' maps $\mathcal{L}$ one-to-one onto $\mathcal{U}_0$, and by the closed graph theorem, then inverse of this operator is also bounded. Assertions 3) and 4) were established as a part of the proof of Lemma 3.1. $\square$

Theorem 3.3. *Let $\Sigma_{dv/s/s} = \left(\begin{bmatrix} A' & B' \\ C' & D' \end{bmatrix}; \mathcal{X}, \mathcal{L}, \mathcal{W} \right)$ be a driving variable representation of a state signal system Σ, and let*

$$\begin{bmatrix} A'_1 & B'_1 \\ C'_1 & D'_1 \end{bmatrix} = \begin{bmatrix} A' & B' \\ C' & D' \end{bmatrix} \begin{bmatrix} 1_{\mathcal{X}} & 0 \\ K' & M' \end{bmatrix} \tag{3.10}$$

where

$$K' \in \mathcal{B}(\mathcal{X}; \mathcal{L}), \ M' \in \mathcal{B}(\mathcal{L}_1; \mathcal{L}), \text{ and } M' \text{ has a bounded inverse,} \tag{3.11}$$

for some Hilbert space $\mathcal{L}_1$. Then $\Sigma^1_{dv/s/s} = \left(\begin{bmatrix} A'_1 & B'_1 \\ C'_1 & D'_1 \end{bmatrix}; \mathcal{X}, \mathcal{L}_1, \mathcal{W} \right)$ is a driving variable representation of Σ. Conversely, every driving variable representation $\Sigma^1_{dv/s/s}$ of Σ may be obtained from formula (3.10) for some operators K' and M' satisfying (3.11). The operators K' and M' are uniquely defined by $\Sigma_{dv/s/s}$ and $\Sigma^1_{dv/s/s}$ via

$$D'K' = C'_1 - C' \text{ and } D'M' = D'_1. \tag{3.12}$$

Proof. Suppose that $\Sigma^1_{dv/s/s} = \left(\begin{bmatrix} A'_1 & B'_1 \\ C'_1 & D'_1 \end{bmatrix}; \mathcal{X}, \mathcal{L}_1, \mathcal{W} \right)$ given by (3.10) for some operators K' and M' satisfying (3.11). It follows from (3.11) that $\begin{bmatrix} 1_{\mathcal{X}} & 0 \\ K' & M' \end{bmatrix}$ maps $\begin{bmatrix} \mathcal{X} \\ \mathcal{L} \end{bmatrix}$ one-to-one onto $\begin{bmatrix} \mathcal{X} \\ \mathcal{L}_1 \end{bmatrix}$. By (3.3) and (3.10),

$$\begin{bmatrix} A'_1 & B'_1 \\ 1_{\mathcal{X}} & 0 \\ C'_1 & D'_1 \end{bmatrix} \begin{bmatrix} \mathcal{X} \\ \mathcal{L}_1 \end{bmatrix} = \begin{bmatrix} A' & B' \\ 1_{\mathcal{X}} & 0 \\ C' & D' \end{bmatrix} \begin{bmatrix} 1_{\mathcal{X}} & 0 \\ K' & M' \end{bmatrix} \begin{bmatrix} \mathcal{X} \\ \mathcal{L}_1 \end{bmatrix} = \begin{bmatrix} A' & B' \\ 1_{\mathcal{X}} & 0 \\ C' & D' \end{bmatrix} \begin{bmatrix} \mathcal{X} \\ \mathcal{L} \end{bmatrix} = V.$$

Furthermore, $D_1' = D'M'$ is injective and has closed range. Thus $\Sigma^1_{dv/s/s}$ is a driving variable representation of Σ.

We next turn to the converse part. By statements 1) and 3) of Proposition 3.2, the operator $\begin{bmatrix} G' & H' \\ K' & M' \end{bmatrix} := \begin{bmatrix} 1_{\mathcal{X}} & 0 \\ C' & D' \end{bmatrix}^{-1} \begin{bmatrix} 1_{\mathcal{X}} & 0 \\ C_1' & D_1' \end{bmatrix}$ is a bounded linear operator mapping $\begin{bmatrix} \mathcal{X} \\ \mathcal{L}_1 \end{bmatrix}$ one-to-one onto $\begin{bmatrix} \mathcal{X} \\ \mathcal{L} \end{bmatrix}$. It follows from the identity $\begin{bmatrix} 1_{\mathcal{X}} & 0 \\ C_1' & D_1' \end{bmatrix} = \begin{bmatrix} 1_{\mathcal{X}} & 0 \\ C' & D' \end{bmatrix} \begin{bmatrix} G' & H' \\ K' & M' \end{bmatrix}$ that $G' = 1_{\mathcal{X}}$ and that $H' = 0$, and the invertibility of $\begin{bmatrix} G' & H' \\ K' & M' \end{bmatrix} = \begin{bmatrix} 1_{\mathcal{X}} & 0 \\ K' & M' \end{bmatrix}$ implies that M' is invertible. Thus, (3.11) and (3.12) hold. By statement 4) of Proposition 3.2,

$$ F = \begin{bmatrix} A' & B' \end{bmatrix} \begin{bmatrix} 1_{\mathcal{X}} & 0 \\ C' & D' \end{bmatrix}^{-1} = \begin{bmatrix} A_1' & B_1' \end{bmatrix} \begin{bmatrix} 1_{\mathcal{X}} & 0 \\ C_1' & D_1' \end{bmatrix}^{-1}, $$

hence $\begin{bmatrix} A_1' & B_1' \end{bmatrix} = \begin{bmatrix} A' & B' \end{bmatrix} \begin{bmatrix} 1_{\mathcal{X}} & 0 \\ K' & M' \end{bmatrix}$. Thus equation (3.10) holds.

Finally, we remark that (3.12) determines K' and M' uniquely since D' is injective. $\qquad\square$

Definition 3.4. Let $\Sigma = (V; \mathcal{X}, \mathcal{W})$ be a state/signal system.

1) By an *externally generated* trajectory of Σ on $[0, n]$ or on $\mathbb{Z}^+$ we mean a trajectory $(x(\cdot), w(\cdot))$ satisfying $x(0) = 0$.
2) The *reachable subspace* $\mathfrak{R}_n$ of Σ *in time* n is the subspace of all the final states $x(n+1)$ of all externally generated trajectories $(x(\cdot), w(\cdot))$ of the system Σ on the interval $[0, n]$.
3) The (approximately) *reachable subspace* $\mathfrak{R}$ of Σ (in infinite time) is the closure in $\mathcal{X}$ of all the possible values of the state components $x(\cdot)$ of all externally generated trajectories $(x(\cdot), w(\cdot))$ of the system Σ on $\mathbb{Z}^+$.
4) The system is (approximately) *controllable* if the reachable subspace is all of $\mathcal{X}$.

Thus,

$$ \mathfrak{R}_n \subset \mathfrak{R}_{n+1}, \qquad \mathfrak{R} = \vee_{n \in \mathbb{Z}^+} \mathfrak{R}_n $$

(we get the first inclusion by taking $x(0) = 0$ and $w(0) = 0$, so that also $x(1) = 0$; for the second inclusion we use part 1) of Proposition 2.2). Observe, in particular, that the subspace $\mathfrak{R}_0$ defined above coincides with the subspace $\mathfrak{R}_0$ defined in (3.5).

The subspaces $\mathfrak{R}_n$ and $\mathfrak{R}$ in Definition 3.4 have the following simple characterizations in terms of an arbitrary driving variable representation of Σ.

Proposition 3.5. *Let* $\Sigma = (V; \mathcal{X}, \mathcal{W})$ *be a state/signal system, with a driving variable representation* $\Sigma_{dv/s/s} = \left(\begin{bmatrix} A' & B' \\ C' & D' \end{bmatrix}; \mathcal{X}, \mathcal{L}, \mathcal{W} \right)$. *Then the subspaces* $\mathfrak{R}_n$ *defined above and the reachable subspace* $\mathfrak{R}$ *are given by*

$$ \mathfrak{R}_n = \operatorname{span}\{\mathcal{R}\left((A')^k B'\right) \mid 0 \le k \le n\}, \quad n \in \mathbb{Z}^+, \tag{3.13} $$

$$ \mathfrak{R} = \vee_{k \in \mathbb{Z}^+} \mathcal{R}\left((A')^k B'\right). \tag{3.14} $$

In particular, Σ *is controllable if and only if*

$$ \mathcal{X} = \vee_{k \in \mathbb{Z}^+} \mathcal{R}\left((A')^k B'\right). \tag{3.15} $$

Proof. Let $(x(\cdot), w(\cdot))$ be an externally generated trajectory of Σ on $[0, n]$. It follows from the representation (3.3) (by induction) that $x(n + 1)$ can be written in the form

$$x(n + 1) = \sum_{k=0}^{n} (A')^k B' \ell(n - k)$$

for some sequence $\{\ell(k)\}_{k=0}^{n}$. Thus, $x(n + 1)$ belongs to the linear span of $\{\mathcal{R}\left((A')^k B'\right)\}_{k=0}^{n}$. Conversely, to each such sequence $\{\ell(k)\}_{k=0}^{n}$ corresponds a trajectory on $[0, n]$ for which $x(n+1)$ is given by the formula above. This proves (3.13). Letting $n \to \infty$ in (3.13) we get (3.14). The final statement follows from (3.14) and the definition of controllability. $\qquad\square$

4. The output nulling representation

In our second representation of the generating subspace V we write V as the kernel of a surjective operator of the following type.

Lemma 4.1. *Let V be a subspace of the product space $\mathfrak{K} := \begin{bmatrix} \mathcal{X} \\ \mathcal{X} \\ \mathcal{W} \end{bmatrix}$, where $\mathcal{X}$ and $\mathcal{W}$ are Hilbert spaces. If there exists a Hilbert space $\mathcal{K}$ and four operators*

$$A'' \in \mathcal{B}(\mathcal{X}), \; B'' \in \mathcal{B}(\mathcal{W}; \mathcal{X}), \; C'' \in \mathcal{B}(\mathcal{X}, \mathcal{K}), \; \text{and } D'' \in \mathcal{B}(\mathcal{W}; \mathcal{K}) \qquad (4.1)$$

where

$$D'' \text{ is surjective} \qquad (4.2)$$

such that

$$V = \mathcal{N}\left(\begin{bmatrix} -1_{\mathcal{X}} & A'' & B'' \\ 0 & C'' & D'' \end{bmatrix}\right) = \left\{ \begin{bmatrix} z \\ x \\ w \end{bmatrix} \in \mathfrak{K} \; \middle| \; \begin{matrix} z = A''x + B''w \\ 0 = C''x + D''w \end{matrix} \right\}, \qquad (4.3)$$

then V has properties (i)–(iv) listed in Definition 2.1, i.e., $(V; \mathcal{X}, \mathcal{W})$ is a state/ signal node. Conversely, if V has properties (i)–(iv) listed in Definition 2.1 then V is given by (4.3) for some Hilbert space $\mathcal{K}$ and some operators A'', B'', C'', and D'' satisfying (4.1) and (4.2).

Proof. Trivially, if V is given by (4.3), then V has property (iii). That (i) holds follows from the fact that V is the kernel of the bounded linear operator $\begin{bmatrix} -1_{\mathcal{X}} & A'' & B'' \\ 0 & C'' & D'' \end{bmatrix}$. Define F as in Lemma 2.3. That (iv) holds follows from the fact that $\mathcal{D}(F)$ is the kernel of the bounded linear operator $\begin{bmatrix} C'' & D'' \end{bmatrix}$. Finally, (ii) holds since the surjectivity of D'' guarantees that for every $x \in \mathcal{X}$ it is possible to find some $w \in \mathcal{W}$ such that $C''x + D''w = 0$, i.e., $\begin{bmatrix} x \\ w \end{bmatrix} \in \mathcal{D}(F)$.

Conversely, suppose that V has properties (i)–(iv). Then the operator F in Lemma 2.3 is bounded and $\mathcal{D}(F)$ is closed. Let $\begin{bmatrix} C'' & D'' \end{bmatrix} \in \mathcal{B}(\begin{bmatrix} \mathcal{X} \\ \mathcal{W} \end{bmatrix}; \mathcal{K})$ be an arbitrary surjective operator with $\mathcal{N}\left(\begin{bmatrix} C'' & D'' \end{bmatrix}\right) = \mathcal{D}(F)$ (e.g., let $\mathcal{K}$ be a complementary subspace to $\mathcal{D}(F)$ in $\left[\begin{smallmatrix}\mathcal{X}\\\mathcal{W}\end{smallmatrix}\right]$ and let $\begin{bmatrix}C'' & D''\end{bmatrix} = P_{\mathcal{K}}^{\mathcal{D}(F)})$. Let $\begin{bmatrix}A'' & B''\end{bmatrix}$ be an arbitrary extension of F to an operator in $\mathcal{B}(\left[\begin{smallmatrix}\mathcal{X}\\\mathcal{W}\end{smallmatrix}\right];\mathcal{X})$ (e.g., take $\begin{bmatrix}A'' & B''\end{bmatrix} = FP_{\mathcal{D}(F)}^{\mathcal{K}}$ with $\mathcal{K}$ chosen as above). Then $\begin{bmatrix}C'' & D''\end{bmatrix}$ is surjective and (4.1) and (4.3) hold.

It remains to show that D'' is surjective, and for this we need property (ii) (which has not yet been used). It follows from (4.3) that (ii) holds if and only if $\mathcal{R}(C'') \subset \mathcal{R}(D'')$. Because of the surjectivity of $\begin{bmatrix}C'' & D''\end{bmatrix}$, this is equivalent to (4.2). $\qquad\square$

We shall call a colligation $\Sigma_{s/s/on} := \left(\begin{bmatrix}A'' & B''\\C'' & D''\end{bmatrix};\mathcal{X},\mathcal{W},\mathcal{K}\right)$, where $\mathcal{K}$ is a Hilbert space and A'', B'', C'', and D'' satisfy (4.1)–(4.3) an *output nulling representation of the state/signal node* $\Sigma = (V;\mathcal{X},\mathcal{W})$. (Output nulling representations are known in the finite-dimensional case from the theory of behaviors; see, e.g., [WT02].) We shall also refer to $\Sigma_{s/s/on}$ as a *signal/state/output nulling node*. By the signal/state/output nulling *system* $\Sigma_{s/s/on}$ we mean the node $\Sigma_{s/s/on}$ itself together with the set of all trajectories generated by this node. However, the notion of a trajectory of such a node differs slightly from the corresponding notions for a state/signal node or a driving-variable/state/signal node. By a trajectory of $\Sigma_{s/s/on}$ on $[n_1, n_2]$ we mean a triple of sequences $(x(\cdot), w(\cdot), e(\cdot))$ which satisfy

$$\begin{aligned}
x(k+1) &= A''x(k) + B''w(k),\\
e(k) &= C''x(k) + D''w(k), \qquad n_1 \le k \le n_2.
\end{aligned} \tag{4.4}$$

Here we interpret w as input data and e as output data. Thus, *not every trajectory of (4.4) corresponds to a trajectory of the corresponding state/signal system* Σ; this is true exactly for those trajectories whose output $e(\cdot)$ is null (i.e., it vanishes identically). We shall refer to e as the *error variable*, and to the space $\mathcal{K}$ as the *error space*.

Output nulling representations have a number of important properties listed below.

Proposition 4.2. *Let* $\Sigma = (V;\mathcal{X},\mathcal{W})$ *be a state/signal node with the output nulling representation* $\Sigma_{s/s/on} = \left(\begin{bmatrix}A'' & B''\\C'' & D''\end{bmatrix};\mathcal{X},\mathcal{W},\mathcal{K}\right)$, *and let* $F\colon \mathcal{D}(F) \to \mathcal{X}$ *be the linear operator defined in Lemma 2.3. Then the following assertions are true.*

1) *The operator* F *is given by*

$$F = \begin{bmatrix}A'' & B''\end{bmatrix}\big|_{\mathcal{D}(F)} \text{ with } \mathcal{D}(F) = \mathcal{N}\left(\begin{bmatrix}C'' & D''\end{bmatrix}\right). \tag{4.5}$$

2) *We have*

$$\mathcal{N}(D'') = \mathcal{U}_0,\ \mathcal{N}(C'') = \mathfrak{U}_0,\ \mathcal{R}(B''|_{\mathcal{U}_0}) = \mathfrak{R}_0, \tag{4.6}$$

where $\mathfrak{R}_0, \mathcal{U}_0,$ *and* $\mathfrak{U}_0$ *are defined in (3.5)–(3.7). Consequently, the range and kernels listed above do not depend on the particular choice of* $\Sigma_{s/s/on}$.

3) *Let* $\mathcal{Y}_0$ *be a direct complement in* $\mathcal{W}$ *to the space* $\mathcal{U}_0$ *defined in (3.6), i.e.,* $\mathcal{W} = \mathcal{Y}_0 \dotplus \mathcal{U}_0$. *Then* $D''|_{\mathcal{Y}_0}$ *maps* $\mathcal{Y}_0$ *one-to-one onto* $\mathcal{K}$ *and* $\begin{bmatrix}1_{\mathcal{X}} & B''|_{\mathcal{Y}_0}\\0 & D''|_{\mathcal{Y}_0}\end{bmatrix}$ *maps*

$\left[\begin{smallmatrix} \mathcal{X} \\ \mathcal{Y}_0 \end{smallmatrix}\right]$ *one-to-one onto* $\left[\begin{smallmatrix} \mathcal{X} \\ \mathcal{K} \end{smallmatrix}\right]$, *and consequently, these operators are boundedly invertible. Moreover,*

$$
\begin{bmatrix} F \\ 0 \end{bmatrix} \begin{bmatrix} 1_{\mathcal{X}} \\ H_{\mathcal{Y}_0} \end{bmatrix} = \begin{bmatrix} A'' & B'' \\ C'' & D'' \end{bmatrix} \begin{bmatrix} 1_{\mathcal{X}} \\ H_{\mathcal{Y}_0} \end{bmatrix},
\tag{4.7}
$$

or equivalently,

$$
\begin{bmatrix} A'' \\ C'' \end{bmatrix} = \begin{bmatrix} 1_{\mathcal{X}} & B''|_{\mathcal{Y}_0} \\ 0 & D''|_{\mathcal{Y}_0} \end{bmatrix} \left(\begin{bmatrix} F \\ 0 \end{bmatrix} \begin{bmatrix} 1_{\mathcal{X}} \\ H_{\mathcal{Y}_0} \end{bmatrix} - \begin{bmatrix} 0 \\ H_{\mathcal{Y}_0} \end{bmatrix} \right),
\tag{4.8}
$$

where $H_{\mathcal{Y}_0} \colon \mathcal{X} \to \mathcal{W}$ *is the operator defined by* $H_{\mathcal{Y}_0} x = w$, *where* w *is the unique element in* $\mathcal{Y}_0$ *such that* $\left[\begin{smallmatrix} x \\ w \end{smallmatrix}\right] \in \mathcal{D}(F)$. *Consequently,* A'' *is determined uniquely by* B'' *and* C'' *is determined uniquely by* D''.

4) *The space* $\mathcal{K}$ *is isomorphic to every direct complement in* $\mathcal{W}$ *to the space* $\mathcal{U}_0$ *defined in* (3.6).

Proof. We leave the straightforward proofs of 1) and 2) to the reader. That the restriction of D'' to any complement $\mathcal{Y}_0$ of $\mathcal{U}_0$ is invertible with a bounded inverse follows from the fact that $\mathcal{N}(D'') = \mathcal{U}_0$. This implies that the restriction of $\left[\begin{smallmatrix} 1_{\mathcal{X}} & B'' \\ 0 & D'' \end{smallmatrix}\right]$ to $\left[\begin{smallmatrix} \mathcal{X} \\ \mathcal{Y}_0 \end{smallmatrix}\right]$ is invertible with a bounded inverse. Formula (4.7) follows from (4.3) and (4.5). Clearly (4.8) is equivalent to (4.7). Finally, 4) follows from the invertibility of $D''|_{\mathcal{Y}_0}$ established in 3). $\qquad\square$

Theorem 4.3. *Let* $\Sigma_{s/s/on} = \left(\left[\begin{smallmatrix} A'' & B'' \\ C'' & D'' \end{smallmatrix}\right]; \mathcal{X}, \mathcal{W}, \mathcal{K} \right)$ *be an output nulling representation of a state/signal system Σ, and let*

$$
\begin{bmatrix} A''_1 & B''_1 \\ C''_1 & D''_1 \end{bmatrix} = \begin{bmatrix} 1_{\mathcal{X}} & K'' \\ 0 & M'' \end{bmatrix} \begin{bmatrix} A'' & B'' \\ C'' & D'' \end{bmatrix},
\tag{4.9}
$$

where

$$
K'' \in \mathcal{B}(\mathcal{K}, \mathcal{X}), \ M'' \in \mathcal{B}(\mathcal{K}, \mathcal{K}_1), \ and \ M'' \ has \ a \ bounded \ inverse,
\tag{4.10}
$$

for some Hilbert space $\mathcal{K}_1$. Then

$$
\Sigma^1_{s/s/on} = \left(\begin{bmatrix} A''_1 & B''_1 \\ C''_1 & D''_1 \end{bmatrix}; \mathcal{X}, \mathcal{W}, \mathcal{K}_1 \right)
$$

is an output nulling representation of Σ. Conversely, every output nulling representation $\Sigma^1_{s/s/on}$ of Σ may be obtained from the formula (4.9) for some operators M'' and K'' satisfying (4.10). The operators M'' and K'' are uniquely defined by $\Sigma_{s/s/on}$ and $\Sigma^1_{s/s/on}$ via

$$
M'' D'' = D''_1 \quad and \quad K'' D'' = B''_1 - B''.
\tag{4.11}
$$

Proof. Suppose that $\Sigma^1_{s/s/on} = \left(\left[\begin{smallmatrix} A''_1 & B''_1 \\ C''_1 & D''_1 \end{smallmatrix}\right]; \mathcal{X}, \mathcal{W}, \mathcal{K}_1 \right)$ is given by (4.9) for some operators K'' and M'' satisfying (4.10). It follows from (4.9) and (4.10) that

$D_1'' = M''D''$ is surjective, that $\begin{bmatrix} 1_{\mathcal{X}} & 0 & 0 \\ 0 & 1_{\mathcal{X}} & K'' \\ 0 & 0 & M'' \end{bmatrix}$ is invertible, and that

$$\mathcal{N}\left(\begin{bmatrix} -1_{\mathcal{X}} & A_1'' & B_1'' \\ 0 & C_1'' & D_1'' \end{bmatrix}\right) = \mathcal{N}\left(\begin{bmatrix} 1_{\mathcal{X}} & 0 & 0 \\ 0 & 1_{\mathcal{X}} & K'' \\ 0 & 0 & M'' \end{bmatrix}\begin{bmatrix} -1_{\mathcal{X}} & A'' & B'' \\ 0 & C'' & D'' \end{bmatrix}\right)$$

$$= \mathcal{N}\left(\begin{bmatrix} -1_{\mathcal{X}} & A'' & B'' \\ 0 & C'' & D'' \end{bmatrix}\right) = V.$$

Thus $\Sigma^1_{s/s/on}$ is an output nulling representation of Σ.

We next turn to the converse part. Let $\mathcal{Y}$ be an arbitrary complement to $\mathcal{D}(F)$. By part 3) of Proposition 4.2, the operator

$$\begin{bmatrix} G'' & K'' \\ H'' & M'' \end{bmatrix} := \begin{bmatrix} 1_{\mathcal{X}} & B_1''|_{\mathcal{Y}_0} \\ 0 & D_1''|_{\mathcal{Y}_0} \end{bmatrix}\begin{bmatrix} 1_{\mathcal{X}} & B''|_{\mathcal{Y}_0} \\ 0 & D''|_{\mathcal{Y}_0} \end{bmatrix}^{-1}$$

is a bounded linear operator mapping $\begin{bmatrix} \mathcal{X} \\ \mathcal{K} \end{bmatrix}$ one-to-one onto $\begin{bmatrix} \mathcal{X} \\ \mathcal{K}_1 \end{bmatrix}$. It follows from the identity $\begin{bmatrix} 1_{\mathcal{X}} & B_1''|_{\mathcal{Y}_0} \\ 0 & D_1''|_{\mathcal{Y}_0} \end{bmatrix} = \begin{bmatrix} G'' & K'' \\ H'' & M'' \end{bmatrix}\begin{bmatrix} 1_{\mathcal{X}} & B''|_{\mathcal{Y}_0} \\ 0 & D''|_{\mathcal{Y}_0} \end{bmatrix}$ that $G'' = 1_{\mathcal{X}}$ and that $H'' = 0$, and the invertibility of $\begin{bmatrix} G'' & K'' \\ H'' & M'' \end{bmatrix} = \begin{bmatrix} 1_{\mathcal{X}} & K'' \\ 0 & M'' \end{bmatrix}$ implies that M'' is invertible. Thus, (4.10) and (4.11) hold. By (4.8), $\begin{bmatrix} 1_{\mathcal{X}} & B_1''|_{\mathcal{Y}_0} \\ 0 & D_1''|_{\mathcal{Y}_0} \end{bmatrix}^{-1}\begin{bmatrix} A_1'' \\ C_1'' \end{bmatrix} = \begin{bmatrix} 1_{\mathcal{X}} & B''|_{\mathcal{Y}_0} \\ 0 & D''|_{\mathcal{Y}_0} \end{bmatrix}^{-1}\begin{bmatrix} A'' \\ C'' \end{bmatrix}$, hence $\begin{bmatrix} A_1'' \\ C_1'' \end{bmatrix} = \begin{bmatrix} 1_{\mathcal{X}} & K'' \\ 0 & M'' \end{bmatrix}\begin{bmatrix} A'' \\ C'' \end{bmatrix}$. Thus equation (4.9) holds.

Finally, we remark that (4.11) determines K'' and M'' uniquely since D'' is surjective. $\qquad\square$

Definition 4.4. Let $\Sigma = (V; \mathcal{X}, \mathcal{W})$ be a state/signal system.

1) By an *unobservable* trajectory of Σ on $[0, n]$ or on $\mathbb{Z}^+$ we mean a trajectory $(x(\cdot), 0)$ (i.e., the signal component of this trajectory is identically zero on $[0, n]$ or on $\mathbb{Z}^+$).
2) The *unobservable subspace* $\mathfrak{U}_n$ of Σ *in time* n is the subspace of the initial states $x(0)$ of all unobservable trajectories $(x(\cdot), 0)$ of Σ on $[0, n]$.
3) The *unobservable subspace* $\mathfrak{U}$ of Σ (in infinite time) is the subspace of the initial states $x(0)$ of all unobservable trajectories $(x(\cdot), 0)$ of Σ on $\mathbb{Z}^+$.
4) The system is (approximately) *observable* if the unobservable subspace is $\{0\}$.

Thus,

$$\mathfrak{U}_{n+1} \subset \mathfrak{U}_n, \qquad \mathfrak{U} = \cap_{n \in \mathbb{Z}^+}\mathfrak{U}_n.$$

Observe, in particular, that the subspace $\mathfrak{U}_0$ defined above coincides with the subspace $\mathfrak{U}_0$ defined in (3.7).

The subspaces $\mathfrak{U}_n$ and $\mathfrak{U}$ in Definition 4.4 have the following simple characterizations in terms of an arbitrary output nulling representation of Σ.

Proposition 4.5. *Let $\Sigma = (V; \mathcal{X}, \mathcal{W})$ be a state/signal system and let $\Sigma_{s/s/on} = \left(\left[\begin{smallmatrix} A'' & B'' \\ C'' & D'' \end{smallmatrix}\right]; \mathcal{X}, \mathcal{W}, \mathcal{K}\right)$ be an output nulling representation of this system. Then*

$$\mathfrak{U}_n = \cap_{0 \leq k \leq n} \mathcal{N}\left(C''(A'')^k\right), \tag{4.12}$$

$$\mathfrak{U} = \cap_{k \in \mathbb{Z}^+} \mathcal{N}\left(C''(A'')^k\right). \tag{4.13}$$

In particular, Σ is observable if and only if

$$\cap_{k \in \mathbb{Z}^+} \mathcal{N}\left(C''(A'')^k\right) = \{0\}. \tag{4.14}$$

Proof. If $x_0 \in \cap_{0 \leq k \leq n} \mathcal{N}\left(C''(A'')^k\right)$, i.e., if $C''(A'')^k x_0 = 0$ for $0 \leq k \leq n$, then it follows from (4.3) that $(x(\cdot), w(\cdot))$, where $x(k) = (A'')^k x_0$ and $w(k) = 0, 0 \leq k \leq n$, is a trajectory of Σ on the interval $[0, n]$. Thus, $x_0 \in \mathfrak{U}_n$ in this case. Conversely, if $(x(\cdot), w(\cdot))$ is a trajectory of Σ on $[0, n]$ with $x(0) = x_0$ and $w(k) = 0, 0 \leq k \leq n$, then by (4.3)

$$x(k+1) = A''x(k)$$

$$0 = C''x(k), \quad 0 \leq k \leq n,$$

which gives $x_0 \in \mathcal{N}\left(C''(A'')^k\right)$ for all k, $0 \leq k \leq n$. Thus (4.12) holds. Letting $n \to \infty$ in (4.12) we get (4.13). The final statement follows from (4.13) and the definition of observability. $\qquad\square$

5. The input/state/output representation

In this section we shall discuss a third type of representation of a state/signal system $\Sigma = (V; \mathcal{X}, \mathcal{W})$ in which trajectories $(x(\cdot), w(\cdot))$ on $\mathbb{Z}^+$ of Σ are described by the usual system of equations (1.2) in the traditional input/state/output theory.

Theorem 5.1. *Let V be a subspace of the product space $\mathfrak{K} := \left[\begin{smallmatrix} \mathcal{X} \\ \mathcal{X} \\ \mathcal{W} \end{smallmatrix}\right]$, where $\mathcal{X}$ and $\mathcal{W}$ are Hilbert spaces, and suppose that $\mathcal{W} = \mathcal{Y} \dotplus \mathcal{U}$ is the direct sum of two complementary closed subspaces $\mathcal{Y}$ and $\mathcal{U}$. If there exists four operators*

$$A \in \mathcal{B}(\mathcal{X}), \; B \in \mathcal{B}(\mathcal{U}; \mathcal{X}), \; C \in \mathcal{B}(\mathcal{X}, \mathcal{Y}), \; and \; D \in \mathcal{B}(\mathcal{U}; \mathcal{Y}), \tag{5.1}$$

such that

$$V = \mathcal{R}\left(\begin{bmatrix} A & B \\ 1_{\mathcal{X}} & 0 \\ C & D \\ 0 & 1_{\mathcal{U}} \end{bmatrix}\right) = \mathcal{N}\left(\begin{bmatrix} -1_{\mathcal{X}} & A & 0 & B \\ 0 & C & -1_{\mathcal{Y}} & D \end{bmatrix}\right)$$

$$= \left\{ \begin{bmatrix} Ax + Bu \\ x \\ Cx + Du + u \end{bmatrix} \;\middle|\; x \in \mathcal{X}, \; u \in \mathcal{U} \right\}, \tag{5.2}$$

then V has properties (i)–(iv) listed in Definition 2.1, i.e., $(V; \mathcal{X}, \mathcal{W})$ is a state/ signal node. Conversely, if V has properties (i)–(iv) listed in Definition 2.1 then V is given by (5.2) for some operators A, B, C, and D satisfying (5.1) for some decomposition $\mathcal{W} = \mathcal{Y} \dotplus \mathcal{U}$. These operators are uniquely defined by V and by the decomposition $\mathcal{W} = \mathcal{Y} \dotplus \mathcal{U}$.

Proof. The representation (5.2) has an obvious interpretation as a driving variable representation of V (take $C' = \left[\begin{smallmatrix} C \\ 0 \end{smallmatrix}\right]$ and $D' = \left[\begin{smallmatrix} D \\ 1_{\mathcal{U}} \end{smallmatrix}\right]$). Thus, by Lemma 3.1, if V is given by (5.2) for some operators A, B, C, and D satisfying (5.1), then V has properties (i)–(iv).

To prove the converse part we start from an arbitrary driving variable representation of V (e.g., from the one constructed in the proof of the converse part of Lemma 3.1), i.e., we let $\mathcal{L}$ be a Hilbert space, and let A', B', C', and D' satisfy (3.1)–(3.3). Then each $\left[\begin{smallmatrix} z \\ x \\ w \end{smallmatrix}\right] \in V$ can be written in the form

$$\begin{bmatrix} z \\ x \\ w \end{bmatrix} = \begin{bmatrix} A' & B' \\ 1_{\mathcal{X}} & 0 \\ C' & D' \end{bmatrix} \begin{bmatrix} x \\ \ell \end{bmatrix},$$

for a unique $\ell \in \mathcal{L}$. Let $\mathcal{W} = \mathcal{Y} \dotplus \mathcal{U}$ be an arbitrary decomposition of $\mathcal{W}$ with the property that $P_{\mathcal{U}}^{\mathcal{Y}} D'$ maps $\mathcal{L}$ one-to-one onto $\mathcal{U}$ (for example, we can take $\mathcal{U} = \mathcal{U}_0$, with $\mathcal{U}_0$ defined as in (3.6), and take $\mathcal{Y}$ to be an arbitrary direct complement to $\mathcal{U}_0$). With respect to this decomposition of $\mathcal{W}$ the vector $\left[\begin{smallmatrix} z \\ x \\ w \end{smallmatrix}\right]$ can be written in the form (where we denote $u = P_{\mathcal{U}}^{\mathcal{Y}} w$ and $y = P_{\mathcal{Y}}^{\mathcal{U}} w$)

$$\begin{bmatrix} z \\ x \\ y \\ u \end{bmatrix} = \begin{bmatrix} A' & B' \\ 1_{\mathcal{X}} & 0 \\ P_{\mathcal{Y}}^{\mathcal{U}} C' & P_{\mathcal{Y}}^{\mathcal{U}} D' \\ P_{\mathcal{U}}^{\mathcal{Y}} C' & P_{\mathcal{U}}^{\mathcal{Y}} D' \end{bmatrix} \begin{bmatrix} x \\ \ell \end{bmatrix}.$$

Since $P_{\mathcal{U}}^{\mathcal{Y}} D'$ is boundedly invertible, we can solve for ℓ to get the equivalent representation

$$\begin{bmatrix} z \\ x \\ y \\ u \end{bmatrix} = \begin{bmatrix} A' & B' \\ 1_{\mathcal{X}} & 0 \\ P_{\mathcal{Y}}^{\mathcal{U}} C' & P_{\mathcal{Y}}^{\mathcal{U}} D' \\ P_{\mathcal{U}}^{\mathcal{Y}} C' & P_{\mathcal{U}}^{\mathcal{Y}} D' \end{bmatrix} \begin{bmatrix} 1_{\mathcal{X}} & 0 \\ P_{\mathcal{U}}^{\mathcal{Y}} C' & P_{\mathcal{U}}^{\mathcal{Y}} D' \end{bmatrix}^{-1} \begin{bmatrix} x \\ u \end{bmatrix}$$

$$= \begin{bmatrix} A' - B'(P_{\mathcal{U}}^{\mathcal{Y}} D')^{-1} P_{\mathcal{U}}^{\mathcal{Y}} C' & B'(P_{\mathcal{U}}^{\mathcal{Y}} D')^{-1} \\ 1_{\mathcal{X}} & 0 \\ P_{\mathcal{Y}}^{\mathcal{U}} C' - P_{\mathcal{Y}}^{\mathcal{U}} D'(P_{\mathcal{U}}^{\mathcal{Y}} D')^{-1} P_{\mathcal{U}}^{\mathcal{Y}} C' & P_{\mathcal{Y}}^{\mathcal{U}} D'(P_{\mathcal{U}}^{\mathcal{Y}} D')^{-1} \\ 0 & 1_{\mathcal{U}} \end{bmatrix} \begin{bmatrix} x \\ u \end{bmatrix}.$$

This representation is of the type (5.2) with

$$\begin{aligned}
\begin{bmatrix} A & B \\ C & D \end{bmatrix} &= \begin{bmatrix} A' & B' \\ P_{\mathcal{Y}}^{\mathcal{U}} C' & P_{\mathcal{Y}}^{\mathcal{U}} D' \end{bmatrix} \begin{bmatrix} 1_{\mathcal{X}} & 0 \\ P_{\mathcal{U}}^{\mathcal{Y}} C' & P_{\mathcal{U}}^{\mathcal{Y}} D' \end{bmatrix}^{-1} \\
&= \begin{bmatrix} A' - B'(P_{\mathcal{U}}^{\mathcal{Y}} D')^{-1} P_{\mathcal{U}}^{\mathcal{Y}} C' & B'(P_{\mathcal{U}}^{\mathcal{Y}} D')^{-1} \\ P_{\mathcal{Y}}^{\mathcal{U}} C' - P_{\mathcal{Y}}^{\mathcal{U}} D'(P_{\mathcal{U}}^{\mathcal{Y}} D')^{-1} P_{\mathcal{U}}^{\mathcal{Y}} C' & P_{\mathcal{Y}}^{\mathcal{U}} D'(P_{\mathcal{U}}^{\mathcal{Y}} D')^{-1} \end{bmatrix}.
\end{aligned} \tag{5.3}$$

134   D.Z. Arov and O.J. Staffans

The uniqueness of $\left[\begin{smallmatrix} A & B \\ C & D \end{smallmatrix}\right]$ follows from the fact that (5.2) is a graph representation of V with respect to the decomposition of $\mathfrak{K}$ into $\mathfrak{K} = \left[\begin{smallmatrix} \mathcal{X} \\ 0 \\ \mathcal{Y} \\ 0 \end{smallmatrix}\right] \dotplus \left[\begin{smallmatrix} 0 \\ \mathcal{X} \\ 0 \\ \mathcal{U} \end{smallmatrix}\right]$, and the operator appearing in this graph representation is unique. $\qquad\square$

We shall call a colligation $\Sigma_{i/s/o} := \left(\left[\begin{smallmatrix} A & B \\ C & D \end{smallmatrix}\right]; \mathcal{X}, \mathcal{U}, \mathcal{Y}\right)$, where $\mathcal{W} = \mathcal{Y} \dotplus \mathcal{U}$ and A, B, C, and D satisfy (5.1) and (5.2) an *input/state/output representation of the state/signal node* $\Sigma = (V; \mathcal{X}, \mathcal{W})$. We shall also refer to $\Sigma_{i/s/o}$ as an *input/state/output node*. By the input/state/output *system* $\Sigma_{i/s/o}$ we mean the node $\Sigma_{i/s/o}$ itself together with the set of all trajectories $(x(\cdot), u(\cdot), y(\cdot))$ generated by this node through the equations

$$x(k+1) = Ax(k) + Bu(k),$$
$$y(k) = Cx(k) + Du(k), \qquad n_1 \leq k \leq n_2. \tag{5.4}$$

The subspace $\mathcal{U}$ considered above is called an *input space*, and the vector $u \in \mathcal{U}$ in (5.2) is called an *input variable*. Analogously, the subspace $\mathcal{Y}$ considered above is called an *output space*, and the vector $y \in \mathcal{Y}$ in (5.2) is called an *output variable*. From each trajectory $(x(\cdot), u(\cdot), y(\cdot))$ of the input/state/output system $\Sigma_{i/s/o}$ we get a trajectory $(x(\cdot), w(\cdot))$ of the state/signal system Σ by taking $w = u + y$, and conversely, from each trajectory $(x(\cdot), w(\cdot))$ of the state/signal system Σ we get a trajectory $(x(\cdot), u(\cdot), y(\cdot))$ of the input/state/output system $\Sigma_{i/s/o}$ by taking $u = P_{\mathcal{U}}^{\mathcal{Y}} w$ and $y = P_{\mathcal{Y}}^{\mathcal{U}} w$.

Remark 5.2. Every input/state/output representation can be interpreted both as a driving variable representation and as an output nulling representation. In both cases we combined u and y into the signal vector $w = \left[\begin{smallmatrix} y \\ u \end{smallmatrix}\right]$. We get a driving variable representation by writing (5.2) in the form

$$z = Ax + Bu,$$
$$\begin{bmatrix} y \\ u \end{bmatrix} = \begin{bmatrix} C \\ 0 \end{bmatrix} x + \begin{bmatrix} D \\ 1_{\mathcal{U}} \end{bmatrix} u,$$

with driving variable space $\mathcal{U}$ (the operator $D' = \left[\begin{smallmatrix} D \\ 1_{\mathcal{U}} \end{smallmatrix}\right]$ is injective and has closed range), and we get an output nulling representation by writing it in the form

$$z = Ax + \begin{bmatrix} 0 & B \end{bmatrix} \begin{bmatrix} y \\ u \end{bmatrix},$$
$$0 = Cx + \begin{bmatrix} -1_{\mathcal{Y}} & D \end{bmatrix} \begin{bmatrix} y \\ u \end{bmatrix},$$

with error space $\mathcal{Y}$ (the operator $D'' = \begin{bmatrix} -1_{\mathcal{Y}} & D \end{bmatrix}$ is surjective).

Remark 5.3. In the standard input/state/output systems theory one considers trajectories $(x(\cdot), u(\cdot), y(\cdot))$ generated by (5.4), but the input space $\mathcal{U}$ and the output space $\mathcal{Y}$ are not required to be complementary subspaces of a given signal space $\mathcal{W}$. Nevertheless, also in this situation it is possible to introduce the product space $\mathcal{W} = \left[\begin{smallmatrix} \mathcal{Y} \\ \mathcal{U} \end{smallmatrix}\right]$ with an appropriate inner product, to identify $\mathcal{Y}$ with the subspace

$\left[\begin{smallmatrix} \mathcal{Y} \\ 0 \end{smallmatrix}\right]$ of $\mathcal{W}$, and to identify $\mathcal{U}$ with the subspace $\left[\begin{smallmatrix} 0 \\ \mathcal{U} \end{smallmatrix}\right]$ of $\mathcal{W}$. Then $\mathcal{W} = \mathcal{Y} \dotplus \mathcal{U}$, the triple $\Sigma = (V; \mathcal{X}, \mathcal{W})$ with V defined by (5.2) is a state/signal node, and the original input/state/output system is an input/state/output representation of this node.

Remark 5.4. Each driving variable representation $\Sigma_{dv/s/s}$ of a state/signal system may be interpreted as an input/state/output system, with the driving variable as input data and the original signal as output data. We can and will therefore apply all notions, notations, and results that we will define or obtain for input/state/output systems to such driving variable representations. In this connection we throughout replace the word "input" by "driving" and the word "output" by "signal". An analogous remark is valid for output nulling representations of state signal systems. When we interpret such representations as input/state/output systems we throughout replace the word "input" by "signal" and the word "output" by "error".

Proposition 5.5. *Let $\Sigma = (V; \mathcal{X}, \mathcal{W})$ be a state/signal system, with an input/state/ output representation $\Sigma_{i/s/o} = \left(\left[\begin{smallmatrix} A & B \\ C & D \end{smallmatrix}\right]; \mathcal{X}, \mathcal{U}, \mathcal{Y}\right)$.*

1) *The reachable subspaces $\mathfrak{R}_n$ in time n and the reachable subspace $\mathfrak{R}$ are given by*

$$\mathfrak{R}_n = \mathrm{span}\{\mathcal{R}\left(A^k B\right) \mid 0 \le k \le n\}, \quad n \in \mathbb{Z}^+, \tag{5.5}$$

$$\mathfrak{R} = \vee_{k \in \mathbb{Z}^+} \mathcal{R}\left(A^k B\right). \tag{5.6}$$

In particular, Σ is controllable if and only if

$$\mathcal{X} = \vee_{k \in \mathbb{Z}^+} \mathcal{R}\left(A^k B\right). \tag{5.7}$$

2) *The unobservable subspaces $\mathfrak{U}_n$ in time n and the unobservable subspace $\mathfrak{U}$ are given by*

$$\mathfrak{U}_n = \cap_{0 \le k \le n} \mathcal{N}\left(CA^k\right), \tag{5.8}$$

$$\mathfrak{U} = \cap_{k \in \mathbb{Z}^+} \mathcal{N}\left(CA^k\right). \tag{5.9}$$

In particular, Σ is observable if and only if

$$\cap_{k \in \mathbb{Z}^+} \mathcal{N}\left(CA^k\right) = \{0\}. \tag{5.10}$$

Proof. This follows from Propositions 3.5 and 4.5 and Remark 5.2. $\qquad\square$

Definition 5.6. Let $\Sigma = (V; \mathcal{X}, \mathcal{W})$ be a state/signal system. We call the ordered direct sum decomposition $\mathcal{W} = \mathcal{Y} \dotplus \mathcal{U}$ (also denoted by $\mathcal{W} = \left[\begin{smallmatrix} \mathcal{Y} \\ \mathcal{U} \end{smallmatrix}\right]$) an *admissible (input/output) decomposition for* Σ if Σ has an input/state/output representation with input space $\mathcal{U}$ and output space $\mathcal{Y}$.

Our following theorem characterizes the set of all admissible input/output decompositions.

Lemma 5.7. *Let $\Sigma = (V; \mathcal{X}, \mathcal{W})$ be a state/signal node, and let $\mathcal{W} = \mathcal{Y} \dotplus \mathcal{U}$ be a direct sum decomposition of $\mathcal{W}$. Define $\mathcal{U}_0$ as in (3.6). Then the following statements are equivalent:*

1) $\mathcal{W} = \mathcal{Y} \dotplus \mathcal{U}$ is an admissible input/output decomposition for Σ.

2) $P_{\mathcal{U}}^{\mathcal{Y}}|_{\mathcal{U}_0}$ maps $\mathcal{U}_0$ one-to-one onto $\mathcal{U}$, i.e., $(P_{\mathcal{U}}^{\mathcal{Y}}|_{\mathcal{U}_0})^{-1} \in \mathcal{B}(\mathcal{U};\mathcal{U}_0)$.

3) The space $\mathcal{U}_0$ has the graph representation

$$\mathcal{U}_0 = \left\{ w = \begin{bmatrix} D \\ 1_{\mathcal{U}} \end{bmatrix} u \mid u \in \mathcal{U} \right\}, \tag{5.11}$$

for some $D \in \mathcal{B}(\mathcal{U};\mathcal{Y})$.

If the decomposition $\mathcal{W} = \mathcal{Y} \dotplus \mathcal{U}$ is admissible for Σ, then the operator D in (5.11) coincides with the operator D in (5.2).

Proof. Proof of 1) $\Rightarrow$ 3): If 1) holds, then the representation (5.2) of V gives us a graph space representation of $\mathcal{U}_0$ (with the same operator D as in (5.2)).

Proof of 3) $\Rightarrow$ 2): If 3) holds, then $P_{\mathcal{U}}^{\mathcal{Y}}$ maps $\mathcal{U}_0$ one-to-one onto $\mathcal{U}$, and $D = P_{\mathcal{Y}}^{\mathcal{U}}(P_{\mathcal{U}}^{\mathcal{Y}}|_{\mathcal{U}_0})^{-1}$.

Proof of 2) $\Rightarrow$ 1): Let $\Sigma_{dv/s/s} = \left(\begin{bmatrix} A' & B' \\ C' & D' \end{bmatrix}; \mathcal{X}, \mathcal{L}, \mathcal{W} \right)$ be an arbitrary driving variable representation of Σ. Then $P_{\mathcal{U}}^{\mathcal{Y}}$ maps $\mathcal{U}_0$ one-to-one onto $\mathcal{U}$ and $P_{\mathcal{U}}^{\mathcal{Y}} D'$ maps $\mathcal{L}$ one-to-one onto $\mathcal{U}$. The proof of Theorem 5.1 provides us with an input/state/output representation of Σ with input space $\mathcal{U}$ and output space $\mathcal{Y}$. $\square$

Remark 5.8. According to Lemma 5.7, if $\mathcal{Y}$ is an arbitrary direct complement to the subspace $\mathcal{U}_0$ in (3.6), then $\mathcal{W} = \mathcal{Y} \dotplus \mathcal{U}_0$ is an admissible decomposition for Σ. For this reason we shall refer to $\mathcal{U}_0$ as the *canonical input space*.

The admissibility of a given decomposition of the signal space of a given state/signal system Σ can also be studied by means of a given driving variable, or output nulling, or input/state/output representation of the given system Σ.

Lemma 5.9. *Let $\Sigma = (V; \mathcal{X}, \mathcal{W})$ be a state/signal node with the driving variable representation $\Sigma_{dv/s/s} = \left(\begin{bmatrix} A' & B' \\ C' & D' \end{bmatrix}; \mathcal{X}, \mathcal{L}, \mathcal{W} \right)$.*

1) $\mathcal{W} = \mathcal{Y} \dotplus \mathcal{U}$ *is an admissible input/output decomposition for Σ if and only if*

$$P_{\mathcal{U}}^{\mathcal{Y}} D' \text{ maps } \mathcal{L} \text{ one-to-one onto } \mathcal{U}, \text{ i.e., } (P_{\mathcal{U}}^{\mathcal{Y}} D')^{-1} \in \mathcal{B}(\mathcal{U};\mathcal{L}). \tag{5.12}$$

2) *If the decomposition $\mathcal{W} = \mathcal{Y} \dotplus \mathcal{U}$ is admissible for Σ, then the corresponding operators A, B, C, and D in (5.2) are given by (5.3).*

Proof. In the proof of Theorem 5.1 we constructed an input/state/output representation of Σ under the assumption that (5.12) holds. Thus, (5.12) is sufficient for admissibility. Conversely, suppose that the decomposition is admissible for Σ. Then by Lemma 5.7, $P_{\mathcal{U}}^{\mathcal{Y}}$ maps the canonical input space $\mathcal{U}_0 = \mathcal{R}(D')$ one-to-one onto $\mathcal{U}$, and D' is injective. Thus, (5.12) is also necessary for admissibility. $\square$

Lemma 5.10. *Let $\Sigma = (V; \mathcal{X}, \mathcal{W})$ be a state/signal node with the output nulling representation $\Sigma_{s/s/on} = \left(\begin{bmatrix} A'' & B'' \\ C'' & D'' \end{bmatrix}; \mathcal{X}, \mathcal{W}, \mathcal{K} \right)$, and let $\mathcal{W} = \mathcal{Y} \dotplus \mathcal{U}$ be a direct sum decomposition of $\mathcal{W}$.*

1) $\mathcal{W} = \mathcal{Y} \dotplus \mathcal{U}$ *is an admissible input/output decomposition for Σ if and only if*

$$D''|_{\mathcal{Y}} \text{ maps } \mathcal{Y} \text{ one-to-one onto } \mathcal{K}, \text{ i.e., } (D''|_{\mathcal{Y}})^{-1} \in \mathcal{B}(\mathcal{K};\mathcal{Y}). \tag{5.13}$$

2) *If the decomposition $W = \mathcal{Y} \dotplus \mathcal{U}$ is admissible for Σ, then the corresponding operators A, B, C, and D in (5.2) are given by*

$$\begin{bmatrix} A & B \\ C & D \end{bmatrix} = \begin{bmatrix} 1_\mathcal{X} & -B''|_\mathcal{Y} \\ 0 & -D''|_\mathcal{Y} \end{bmatrix}^{-1} \begin{bmatrix} A'' & B''|_\mathcal{U} \\ C'' & D''|_\mathcal{U} \end{bmatrix}$$
$$= \begin{bmatrix} A'' - B''|_\mathcal{Y}(D''|_\mathcal{Y})^{-1}C'' & B''|_\mathcal{U} - B''|_\mathcal{Y}(D''|_\mathcal{Y})^{-1}D''|_\mathcal{U} \\ -(D''|_\mathcal{Y})^{-1}C'' & -(D''|_\mathcal{Y})^{-1}D''|_\mathcal{U} \end{bmatrix}. \tag{5.14}$$

Proof. Take an arbitrary $\begin{bmatrix} z \\ x \\ w \end{bmatrix} \in \mathfrak{K}$. By (4.3), $\begin{bmatrix} z \\ x \\ w \end{bmatrix} \in V$ if and only if

$$\begin{bmatrix} z \\ 0 \end{bmatrix} = \begin{bmatrix} A'' & B'' \\ C'' & D'' \end{bmatrix} \begin{bmatrix} x \\ w \end{bmatrix}.$$

With $u = P_\mathcal{U}^\mathcal{Y} w$ and $y = P_\mathcal{Y}^\mathcal{U} w$ this can be written in the equivalent form

$$\begin{bmatrix} z \\ 0 \end{bmatrix} = \begin{bmatrix} A'' & B''|_\mathcal{Y} & B''|_\mathcal{U} \\ C'' & D''|_\mathcal{Y} & D''|_\mathcal{U} \end{bmatrix} \begin{bmatrix} x \\ y \\ u \end{bmatrix}. \tag{5.15}$$

If the decomposition $W = \mathcal{Y} \dotplus \mathcal{U}$ is admissible for Σ, then the condition $\begin{bmatrix} z \\ x \\ w \end{bmatrix} \in V$ determines y uniquely as a continuous function of x and u (by (5.2), $y = Cx + Du$), and therefore the operator $D''|_\mathcal{Y}$ in (5.15) must map $\mathcal{Y}$ one-to-one onto $\mathcal{K}$ (recall that the range of D'' is all of $\mathcal{K}$). Thus (5.13) is a necessary condition for admissibility. Conversely, suppose that (5.13) holds. Then (5.15) can be written in the equivalent form

$$\begin{bmatrix} z \\ y \end{bmatrix} = \begin{bmatrix} 1_\mathcal{X} & -B''|_\mathcal{Y} \\ 0 & -D''|_\mathcal{Y} \end{bmatrix}^{-1} \begin{bmatrix} A'' & B''|_\mathcal{U} \\ C'' & D''|_\mathcal{U} \end{bmatrix} \begin{bmatrix} x \\ u \end{bmatrix}$$
$$= \begin{bmatrix} A'' - B''|_\mathcal{Y}(D''|_\mathcal{Y})^{-1}C'' & B''|_\mathcal{U} - B''|_\mathcal{Y}(D''|_\mathcal{Y})^{-1}D''|_\mathcal{U} \\ -(D''|_\mathcal{Y})^{-1}C'' & -(D''|_\mathcal{Y})^{-1}D''|_\mathcal{U} \end{bmatrix} \begin{bmatrix} x \\ u \end{bmatrix}.$$

This is an input/state/output representation with A'', B'', C'', and D'' given by (5.14). Thus, (5.13) is also sufficient for the admissibility of the decomposition $W = \mathcal{Y} \dotplus \mathcal{U}$. $\square$

Theorem 5.11. *Let $\Sigma = (V; \mathcal{X}, \mathcal{W})$ be a state/signal node with the input/state/output representation $\Sigma_{i/s/o} = \left(\begin{bmatrix} A & B \\ C & D \end{bmatrix}; \mathcal{X}, \mathcal{U}, \mathcal{Y}\right)$. Let $W = \mathcal{Y}_1 \dotplus \mathcal{U}_1$ be a direct sum decomposition of W, and define $\Theta \in \mathcal{B}(\begin{bmatrix} \mathcal{Y} \\ \mathcal{U} \end{bmatrix}; \begin{bmatrix} \mathcal{Y}_1 \\ \mathcal{U}_1 \end{bmatrix})$ by (1.6).*

1) *$W = \mathcal{Y}_1 \dotplus \mathcal{U}_1$ is an admissible input/output decomposition for Σ if and only if*

$$\Theta_{21}D + \Theta_{22} \text{ maps } \mathcal{U} \text{ one-to-one onto } \mathcal{U}_1, \text{ i.e.,}$$
$$(\Theta_{21}D + \Theta_{22})^{-1} \in \mathcal{B}(\mathcal{U}_1; \mathcal{U}). \tag{5.16}$$

2) *If the decomposition $W = \mathcal{Y}_1 \dotplus \mathcal{U}_1$ is admissible for Σ, then the corresponding operators A_1, B_1, C_1, and D_1 are given by*

$$\begin{bmatrix} A_1 & B_1 \\ C_1 & D_1 \end{bmatrix} = \begin{bmatrix} A & B \\ \Theta_{11}C & \Theta_{11}D + \Theta_{12} \end{bmatrix} \begin{bmatrix} 1_\mathcal{X} & 0 \\ \Theta_{21}C & \Theta_{21}D + \Theta_{22} \end{bmatrix}^{-1}, \tag{5.17}$$

or equivalently,

$$\begin{aligned}
A_1 &= A - B(\Theta_{21}D + \Theta_{22})^{-1}\Theta_{21}C, \\
B_1 &= B(\Theta_{21}D + \Theta_{22})^{-1}, \\
C_1 &= \Theta_{11}C - (\Theta_{11}D + \Theta_{12})(\Theta_{21}D + \Theta_{22})^{-1}\Theta_{21}C, \\
D_1 &= (\Theta_{11}D + \Theta_{12})(\Theta_{21}D + \Theta_{22})^{-1}.
\end{aligned} \tag{5.18}$$

Proof. This follows from Remark 5.2 and Lemma 5.9. $\qquad\square$

Theorem 5.12. *Let* $\Sigma = (V; \mathcal{X}, \mathcal{W})$ *be a state/signal node with the input/state/output representation* $\Sigma_{i/s/o} = ([\begin{smallmatrix} A & B \\ C & D \end{smallmatrix}]; \mathcal{X}, \mathcal{U}, \mathcal{Y})$, *and let* $\mathcal{W} = \mathcal{Y}_1 \dotplus \mathcal{U}_1$ *be a direct sum decomposition of* $\mathcal{W}$. *Define* $\widetilde{\Theta} \in \mathcal{B}([\begin{smallmatrix} \mathcal{Y}_1 \\ \mathcal{U}_1 \end{smallmatrix}]; [\begin{smallmatrix} \mathcal{Y} \\ \mathcal{U} \end{smallmatrix}])$ *by*

$$\widetilde{\Theta} = \begin{bmatrix} \widetilde{\Theta}_{11} & \widetilde{\Theta}_{12} \\ \widetilde{\Theta}_{21} & \widetilde{\Theta}_{22} \end{bmatrix} = \begin{bmatrix} P_{\mathcal{Y}}^{\mathcal{U}}|_{\mathcal{Y}_1} & P_{\mathcal{Y}}^{\mathcal{U}}|_{\mathcal{U}_1} \\ P_{\mathcal{U}}^{\mathcal{Y}}|_{\mathcal{Y}_1} & P_{\mathcal{U}}^{\mathcal{Y}}|_{\mathcal{U}_1} \end{bmatrix}. \tag{5.19}$$

1) $\mathcal{W} = \mathcal{Y}_1 \dotplus \mathcal{U}_1$ *is an admissible input/output decomposition for* Σ *if and only if*

$$\widetilde{\Theta}_{11} - D\widetilde{\Theta}_{21} \text{ maps } \mathcal{Y}_1 \text{ one-to-one onto } \mathcal{Y}. \tag{5.20}$$

2) *If the decomposition* $\mathcal{W} = \mathcal{Y}_1 \dotplus \mathcal{U}_1$ *is admissible for* Σ, *then the corresponding operators* A_1, B_1, C_1, *and* D_1 *are given by*

$$\begin{bmatrix} A_1 & B_1 \\ C_1 & D_1 \end{bmatrix} = \begin{bmatrix} 1_{\mathcal{X}} & -B\widetilde{\Theta}_{21} \\ 0 & \widetilde{\Theta}_{11} - D\widetilde{\Theta}_{21} \end{bmatrix}^{-1} \begin{bmatrix} A & B\widetilde{\Theta}_{22} \\ C & -\widetilde{\Theta}_{12} + D\widetilde{\Theta}_{22} \end{bmatrix}, \tag{5.21}$$

or equivalently,

$$\begin{aligned}
A_1 &= A + B\widetilde{\Theta}_{21}(\widetilde{\Theta}_{11} - D\widetilde{\Theta}_{21})^{-1}C, \\
B_1 &= B\widetilde{\Theta}_{22} + B\widetilde{\Theta}_{21}(\widetilde{\Theta}_{11} - D\widetilde{\Theta}_{21})^{-1}(-\widetilde{\Theta}_{12} + D\widetilde{\Theta}_{22}), \\
C_1 &= (\widetilde{\Theta}_{11} - D\widetilde{\Theta}_{21})^{-1}C, \\
D_1 &= (\widetilde{\Theta}_{11} - D\widetilde{\Theta}_{21})^{-1}(-\widetilde{\Theta}_{12} + D\widetilde{\Theta}_{22}).
\end{aligned} \tag{5.22}$$

Proof. This follows from Remark 5.2 and Lemma 5.10. $\qquad\square$

6. Transfer functions

The (input-output) transfer function of discrete time input/state/output system $\Sigma_{i/s/o} = ([\begin{smallmatrix} A & B \\ C & D \end{smallmatrix}]; \mathcal{X}, \mathcal{U}, \mathcal{Y})$ is defined by the formula

$$\mathfrak{D}(z) = D + zC(1_{\mathcal{X}} - zA)^{-1}B, \quad z \in \Lambda_A, \tag{6.1}$$

where Λ_A is the set of points $z \in \mathbb{C}$ for which $(1_{\mathcal{X}} - zA)$ has a bounded inverse, plus the point at infinity if A is boundedly invertible. The set Λ_A is the maximal domain of analyticity of the function $z\mathfrak{A}(z)$, where $\mathfrak{A}$ is the (Fredholm) resolvent of A, i.e.,

$$\mathfrak{A}(z) = (1_{\mathcal{X}} - zA)^{-1}, \quad z \in \Lambda_A. \tag{6.2}$$

Thus, both $\mathfrak{D}$ and $\mathfrak{A}$ will be defined on the same subset Λ_A of the extended complex plane. The resolvent $\mathfrak{A}$ may have an analytic extension to the point at infinity even if A does not have a bounded inverse, and the transfer function $\mathfrak{D}$ may have an analytic extension to a larger domain, but in this paper we shall not make any use of such extensions. Note that $\mathfrak{D}(z) = D + zC\mathfrak{A}(z)B$, that $\mathfrak{D}(0) = D$ and that $\mathfrak{D}(\infty) = D - CA^{-1}B$ (if A is boundedly invertible).

The function $\mathfrak{D}$ arises in a natural way when one studies the Z-transform of a trajectory $(x(\cdot), u(\cdot), y(\cdot))$ of $\Sigma_{i/s/o}$ on $\mathbb{Z}^+$. Let us denote the formal power series induced by the sequences $\{x(n)\}_{n=0}^\infty$, $\{y(n)\}_{n=0}^\infty$, and $\{u(n)\}_{n=0}^\infty$ by[2]

$$\hat{x}(z) = \sum_{n=0}^\infty x(n)z^n, \quad \hat{y}(z) = \sum_{n=0}^\infty y(n)z^n, \quad \hat{u}(z) = \sum_{n=0}^\infty u(n)z^n.$$

The system of equations (1.2) is then equivalent to the following system of equations for formal power series:

$$\begin{aligned}
\hat{x}(z) &= x(0) + zA\hat{x}(z) + zB\hat{u}(z), \\
\hat{y}(z) &= C\hat{x}(z) + D\hat{u}(z).
\end{aligned} \tag{6.3}$$

Solving these equations for $\hat{x}$ and $\hat{y}$ in terms of $x(0)$ and $\hat{u}$ we get the more explicit formula

$$\begin{bmatrix} \hat{x}(z) \\ \hat{y}(z) \end{bmatrix} = \begin{bmatrix} \mathfrak{A}(z) \\ \mathfrak{C}(z) \end{bmatrix} x(0) + \begin{bmatrix} \mathfrak{B}(z) \\ \mathfrak{D}(z) \end{bmatrix} \hat{u}(z), \tag{6.4}$$

where the right-hand side should be interpreted as sums and products of (formal) power series of the following type: $x(0)$ is just a constant, $\hat{u}(z)$ is the formal power series induced by the sequence $\{u(n)\}_{n=0}^\infty$, and the multipliers $\mathfrak{A}(z)$, $\mathfrak{B}(z)$, $\mathfrak{C}(z)$, and $\mathfrak{D}(z)$, represent the MacLaurin series of the corresponding functions defined by (6.1), (6.2), and by

$$\begin{aligned}
\mathfrak{B}(z) &= z(1_{\mathcal{X}} - zA)^{-1}B = z\mathfrak{A}(z)B, & z \in \Lambda_A, \\
\mathfrak{C}(z) &= C(1_{\mathcal{X}} - zA)^{-1} = C\mathfrak{A}(z), & z \in \Lambda_A,
\end{aligned} \tag{6.5}$$

that is,

$$\begin{aligned}
\mathfrak{A}(z) &= \sum_{n=0}^\infty A^n z^n, & \mathfrak{B}(z) &= \sum_{n=0}^\infty A^n B z^{n+1}, \\
\mathfrak{C}(z) &= \sum_{n=0}^\infty CA^n z^n, & \mathfrak{D}(z) &= D + \sum_{n=0}^\infty CA^n B z^{n+1}.
\end{aligned} \tag{6.6}$$

---

[2]The alternative transform where z is replaced by $1/z$ is also frequently used. The corresponding transfer function is then given by $D + C(z - A)^{-1}B$, defined on the resolvent set of A, including the point at infinity.

The corresponding time-domain formulas are

$$
x(n) = A^n x(0) + \sum_{k=0}^{n-1} A^k B u(n - k - 1),
$$

$$
y(n) = C A^n x(0) + D u(n) + \sum_{k=0}^{n-1} C A^k B u(n - k - 1), \qquad n \in \mathbb{Z}^+ \tag{6.7}
$$

(where we interpret an empty sum as zero). From time to time we shall need to refer to the different maps in (6.7), and therefore introduce the following terminology. We define the *state-to-state map* $\check{\mathfrak{A}} \colon \mathcal{X} \to \mathcal{X}^{\mathbb{Z}^+}$, the *input-to-state map* $\check{\mathfrak{B}} \colon \mathcal{U}^{\mathbb{Z}^+} \to \mathcal{X}^{\mathbb{Z}^+}$, the *state-to-output map* $\check{\mathfrak{C}} \colon \mathcal{X} \to \mathcal{Y}^{\mathbb{Z}^+}$, and the *input-to-output map* $\check{\mathfrak{D}} \colon \mathcal{U}^{\mathbb{Z}^+} \to \mathcal{U}^{\mathbb{Z}^+}$ by

$$
(\check{\mathfrak{A}}x)(n) = A^n x, \qquad\qquad\qquad n \in \mathbb{Z}^+,
$$

$$
(\check{\mathfrak{B}}u)(n) = \sum_{k=0}^{n-1} A^k B u(n - k - 1), \qquad n \in \mathbb{Z}^+,
$$

$$
(\check{\mathfrak{C}}x)(n) = C A^n x, \qquad\qquad\qquad n \in \mathbb{Z}^+, \tag{6.8}
$$

$$
(\check{\mathfrak{D}}u)(n) = D + \sum_{k=0}^{n-1} C A^k B u(n - k - 1), \quad n \in \mathbb{Z}^+.
$$

It is frequently possible to interpret the above equations as equations between analytic functions defined in a neighborhood of zero rather than formal power series. It suffices to assume that the (formal) power series defining $\hat{u}$ has a strictly positive radius of convergence. This implies that also the series defining $\hat{x}$ and $\hat{y}$ have a positive radius of convergence, that $\hat{u}$, $\hat{z}$, and $\hat{y}$ are analytic functions defined in a neighborhood of zero, and that (6.4) holds with $\mathfrak{A}(z)$, $\mathfrak{B}(z)$, $\mathfrak{C}(z)$, and $\mathfrak{D}(z)$ defined by (6.1), (6.2), and (6.5). In particular, if $x(0) = 0$, then $\hat{y}(z) = \mathfrak{D}(z)\hat{u}(z)$ in a neighborhood of zero, and this explains why the function $\mathfrak{D}$ is called the input-output transfer function. Similar interpretations are valid for the transfer functions $\mathfrak{A}$ (state to state), $\mathfrak{B}$ (input to state), and $\mathfrak{C}$ (state to output).

A more compact way of writing (6.1), (6.2), and (6.5) is

$$
\begin{bmatrix} z\mathfrak{A}(z) & \mathfrak{B}(z) \\ z\mathfrak{C}(z) & \mathfrak{D}(z) \end{bmatrix} = \begin{bmatrix} (1/z - A)^{-1} & (1/z - A)^{-1}B \\ C(1/z - A)^{-1} & D + C(1/z - A)^{-1}B \end{bmatrix}
$$

$$
= \begin{bmatrix} 1_{\mathcal{X}} & 0 \\ C & D \end{bmatrix} \begin{bmatrix} 1/z - A & -B \\ 0 & 1_{\mathcal{U}} \end{bmatrix}^{-1} \tag{6.9}
$$

$$
= \begin{bmatrix} 1/z - A & 0 \\ -C & 1_{\mathcal{Y}} \end{bmatrix}^{-1} \begin{bmatrix} 1_{\mathcal{X}} & B \\ 0 & D \end{bmatrix}, \qquad z \in \Lambda_A, \quad z \neq 0
$$

(the value at infinity is obtained by taking limits as $z \to \infty$, and the corresponding formula for $z = 0$ is trivial).

We shall call

$$\mathfrak{V}(z) := \begin{bmatrix} \mathfrak{A}(z) & \mathfrak{B}(z) \\ \mathfrak{C}(z) & \mathfrak{D}(z) \end{bmatrix}$$

the *four block input/state/output* transfer function of the system $\Sigma_{i/s/o}$.

A driving-variable/state/signal system $\Sigma_{dv/s/s} = \left(\left[\begin{smallmatrix} A' & B' \\ C' & D' \end{smallmatrix}\right]; \mathcal{X}, \mathcal{L}, \mathcal{W}\right)$ may be interpreted as an input/state/output system with $\mathcal{L}$ as input space, $\mathcal{X}$ as state space, and $\mathcal{W}$ as output space. The Z-transform $(\hat{x}, \hat{\ell}, \hat{w})$ of a trajectory $(x(\cdot), \ell(\cdot), w(\cdot))$ of this system on $\mathbb{Z}^+$ therefore satisfies

$$\begin{bmatrix} \hat{x}(z) \\ \hat{w}(z) \end{bmatrix} = \mathfrak{V}'(z) \begin{bmatrix} x(0) \\ \hat{\ell}(z) \end{bmatrix} := \begin{bmatrix} \mathfrak{A}'(z) & \mathfrak{B}'(z) \\ \mathfrak{C}'(z) & \mathfrak{D}'(z) \end{bmatrix} \begin{bmatrix} x(0) \\ \hat{\ell}(z) \end{bmatrix}, \qquad (6.10)$$

where $\mathfrak{A}'$, $\mathfrak{B}'$, $\mathfrak{C}'$, and $\mathfrak{D}'$ are given by (6.9) with A, B, C, and D replaced by A', B', C', and D'. We shall call $\mathfrak{V}'$ the *four block driving-variable/state/signal* transfer function of the system $\Sigma_{dv/s/s}$. Analogously, the Z-transform $(\hat{x}, \hat{w}, \hat{e})$ of a trajectory $(x(\cdot), w(\cdot), e(\cdot))$ of a signal/state/output nulling system $\Sigma_{s/s/on} = \left(\left[\begin{smallmatrix} A'' & B'' \\ C'' & D'' \end{smallmatrix}\right]; \mathcal{X}, \mathcal{W}, \mathcal{K}\right)$ on $\mathbb{Z}^+$ therefore satisfies

$$\begin{bmatrix} \hat{x}(z) \\ \hat{e}(z) \end{bmatrix} = \mathfrak{V}''(z) \begin{bmatrix} x(0) \\ \hat{w}(z) \end{bmatrix} := \begin{bmatrix} \mathfrak{A}''(z) & \mathfrak{B}''(z) \\ \mathfrak{C}''(z) & \mathfrak{D}''(z) \end{bmatrix} \begin{bmatrix} x(0) \\ \hat{w}(z) \end{bmatrix}, \qquad (6.11)$$

where $\mathfrak{A}''$, $\mathfrak{B}''$, $\mathfrak{C}''$, and $\mathfrak{D}''$ are given by (6.9) with A, B, C, and D replaced by A'', B'', C'', and D''. We shall call $\mathfrak{V}''$ the *four block signal/state/error* transfer function of the system $\Sigma_{s/s/on}$.

Below we shall study relations between the four block transfer functions $\mathfrak{V}$, $\mathfrak{V}'$, and $\mathfrak{V}''$ that correspond to the three types of representations (input/state/output, driving variable, or output nulling, respectively) of a given state/signal system $\Sigma = (V; \mathcal{X}, \mathcal{W})$.

First we will consider the relationships between the four block driving variable transfer function of two driving-variable representations of a state/signal system.

Theorem 6.1. *Let*

$$\Sigma_{dv/s/s} = \left(\left[\begin{smallmatrix} A' & B' \\ C' & D' \end{smallmatrix}\right]; \mathcal{X}, \mathcal{L}, \mathcal{W}\right) \quad and \quad \Sigma^1_{dv/s/s} = \left(\left[\begin{smallmatrix} A'_1 & B'_1 \\ C'_1 & D'_1 \end{smallmatrix}\right]; \mathcal{X}, \mathcal{L}_1, \mathcal{W}\right)$$

be two driving variable representations of the state/signal system $\Sigma = (V; \mathcal{X}, \mathcal{W})$. Denote the four block transfer functions of $\Sigma_{dv/s/s}$ and $\Sigma^1_{dv/s/s}$ by $\begin{bmatrix} \mathfrak{A}'(z) & \mathfrak{B}'(z) \\ \mathfrak{C}'(z) & \mathfrak{D}'(z) \end{bmatrix}$ and $\begin{bmatrix} \mathfrak{A}'_1(z) & \mathfrak{B}'_1(z) \\ \mathfrak{C}'_1(z) & \mathfrak{D}'_1(z) \end{bmatrix}$, respectively, and let $K' \in \mathcal{B}(\mathcal{X}; \mathcal{L})$ and $M' \in \mathcal{B}(\mathcal{L}_1; \mathcal{L})$ be the operators in Theorem 3.3, uniquely determined by (3.12).

1) *The operator $1_{\mathcal{L}} - K'\mathfrak{B}'(z)$ (defined on $\Lambda_{A'}$) has a bounded inverse if and only if $z \in \Lambda_{A'} \cap \Lambda_{A'_1}$.*
2) *For all $z \in \Lambda_{A'} \cap \Lambda_{A'_1}$,*

$$\begin{bmatrix} \mathfrak{A}'_1(z) & \mathfrak{B}'_1(z) \\ \mathfrak{C}'_1(z) & \mathfrak{D}'_1(z) \end{bmatrix} = \begin{bmatrix} \mathfrak{A}'(z) & \mathfrak{B}'(z) \\ \mathfrak{C}'(z) & \mathfrak{D}'(z) \end{bmatrix} \begin{bmatrix} 1_{\mathcal{X}} & 0 \\ -K'\mathfrak{A}'(z) & 1_{\mathcal{L}} - K'\mathfrak{B}'(z) \end{bmatrix}^{-1} \begin{bmatrix} 1_{\mathcal{X}} & 0 \\ 0 & M' \end{bmatrix},$$
$$(6.12)$$

or equivalently,[3]

$$\mathfrak{A}'_1(z) = (1_{\mathcal{X}} - \mathfrak{B}'(z)K')^{-1}\mathfrak{A}'(z),$$
$$\mathfrak{B}'_1(z) = (1_{\mathcal{X}} - \mathfrak{B}'(z)K')^{-1}\mathfrak{B}'(z)M',$$
$$\mathfrak{C}'_1(z) = \mathfrak{C}'(z) + \mathfrak{D}'(z)K'(1_{\mathcal{X}} - \mathfrak{B}'(z)K')^{-1}\mathfrak{A}'(z), \tag{6.13}$$
$$\mathfrak{D}'_1(z) = \mathfrak{D}'(z)(1_{\mathcal{L}} - K'\mathfrak{B}'(z))^{-1}M'.$$

Proof. The case where $z = 0$ is trivial, so in the sequel we assume that $z \neq 0$.

Assume first that $z \in \Lambda_{A'} \cap \Lambda_{A'_1}$, with $z \neq 0$. Since $z \in \Lambda_{A'_1}$, we get from (6.9),

$$\begin{bmatrix} z\mathfrak{A}'_1(z) & \mathfrak{B}'_1(z) \\ z\mathfrak{C}'_1(z) & \mathfrak{D}'_1(z) \end{bmatrix} = \begin{bmatrix} 1_{\mathcal{X}} & 0 \\ C'_1 & D'_1 \end{bmatrix} \begin{bmatrix} 1/z - A'_1 & -B'_1 \\ 0 & 1_{\mathcal{L}_1} \end{bmatrix}^{-1}$$

$$= \begin{bmatrix} 1_{\mathcal{X}} & 0 \\ C'_1 & D'_1 \end{bmatrix} \begin{bmatrix} 1_{\mathcal{X}} & 0 \\ K' & M' \end{bmatrix}^{-1} \left(\begin{bmatrix} 1/z - A'_1 & -B'_1 \\ 0 & 1_{\mathcal{L}_1} \end{bmatrix} \begin{bmatrix} 1_{\mathcal{X}} & 0 \\ K' & M' \end{bmatrix}^{-1} \right)^{-1}$$

$$= \begin{bmatrix} 1_{\mathcal{X}} & 0 \\ C' & D' \end{bmatrix} \begin{bmatrix} 1/z - A' & -B' \\ -(M')^{-1}K' & (M')^{-1} \end{bmatrix}^{-1}.$$

Observe, in particular, that the last block matrix above is boundedly invertible. Since also $z \in \Lambda_{A'}$, we can factor

$$\begin{bmatrix} 1/z - A' & -B' \\ -(M')^{-1}K' & (M')^{-1} \end{bmatrix} = \begin{bmatrix} 1_{\mathcal{X}} & 0 \\ -(M')^{-1}K'z\mathfrak{A}'(z) & (M')^{-1}(1_{\mathcal{L}} - K'\mathfrak{B}'(z)) \end{bmatrix}$$
$$\times \begin{bmatrix} 1/z - A' & -B' \\ 0 & 1_{\mathcal{L}} \end{bmatrix}. \tag{6.14}$$

As we noticed above, the left-hand side in boundedly invertible, and hence also the operator $1_{\mathcal{L}} - K'\mathfrak{B}'(z)$ must be boundedly invertible. Substituting this factorization into the formula above we get

$$\begin{bmatrix} z\mathfrak{A}'_1(z) & \mathfrak{B}'_1(z) \\ z\mathfrak{C}'_1(z) & \mathfrak{D}'_1(z) \end{bmatrix} = \begin{bmatrix} 1_{\mathcal{X}} & 0 \\ C' & D' \end{bmatrix} \begin{bmatrix} 1/z - A' & -B' \\ 0 & 1_{\mathcal{L}} \end{bmatrix}^{-1}$$

$$\times \begin{bmatrix} 1_{\mathcal{X}} & 0 \\ -(M')^{-1}K'z\mathfrak{A}'(z) & (M')^{-1}(1_{\mathcal{L}} - K'\mathfrak{B}'(z)) \end{bmatrix}^{-1}$$

$$= \begin{bmatrix} z\mathfrak{A}'(z) & \mathfrak{B}'(z) \\ z\mathfrak{C}'(z) & \mathfrak{D}'(z) \end{bmatrix} \begin{bmatrix} 1_{\mathcal{X}} & 0 \\ -K'z\mathfrak{A}'(z) & 1_{\mathcal{L}} - K'\mathfrak{B}'(z) \end{bmatrix}^{-1} \begin{bmatrix} 1_{\mathcal{X}} & 0 \\ 0 & M' \end{bmatrix}.$$

Multiplying this identity to the right by $\begin{bmatrix} 1/z & 0 \\ 0 & 1 \end{bmatrix}$ we get (6.12). We have now proved assertion 2) and one half of assertion 1).

To prove the other half of assertion 1) we assume that $z \in \Lambda_{A'}$, $z \neq 0$, and that $1_{\mathcal{L}} - K'\mathfrak{B}'(z)$ is boundedly invertible. Then the block operator matrix on the left-hand side of (6.14) is also boundedly invertible. As we noticed above,

---

[3]Note that, by Lemma 10.1, $1_{\mathcal{L}} - K'\mathfrak{B}'(z)$ has a bounded inverse if and only if $1_{\mathcal{X}} - \mathfrak{B}'(z)K'$ has a bounded inverse.

this matrix factors into $\begin{bmatrix} 1/z - A_1' & -B_1' \\ 0 & 1_{\mathcal{L}_1} \end{bmatrix} \begin{bmatrix} 1_{\mathcal{X}} & 0 \\ K' & M' \end{bmatrix}^{-1}$, and hence $1/z - A_1'$ must be boundedly invertible, i.e., $z \in \Lambda_{A_1'}$.

Theorem 6.2. *Let*

$$\Sigma_{s/s/on} = \left(\begin{bmatrix} A'' & B'' \\ C'' & D'' \end{bmatrix} ; \mathcal{X}, \mathcal{W}, \mathcal{K} \right) \quad and \quad \Sigma^1_{s/s/on} = \left(\begin{bmatrix} A_1'' & B_1'' \\ C_1'' & D_1'' \end{bmatrix} ; \mathcal{X}, \mathcal{W}, \mathcal{K}_1 \right)$$

be two output nulling representations of the state/signal system $\Sigma = (V; \mathcal{X}, \mathcal{W})$. Denote the four block transfer functions of $\Sigma_{s/s/on}$ and $\Sigma^1_{s/s/on}$ by $\begin{bmatrix} \mathfrak{A}''(z) & \mathfrak{B}''(z) \\ \mathfrak{C}''(z) & \mathfrak{D}''(z) \end{bmatrix}$ and $\begin{bmatrix} \mathfrak{A}_1''(z) & \mathfrak{B}_1''(z) \\ \mathfrak{C}_1''(z) & \mathfrak{D}_1''(z) \end{bmatrix}$, respectively, and let K'' and M'' be the operators in Theorem 4.3, uniquely determined by (4.11).

1) *The operator $1_{\mathcal{K}} - z\mathfrak{C}''(z)K''$ (defined on $\Lambda_{A''}$) has a bounded inverse if and only if $z \in \Lambda_{A''} \cap \Lambda_{A_1''}$.*
2) *For all $z \in \Lambda_{A''} \cap \Lambda_{A_1''}$,*

$$\begin{bmatrix} \mathfrak{A}_1''(z) & \mathfrak{B}_1''(z) \\ \mathfrak{C}_1''(z) & \mathfrak{D}_1''(z) \end{bmatrix} = \begin{bmatrix} 1_{\mathcal{X}} & 0 \\ 0 & M'' \end{bmatrix} \begin{bmatrix} 1_{\mathcal{X}} & -z\mathfrak{A}''(z)K'' \\ 0 & 1_{\mathcal{K}} - z\mathfrak{C}''(z)K'' \end{bmatrix}^{-1} \begin{bmatrix} \mathfrak{A}''(z) & \mathfrak{B}''(z) \\ \mathfrak{C}''(z) & \mathfrak{D}''(z) \end{bmatrix},$$

(6.15)

or equivalently,[4]

$$\begin{aligned}
\mathfrak{A}_1''(z) &= \mathfrak{A}''(z)(1_{\mathcal{X}} - zK''\mathfrak{C}''(z))^{-1}\mathfrak{C}''(z), \\
\mathfrak{B}_1''(z) &= \mathfrak{B}''(z) + z\mathfrak{A}''(z)(1_{\mathcal{X}} - zK''\mathfrak{C}''(z))^{-1}K''\mathfrak{D}''(z), \\
\mathfrak{C}_1''(z) &= M''\mathfrak{C}''(z)(1_{\mathcal{X}} - zK''\mathfrak{C}''(z))^{-1}, \\
\mathfrak{D}_1''(z) &= M''(1_{\mathcal{K}} - z\mathfrak{C}''(z)K'')^{-1}\mathfrak{D}''(z).
\end{aligned}$$

(6.16)

The proof of this theorem is similar to the proof of Theorem 6.1, and we leave it to the reader.

Lemma 6.3. *Let $\Sigma_{i/s/o} = \left(\begin{bmatrix} A & B \\ C & D \end{bmatrix} ; \mathcal{X}, \mathcal{U}, \mathcal{Y} \right)$ and $\Sigma_{dv/s/s} = \left(\begin{bmatrix} A' & B' \\ C' & D' \end{bmatrix} ; \mathcal{X}, \mathcal{L}, \mathcal{W} \right)$ be an input/state/output and a driving variable representation, respectively, of the state/signal system $\Sigma = (V; \mathcal{X}, \mathcal{W})$. Denote the four block transfer functions of $\Sigma_{i/s/o}$ and $\Sigma_{dv/s/s}$ by $\begin{bmatrix} \mathfrak{A}(z) & \mathfrak{B}(z) \\ \mathfrak{C}(z) & \mathfrak{D}(z) \end{bmatrix}$ and $\begin{bmatrix} \mathfrak{A}'(z) & \mathfrak{B}'(z) \\ \mathfrak{C}'(z) & \mathfrak{D}'(z) \end{bmatrix}$, respectively.*

1) *The operator $P_{\mathcal{U}}^{\mathcal{Y}}\mathfrak{D}'(z)$ (defined on $\Lambda_{A'}$) has a bounded inverse if and only if $z \in \Lambda_A \cap \Lambda_{A'}$.*
2) *For all $z \in \Lambda_A \cap \Lambda_{A'}$,*

$$\begin{bmatrix} \mathfrak{A}(z) & \mathfrak{B}(z) \\ \mathfrak{C}(z) & \mathfrak{D}(z) \end{bmatrix} = \begin{bmatrix} \mathfrak{A}'(z) & \mathfrak{B}'(z) \\ P_{\mathcal{Y}}^{\mathcal{U}}\mathfrak{C}'(z) & P_{\mathcal{Y}}^{\mathcal{U}}\mathfrak{D}'(z) \end{bmatrix} \begin{bmatrix} 1_{\mathcal{X}} & 0 \\ P_{\mathcal{U}}^{\mathcal{Y}}\mathfrak{C}'(z) & P_{\mathcal{U}}^{\mathcal{Y}}\mathfrak{D}'(z) \end{bmatrix}^{-1}, \qquad (6.17)$$

---

[4]Note that, by Lemma 10.1, $1_{\mathcal{K}} - z\mathfrak{C}''(z)K''$ has a bounded inverse if and only if $1_{\mathcal{X}} - zK''\mathfrak{C}''(z)$ has a bounded inverse.

or equivalently,

$$\begin{aligned}
\mathfrak{A}(z) &= \mathfrak{A}'(z) - \mathfrak{B}'(z)(P_{\mathcal{U}}^{\mathcal{Y}}\mathfrak{D}'(z))^{-1}P_{\mathcal{U}}^{\mathcal{Y}}\mathfrak{C}'(z) \\
\mathfrak{B}(z) &= \mathfrak{B}'(z)(P_{\mathcal{U}}^{\mathcal{Y}}\mathfrak{D}'(z))^{-1} \\
\mathfrak{C}(z) &= P_{\mathcal{Y}}^{\mathcal{U}}\mathfrak{C}'(z) - P_{\mathcal{Y}}^{\mathcal{U}}\mathfrak{D}'(z)(P_{\mathcal{U}}^{\mathcal{Y}}\mathfrak{D}'(z))^{-1}P_{\mathcal{U}}^{\mathcal{Y}}\mathfrak{C}'(z) \\
\mathfrak{D}(z) &= P_{\mathcal{Y}}^{\mathcal{U}}\mathfrak{D}'(z)(P_{\mathcal{U}}^{\mathcal{Y}}\mathfrak{D}'(z))^{-1}.
\end{aligned} \tag{6.18}$$

Proof. We interpret $\Sigma_{i/s/o}$ as a driving variable representation

$$\Sigma_{dv/s/s}^1 = \left(\begin{bmatrix} A_1' & B_1' \\ C_1' & D_1' \end{bmatrix}; \mathcal{X}, \mathcal{L}_1, \mathcal{W}\right)$$

with $\mathcal{L}_1 = \mathcal{U}$ and

$$\left[\begin{array}{c|c} A_1' & B_1' \\ \hline C_1' & D_1' \end{array}\right] = \left[\begin{array}{c|c} A & B \\ \hline C & D \\ 0 & 1_{\mathcal{U}} \end{array}\right];$$

see Remark 5.2. The corresponding block decomposition of $\Sigma_{dv/s/s}$ is given by

$$\left[\begin{array}{c|c} A' & B' \\ \hline C' & D' \end{array}\right] = \left[\begin{array}{c|c} A' & B' \\ \hline P_{\mathcal{Y}}^{\mathcal{U}}C' & P_{\mathcal{Y}}^{\mathcal{U}}D' \\ P_{\mathcal{U}}^{\mathcal{Y}}C' & P_{\mathcal{U}}^{\mathcal{Y}}D' \end{array}\right].$$

To these two driving variable representations we apply Theorem 6.1. By comparing the two representations to each other we find that the operators $K' \in \mathcal{B}(\mathcal{X}; \mathcal{L})$ and $M' \in \mathcal{B}(\mathcal{U}; \mathcal{L})$ are given by

$$M' = [P_{\mathcal{U}}^{\mathcal{Y}}D']^{-1}, \qquad K' = -[P_{\mathcal{U}}^{\mathcal{Y}}D']^{-1}P_{\mathcal{U}}^{\mathcal{Y}}C'.$$

The operator $1_{\mathcal{L}} - K'\mathfrak{B}'(z)$ in part 1) Theorem 6.1 is given by

$$\begin{aligned}
1_{\mathcal{L}} - K'\mathfrak{B}'(z) &= 1_{\mathcal{L}} + [P_{\mathcal{U}}^{\mathcal{Y}}D']^{-1}P_{\mathcal{U}}^{\mathcal{Y}}C'\mathfrak{B}'(z) \\
&= [P_{\mathcal{U}}^{\mathcal{Y}}D']^{-1}(P_{\mathcal{U}}^{\mathcal{Y}}D' + P_{\mathcal{U}}^{\mathcal{Y}}C'\mathfrak{B}'(z)) \\
&= [P_{\mathcal{U}}^{\mathcal{Y}}D']^{-1}P_{\mathcal{U}}^{\mathcal{Y}}\mathfrak{D}(z),
\end{aligned}$$

and it is boundedly invertible if and only if $P_{\mathcal{U}}^{\mathcal{Y}}\mathfrak{D}(z)$ is boundedly invertible. Substituting the above values into (6.12) we get (6.17). $\quad\square$

Lemma 6.4. *Let* $\Sigma_{i/s/o} = \left(\begin{bmatrix} A & B \\ C & D \end{bmatrix}; \mathcal{X}, \mathcal{U}, \mathcal{Y}\right)$ *and* $\Sigma_{s/s/on} = \left(\begin{bmatrix} A'' & B'' \\ C'' & D'' \end{bmatrix}; \mathcal{X}, \mathcal{W}, \mathcal{K}\right)$ *be an input/state/output and a output nulling representation, respectively, of the state/signal system* $\Sigma = (V; \mathcal{X}, \mathcal{W})$. *Denote the four block transfer functions of* $\Sigma_{i/s/o}$ *and* $\Sigma_{s/s/on}$ *by* $\begin{bmatrix} \mathfrak{A}(z) & \mathfrak{B}(z) \\ \mathfrak{C}(z) & \mathfrak{D}(z) \end{bmatrix}$ *and* $\begin{bmatrix} \mathfrak{A}''(z) & \mathfrak{B}''(z) \\ \mathfrak{C}''(z) & \mathfrak{D}''(z) \end{bmatrix}$, *respectively.*

1) *The operator* $\mathfrak{D}''(z)|_{\mathcal{Y}}$ *(defined on* $\Lambda_{A''}$*) has a bounded inverse if and only if* $z \in \Lambda_A \cap \Lambda_{A''}$.

2) *For all* $z \in \Lambda_A \cap \Lambda_{A''}$,

$$\begin{bmatrix} \mathfrak{A}(z) & \mathfrak{B}(z) \\ \mathfrak{C}(z) & \mathfrak{D}(z) \end{bmatrix} = \begin{bmatrix} 1_{\mathcal{X}} & -\mathfrak{B}''(z)|_{\mathcal{Y}} \\ 0 & -\mathfrak{D}''(z)|_{\mathcal{Y}} \end{bmatrix}^{-1} \begin{bmatrix} \mathfrak{A}''(z) & \mathfrak{B}''(z)|_{\mathcal{U}} \\ \mathfrak{C}''(z) & \mathfrak{D}''(z)|_{\mathcal{U}} \end{bmatrix}, \tag{6.19}$$

or equivalently

$$\begin{aligned}
\mathfrak{A}(z) &= \mathfrak{A}''(z) - \mathfrak{B}''(z)|_{\mathcal{Y}}(\mathfrak{D}''(z)|_{\mathcal{Y}})^{-1}\mathfrak{C}''(z), \\
\mathfrak{B}(z) &= \mathfrak{B}''(z)|_{\mathcal{U}} - \mathfrak{B}''(z)|_{\mathcal{Y}}(\mathfrak{D}''(z)|_{\mathcal{Y}})^{-1}\mathfrak{D}''(z)|_{\mathcal{U}}, \\
\mathfrak{C}(z) &= -(\mathfrak{D}''(z)|_{\mathcal{Y}})^{-1}\mathfrak{C}''(z), \\
\mathfrak{D}(z) &= -(\mathfrak{D}''(z)|_{\mathcal{Y}})^{-1}\mathfrak{D}''(z)|_{\mathcal{U}}.
\end{aligned} \tag{6.20}$$

Proof. This lemma is proved in the same way as Lemma 6.3, but this time we interpret $\Sigma_{i/s/o}$ as an output nulling representation of Σ (as in Remark 5.2) and use Theorem 6.2 instead of Theorem 6.1. $\qquad\square$

Theorem 6.5. *Let* $\Sigma_{i/s/o} = \left(\left[\begin{smallmatrix} A & B \\ C & D \end{smallmatrix}\right]; \mathcal{X}, \mathcal{U}, \mathcal{Y}\right)$ *and* $\Sigma^1_{i/s/o} = \left(\left[\begin{smallmatrix} A_1 & B_1 \\ C_1 & D_1 \end{smallmatrix}\right]; \mathcal{X}, \mathcal{U}_1, \mathcal{Y}_1\right)$ *be two input/state/output representations of the state/signal system* $\Sigma = (V; \mathcal{X}, \mathcal{W})$. *Denote the four block transfer functions of* $\Sigma_{i/s/o}$ *and* $\Sigma^1_{i/s/o}$ *by* $\left[\begin{smallmatrix} \mathfrak{A}(z) & \mathfrak{B}(z) \\ \mathfrak{C}(z) & \mathfrak{D}(z) \end{smallmatrix}\right]$ *and* $\left[\begin{smallmatrix} \mathfrak{A}_1(z) & \mathfrak{B}_1(z) \\ \mathfrak{C}_1(z) & \mathfrak{D}_1(z) \end{smallmatrix}\right]$, *respectively. Define* $\Theta \in \mathcal{B}\left(\left[\begin{smallmatrix}\mathcal{Y}\\\mathcal{U}\end{smallmatrix}\right]; \left[\begin{smallmatrix}\mathcal{Y}_1\\\mathcal{U}_1\end{smallmatrix}\right]\right)$ *and* $\widetilde{\Theta} \in \mathcal{B}\left(\left[\begin{smallmatrix}\mathcal{Y}_1\\\mathcal{U}_1\end{smallmatrix}\right]; \left[\begin{smallmatrix}\mathcal{Y}\\\mathcal{U}\end{smallmatrix}\right]\right)$ *by* (1.6) *and* (5.19), *respectively.*

1) *For each* $z \in \Lambda_A$ *the following conditions are equivalent:*
   (a) $z \in \Lambda_{A_1}$.
   (b) *The operator* $\Theta_{21}\mathfrak{D}(z) + \Theta_{22}$ *has a bounded inverse.*
   (c) *The operator* $\widetilde{\Theta}_{11} - \mathfrak{D}(z)\widetilde{\Theta}_{21}$ *has a bounded inverse.*

2) *For all* $z \in \Lambda_A \cap \Lambda_{A_1}$,

$$\begin{bmatrix} \mathfrak{A}_1(z) & \mathfrak{B}_1(z) \\ \mathfrak{C}_1(z) & \mathfrak{D}_1(z) \end{bmatrix} = \begin{bmatrix} \mathfrak{A}(z) & \mathfrak{B}(z) \\ \Theta_{11}\mathfrak{C}(z) & \Theta_{11}\mathfrak{D}(z) + \Theta_{12} \end{bmatrix}\begin{bmatrix} 1_{\mathcal{X}} & 0 \\ \Theta_{21}\mathfrak{C}(z) & \Theta_{21}\mathfrak{D}(z) + \Theta_{22} \end{bmatrix}^{-1}, \tag{6.21}$$

   or equivalently,

$$\begin{aligned}
\mathfrak{A}_1(z) &= \mathfrak{A}(z) - \mathfrak{B}(z)(\Theta_{21}\mathfrak{D}(z) + \Theta_{22})^{-1}\Theta_{21}\mathfrak{C}(z), \\
\mathfrak{B}_1(z) &= \mathfrak{B}(z)(\Theta_{21}\mathfrak{D}(z) + \Theta_{22})^{-1}, \\
\mathfrak{C}_1(z) &= \Theta_{11}\mathfrak{C}(z) - (\Theta_{11}\mathfrak{D}(z) + \Theta_{12})(\Theta_{21}\mathfrak{D}(z) + \Theta_{22})^{-1}\Theta_{21}\mathfrak{C}(z), \\
\mathfrak{D}_1(z) &= (\Theta_{11}\mathfrak{D}(z) + \Theta_{12})(\Theta_{21}\mathfrak{D}(z) + \Theta_{22})^{-1}.
\end{aligned} \tag{6.22}$$

3) *For all* $z \in \Lambda_A \cap \Lambda_{A_1}$,

$$\begin{bmatrix} \mathfrak{A}_1(z) & \mathfrak{B}_1(z) \\ \mathfrak{C}_1(z) & \mathfrak{D}_1(z) \end{bmatrix} = \begin{bmatrix} 1_{\mathcal{X}} & -\mathfrak{B}(z)\widetilde{\Theta}_{21} \\ 0 & \widetilde{\Theta}_{11} - \mathfrak{D}(z)\widetilde{\Theta}_{21} \end{bmatrix}^{-1}\begin{bmatrix} \mathfrak{A}(z) & \mathfrak{B}(z)\widetilde{\Theta}_{22} \\ \mathfrak{C}(z) & -\widetilde{\Theta}_{12} + \mathfrak{D}(z)\widetilde{\Theta}_{22} \end{bmatrix}, \tag{6.23}$$

   or equivalently,

$$\begin{aligned}
\mathfrak{A}_1(z) &= \mathfrak{A}(z) + \mathfrak{B}(z)\widetilde{\Theta}_{21}(\widetilde{\Theta}_{11} - \mathfrak{D}(z)\widetilde{\Theta}_{21})^{-1}\mathfrak{C}(z), \\
\mathfrak{B}_1(z) &= \mathfrak{B}(z)\widetilde{\Theta}_{22} + \mathfrak{B}(z)\widetilde{\Theta}_{21}(\widetilde{\Theta}_{11} - \mathfrak{D}(z)\widetilde{\Theta}_{21})^{-1}(-\widetilde{\Theta}_{12} + \mathfrak{D}(z)\widetilde{\Theta}_{22}), \\
\mathfrak{C}_1(z) &= (\widetilde{\Theta}_{11} - \mathfrak{D}(z)\widetilde{\Theta}_{21})^{-1}\mathfrak{C}(z), \\
\mathfrak{D}_1(z) &= (\widetilde{\Theta}_{11} - \mathfrak{D}(z)\widetilde{\Theta}_{21})^{-1}(-\widetilde{\Theta}_{12} + \mathfrak{D}(z)\widetilde{\Theta}_{22}).
\end{aligned} \tag{6.24}$$

Proof. Assertion 2) follows from Lemma 6.3, assertion 3) from Lemma 6.4, and for assertion 1) we need both of these lemmas. For the proof of 2) we interpret $\Sigma^1_{i/s/o}$ as a driving variable representation, and for the proof of 3) we interpret $\Sigma^1_{i/s/o}$ as an output nulling representation, as explained in Remark 5.2. $\qquad\square$

7. Signal behaviors, external equivalence, and similarity

The behavioral approach to systems theory was introduced by Willems, and has been developed extensively by him and others (see, e.g., [PW98] for a recent presentation of behavioral theory). The vast majority of the literature on behaviors deals with finite-dimensional systems, and the existing extensions to the infinite-dimensional case seem to ignore state space representations of the type that we have introduced above. Below we shall consider the problem of realization of a given behavior on a Hilbert space $\mathcal{W}$ by a state/signal system $\Sigma = (V; \mathcal{X}, \mathcal{W})$.

In order to motivate out definition of a signal behavior we first take a closer look at the signal parts of all externally generated trajectories of a state/signal system $\Sigma = (V; \mathcal{X}, \mathcal{W})$. Let $\mathfrak{W}$ be the set of all the signal sequences $w(\cdot)$, defined on $\mathbb{Z}^+$ with values in $\mathcal{W}$, that are the signal components of externally generated trajectories $(x(\cdot), w(\cdot))$ of Σ on $\mathbb{Z}^+$. It is easy to see that this set $\mathfrak{W}$ is a closed right-shift invariant subspace of the Fréchet space $\mathcal{W}^{\mathbb{Z}^+}$ of all $\mathcal{W}$-valued sequences on $\mathbb{Z}^+$.

We now turn the above property into a definition.

Definition 7.1. Let $\mathcal{W}$ be a Hilbert space.[5] By a (causal signal) *behavior* on the signal space $\mathcal{W}$ we mean a closed right-shift invariant subspace of $\mathcal{W}^{\mathbb{Z}^+}$.

This is a special case of a "manifest behavior", as described, e.g., in [PW98, Definition 1.2.9], but our choice of this particular subclass of behaviors is not a standard one. A similar definition was used by Ball and Staffans [BS05] in continuous time (with an extra growth restriction at infinity that was appropriate in their setting).

A behavior that is induced by a state/signal system $\Sigma = (V; \mathcal{X}, \mathcal{W})$ as explained above is called *realizable*, and the state/signal system Σ that induces this behavior is called a *realization* of the behavior $\mathfrak{W}$.

Definition 7.2. Two state/signal systems with the same signal space are called *externally equivalent* if they induce the same behavior.

A behavior induced by a state/signal system has both an *image representation* and a *kernel representation* of the following type:

Lemma 7.3. *Let $\mathfrak{W}$ be the behavior induced by a state/signal system $\Sigma = (V; \mathcal{X}; \mathcal{W})$. Then*

---

[5] We make only indirect use of the fact that $\mathcal{W}$ is a Hilbert space. See the footnote to Definition 2.1.

1) $\mathfrak{W}$ *is the range of the driving-to-signal map* $\check{\mathfrak{D}}'$ *of every driving variable representation of* Σ, *and*
2) $\mathfrak{W}$ *is the kernel of the signal-to-error map* $\check{\mathfrak{D}}''$ *of every output nulling representation of* Σ.

We leave the easy proof to the reader.

After introducing the above notions we face the following tasks:

1) find criteria of realizability of a given behavior on $\mathcal{W}$;
2) find criteria of external equivalence between two state/signal systems with the same signal space.

The solutions of these problems will be given in this section. These solutions involve some additional notation. If $\mathfrak{W}$ is a behavior on $\mathcal{W}$, then the set

$$\mathfrak{W}(0) = \{w(0) \mid w \in \mathfrak{W}\}. \tag{7.1}$$

is a closed subspace of $\mathcal{W}$. We call this subspace the *zero section* of $\mathfrak{W}$. Observe that, if $\mathfrak{W}$ is induced by a state/signal system, then $\mathfrak{W}(0)$ coincides with the canonical input space $\mathcal{U}_0$ in (3.6).

Definition 7.4. Let $\mathfrak{W}$ be a behavior on $\mathcal{W}$. An ordered direct sum decomposition $\mathcal{W} = \mathcal{Y} \dotplus \mathcal{U}$ (also denoted by $\mathcal{W} = \left[\begin{smallmatrix}\mathcal{Y}\\\mathcal{U}\end{smallmatrix}\right]$) is called an *admissible (input/output) decomposition for* $\mathfrak{W}$ if it has the following two properties:

1) For any sequence $u(\cdot) \in \mathcal{U}^{\mathbb{Z}^+}$ there exists at least one sequence $w(\cdot) \in \mathfrak{W}$ such that $u(n) = P_{\mathcal{U}}^{\mathcal{Y}} w(n)$ for all $n \in \mathbb{Z}^+$ (that is, the projection of $\mathfrak{W}$ onto $\mathcal{U}^{\mathbb{Z}^+}$ along $\mathcal{Y}^{\mathbb{Z}^+}$ is surjective).
2) There exists positive constants M and r such that

$$\sum_{n=0}^{T} \|r^n w(n)\|^2 \leq M^2 \sum_{n=0}^{T} \|r^n P_{\mathcal{U}}^{\mathcal{Y}} w(n)\|^2 \tag{7.2}$$

for all $w(\cdot) \in \mathfrak{W}$ and all $T \in \mathbb{Z}^+$.

Theorem 7.5. *Let* $\mathfrak{W}$ *be a behavior on* $\mathcal{W}$.

1) *The following conditions are equivalent:*
   (a) *The behavior* $\mathfrak{W}$ *is realizable by a state/signal system.*
   (b) *There exists at least one admissible input/output decomposition* $\mathcal{W} = \mathcal{Y} \dotplus \mathcal{U}$ *for* $\mathfrak{W}$.
   (c) *For some direct complement* $\mathcal{Y}_0$ *to the zero section* $\mathfrak{W}(0)$ *the decomposition* $\mathcal{W} = \mathcal{Y}_0 \dotplus \mathfrak{W}(0)$ *is admissible for* $\mathfrak{W}$.
   (d) *For every direct complement* $\mathcal{Y}_0$ *to the zero section* $\mathfrak{W}(0)$ *the decomposition* $\mathcal{W} = \mathcal{Y}_0 \dotplus \mathfrak{W}(0)$ *is admissible for* $\mathfrak{W}$.
2) *Assume that* $\mathfrak{W}$ *is realizable by the state/signal system* $\Sigma = (V; \mathcal{X}, \mathcal{W})$. *Then a direct sum decomposition* $\mathcal{W} = \mathcal{Y} \dotplus \mathcal{U}$ *is admissible for* $\mathfrak{W}$ *if and only if it is admissible for* Σ.

Proof. We begin by proving one half of assertion 2). Suppose first that the behavior $\mathfrak{W}$ is realized by the state/signal system $\Sigma = (V; \mathcal{X}, \mathcal{W})$. Consider some admissible input/output decomposition $\mathcal{W} = \mathcal{Y} \dotplus \mathcal{U}$ for the state/signal system Σ. Let $\Sigma_{i/s/o} = ([\begin{smallmatrix} A & B \\ C & D \end{smallmatrix}]; \mathcal{X}, \mathcal{U}, \mathcal{Y})$ be the input/state/output representation of Σ corresponding to this decomposition. Then, for every externally generated trajectory $(x(\cdot), w(\cdot))$ of Σ on $\mathbb{Z}^+$ we have $w(n) = y(n) + u(n)$, where $u(n) = P_{\mathcal{U}}^{\mathcal{Y}} w(n)$ and $y(n) = P_{\mathcal{Y}}^{\mathcal{U}} w(n)$. Clearly, the projection of $\mathcal{W}$ onto $\mathcal{U}^{\mathbb{Z}^+}$ is surjective (this is the first requirement of an admissible input/output decomposition for $\mathfrak{W}$). To prove that also (7.2) holds we choose some $r > 0$ and rewrite (1.2) in the form

$$x_r(n+1) = rAx_r(n) + rBu_r(n),$$
$$y_r(n) = Cx_r(n) + Du_r(n), \qquad n \in \mathbb{Z}^+, \tag{7.3}$$
$$x(0) = 0,$$

where $x_r(n) = r^n x(n)$, $u_r(n) = r^n u(n)$, and $y_r(n) = r^n y(n)$. Choose r so small that $\|rA\| < 1$. By (6.7) and by the standard fact that the convolution of an ℓ^1-sequence and an ℓ^2-sequence belongs to ℓ^2,

$$\sum_{n=0}^{T} \|y_r(n)\|^2 \leq M_1^2 \sum_{n=0}^{T} \|u_r(n)\|^2,$$

where $M_1 = \|D\| + \|C\|(1 - \|rA\|)^{-1}\|B\|$. Clearly this implies (7.2) with a larger constant M (which depends, among others, on the norms of $P_{\mathcal{Y}}^{\mathcal{U}}$). Thus, the decomposition $\mathcal{W} = \mathcal{Y} \dotplus \mathcal{U}$ is admissible for $\mathfrak{W}$, and we have proved one direction of assertion 2). In addition, we have proved the implication (a) $\Rightarrow$ (d), since the decomposition in (d) is admissible for Σ (see Lemma 5.7). Trivially (d) $\Rightarrow$ (c) and (c) $\Rightarrow$ (b). Thus, it remains to prove the other half of assertion 2) and the implication (b) $\Rightarrow$ (a).

Suppose now that $\mathcal{W} = \mathcal{Y} \dotplus \mathcal{U}$ is an admissible decomposition for the behavior $\mathfrak{W}$. Let r and M be the constants in (7.2). For each $w(\cdot) \in \mathfrak{W}$ we define $w_r(n) = r^n w(n)$, $u_r(n) = r^n P_{\mathcal{U}}^{\mathcal{Y}} w(n)$, and $y_r(n) = r^n P_{\mathcal{Y}}^{\mathcal{U}} w$, $n \in \mathbb{Z}^+$. Then (7.2) implies that the mapping from u_r to y_r is a continuous right-shift invariant mapping from $\ell^2(\mathbb{Z}^+; \mathcal{U})$ to $\ell^2(\mathbb{Z}^+; \mathcal{Y})$. As is well known, this implies that this mapping has a multiplier representation given in terms of Z-transforms by

$$\hat{y}_r(z) = \mathfrak{D}_r(z)\hat{u}_r(z)$$

for some bounded holomorphic $\mathcal{B}(\mathcal{U}; \mathcal{Y})$-valued function in the unit disk $\mathbb{D}$, satisfying $\sup_{z \in \mathbb{D}} \|\mathfrak{D}_r(z)\| \leq M$. This function $\mathfrak{D}_r$ can be realized as the input/output transfer function of an input/state/output system $\Sigma_r = ([\begin{smallmatrix} A_r & B_r \\ C_r & D_r \end{smallmatrix}]; \mathcal{X}, \mathcal{U}, \mathcal{Y})$; see [Aro74, Theorem 3], [Fuh74], or [Hel74, Theorem 3c.1]. We then define

$$\Sigma_{i/s/o} = \left(\begin{bmatrix} r^{-1}A_r & r^{-1}B_r \\ C_r & D_r \end{bmatrix}; \mathcal{X}, \mathcal{U}, \mathcal{Y} \right).$$

This system is an input/state/output representation of a state/signal system $\Sigma = (V; \mathcal{X}, \mathcal{W})$, and the decomposition $\mathcal{W} = \mathcal{Y} \dotplus \mathcal{U}$ is admissible for this system. The system Σ is a state/signal realization of the given behavior $\mathfrak{W}$. This proves the implication (b) $\Rightarrow$ (a), and completes the proof of assertion 1).

It only remains to prove the second half of the assertion 2), namely that every decomposition $W = \mathcal{Y} \dotplus \mathcal{U}$ that is admissible for the behavior $\mathfrak{W}$ is also admissible for its realization Σ. To do this we use the characterization given in Lemma 5.7. Let $u_0 \in \mathcal{U}$, and take some arbitrary $u(\cdot) \in \mathcal{U}^{\mathbb{Z}^+}$ with $u(0) = u_0$. Then there is a corresponding signal $w(\cdot) \in \mathfrak{W}$ such that $P_{\mathcal{U}}^{\mathcal{Y}} w(\cdot) = u(\cdot)$. In particular, $u_0 = P_{\mathcal{U}}^{\mathcal{Y}} w(0)$, where $w(0) \in \mathfrak{W}(0) = \mathcal{U}_0$. Thus $P_{\mathcal{U}}^{\mathcal{Y}}$ maps $\mathcal{U}_0$ onto $\mathcal{U}$. That $P_{\mathcal{U}}^{\mathcal{Y}}|_{\mathcal{U}_0}$ is injective follows from (7.2). By Lemma 5.7, the decomposition $W = \mathcal{Y} \dotplus \mathcal{U}$ is admissible for Σ. $\qquad\square$

Proposition 7.6. *Let $\mathfrak{W}$ be a realizable behavior on W, let $W = \mathcal{Y} \dotplus \mathcal{U}$ be a direct sum decomposition of W. Then the following conditions are equivalent.*

1) *$W = \mathcal{Y} \dotplus \mathcal{U}$ is an admissible input/output decomposition for $\mathfrak{W}$.*
2) *$P_{\mathcal{U}}^{\mathcal{Y}}$ maps $\mathfrak{W}(0)$ one-to-one onto $\mathcal{U}$, i.e., $(P_{\mathcal{U}}^{\mathcal{Y}})^{-1} \in \mathcal{B}(\mathcal{U}; \mathfrak{W}(0))$.*
3) *The space $\mathfrak{W}(0)$ has the graph representation*

$$\mathfrak{W}(0) = \left\{ w = \left[\begin{smallmatrix} D \\ 1_{\mathcal{U}} \end{smallmatrix} \right] u \mid u \in \mathcal{U} \right\}, \tag{7.4}$$

   for some $D \in \mathcal{B}(\mathcal{U}; \mathcal{Y})$.

If the decomposition is admissible, then the operator D in (7.4) is the feedthrough operator of every input/state/output realization of $\mathfrak{W}$ with $W = \mathcal{Y} \dotplus \mathcal{U}$.

This follows from Lemma 5.7 and part 2) of Theorem 7.5 (recall that $\mathfrak{W}(0) = \mathcal{U}_0$).

Theorem 7.7. *Let Σ and Σ^1 be two state/signal systems with the common signal space W.*

1) *If Σ and Σ^1 have a common admissible input/output decomposition $W = \mathcal{Y} \dotplus \mathcal{U}$ and the corresponding input/output transfer functions coincide in a neighborhood of zero, then the two systems are externally equivalent.*
2) *Conversely, if Σ and Σ^1 are externally equivalent, then any direct sum decomposition $W = \mathcal{Y} \dotplus \mathcal{U}$ is admissible for Σ if and only if it is admissible for Σ^1, and the corresponding input/output transfer functions coincide in the (connected) component of $\Lambda_A \cap \Lambda_{A_1}$ which contains zero. In particular, the feedthrough operators also coincide.*

Proof. Proof of 1): We denote the input/state/output representations of Σ and Σ_1 by $\Sigma_{i/s/o} = \left(\left[\begin{smallmatrix} A & B \\ C & D \end{smallmatrix} \right]; \mathcal{X}, \mathcal{U}, \mathcal{Y} \right)$, respectively, $\Sigma_{i/s/o}^1 = \left(\left[\begin{smallmatrix} A_1 & B_1 \\ C_1 & D_1 \end{smallmatrix} \right]; \mathcal{X}, \mathcal{U}, \mathcal{Y} \right)$, and the behaviors induced by Σ and Σ_1 by $\mathfrak{W}$, respectively, $\mathfrak{W}_1$. Let $w(\cdot) \in \mathfrak{W}$. Then there exists a sequence $x(\cdot)$ with $x(0) = 0$ such that $(x(\cdot), w(\cdot))$ is a trajectory of Σ on $\mathbb{Z}^+$. Equivalently, $(x(\cdot), u(\cdot), y(\cdot))$, with $u(\cdot) = P_{\mathcal{U}}^{\mathcal{Y}} w(\cdot)$ and $y(\cdot) = P_{\mathcal{Y}}^{\mathcal{U}} w(\cdot)$ is a trajectory of $\Sigma_{i/s/o}$ on $\mathbb{Z}^+$ with $x(0) = 0$. Let $(x_1(\cdot), u(\cdot), y_1(\cdot))$ be the trajectory of $\Sigma_{i/s/o}^1$ on $\mathbb{Z}^+$ which has $x_1(0) = 0$ and the same input sequence u as above. We claim that $y_1(\cdot) = y(\cdot)$. To prove this is suffices to show that the two input-to-output map (the map $\check{\mathfrak{D}}$ in (6.8)) are the same for the two systems $\Sigma_{i/s/o}$ and $\Sigma_{i/s/o}^1$, i.e., that $D = D_1$ and that $CA^k B = C_1 A_1^k B$ for all $k \in \mathbb{Z}^+$. However, these are the Taylor coefficients of the corresponding transfer functions $\mathfrak{D}$ and $\mathfrak{D}_1$ at the origin, and since we assume that the two transfer functions coincide in a neighborhood of the origin, these Taylor coefficients are the same, too. Thus, $y(\cdot) = y_1(\cdot)$, as claimed. This means that $(x_1(\cdot), w(\cdot))$ is an externally generated trajectory of Σ_1 on $\mathbb{Z}^+$. The above argument shows that $\mathfrak{W} \subset \mathfrak{W}_1$. By interchanging the roles of the two systems Σ and Σ_1 we conclude by the same argument that $\mathfrak{W}_1 \subset \mathfrak{W}$. Thus, the two systems Σ and Σ_1 are externally equivalent.

Proof of 2). Suppose that Σ and Σ^1 are externally equivalent. Then they induce the same behavior $\mathfrak{W}$. By part 2) of Theorem 7.5, the decomposition $W = \mathcal{Y} \dotplus \mathcal{U}$ is admissible for Σ if and only if it is admissible for $\mathfrak{W}$, and this is true if and only if it is admissible for Σ^1. Assume that the decomposition is admissible (for both systems), and denote the corresponding transfer functions by $\mathfrak{D}$, respectively, $\mathfrak{D}_1$. Let $u(\cdot) \in \mathcal{U}^{\mathbb{Z}^+}$, and suppose that the Z-transform of $u(\cdot)$ has a nonzero radius of convergence. Choose some $w(\cdot) \in \mathfrak{W}$ such that $P_{\mathcal{U}}^{\mathcal{Y}} w(\cdot) = u(\cdot)$. Define $y(\cdot) = P_{\mathcal{Y}}^{\mathcal{U}} w(\cdot)$. Then we have in some (possibly smaller) neighborhood of zero,

$$\hat{y}(z) = \mathfrak{D}(z)\hat{u}(z) = \mathfrak{D}_1 \hat{u}(z).$$

This being true for all $u(\cdot) \in \mathcal{U}^{\mathbb{Z}^+}$ whose Z-transform of $u(\cdot)$ has a nonzero radius of convergence, this implies that $\mathfrak{D}(z) = \mathfrak{D}_1(z)$ in some neighborhood of zero. By analytic extension, these two transfer functions must coincide in the connected component of $\Lambda_A \cap \Lambda_{A_1}$ which contains zero. That the feedthrough operators coincide follows from the fact that they are the values of the transfer functions at zero. $\qquad\square$

Instead of testing the external equivalence of two state/signal systems by using input/state/output representations of these systems it is also possible to use driving variable or output nulling representations.

Proposition 7.8. *Let Σ and Σ^1 be two state/signal systems with the common signal space W. Let $\Sigma_{i/s/o}$ and $\Sigma^1_{i/s/o}$ be two input/state/output representations of Σ, respectively, Σ_1 corresponding to the same admissible decomposition $W = \mathcal{Y} \dotplus \mathcal{U}$, let $\Sigma_{dv/s/s}$ and $\Sigma^1_{dv/s/s}$ be two driving variable representations of Σ, respectively, Σ_1, and let $\Sigma_{s/s/on}$ and $\Sigma^1_{s/s/on}$ be two output nulling variable representations of Σ, respectively, Σ_1. Then the following conditions are equivalent:*

1) *Σ and Σ^1 are externally equivalent.*
2) *The input-to-output maps $\check{\mathfrak{D}}$ and $\check{\mathfrak{D}}_1$ of $\Sigma_{i/s/o}$, respectively, $\Sigma^1_{i/s/o}$ coincide.*
3) *The driving-to-signal maps $\check{\mathfrak{D}}'$ and $\check{\mathfrak{D}}'_1$ of $\Sigma_{dv/s/s}$, respectively, $\Sigma^1_{dv/s/s}$ have the same ranges.*
4) *The signal-to-error maps $\check{\mathfrak{D}}''$ and $\check{\mathfrak{D}}''_1$ of $\Sigma_{s/s/on}$, respectively, $\Sigma^1_{s/s/on}$ have the same kernels.*

Proof. This follows from Lemma 7.3, Theorem 7.7, and the fact that the input/output transfer function determines the input-to-output map uniquely. $\qquad\square$

The rest of this section is devoted to a study of similarity and pseudo-similarity of state/signal systems.

Definition 7.9. Two state/signal systems $\Sigma = (V; \mathcal{X}, \mathcal{W})$ and $\Sigma_1 = (V_1; \mathcal{X}_1, \mathcal{W})$ with the same signal space $\mathcal{W}$ are *similar* if there exists a boundedly invertible operator $R \in \mathcal{B}(\mathcal{X}; \mathcal{X}_1)$, called the *similarity operator*, such that $(x(\cdot), w(\cdot))$ is a trajectory of Σ if and only if $(x_1(\cdot), w(\cdot)) = (Rx(\cdot), w(\cdot))$ is a trajectory of Σ_1.

From this definition follows that two similar state/signal systems are externally equivalent.

The corresponding similarity notion is well known for input/state/output systems. Two input/state/output systems $\Sigma_{i/s/o} = \left(\left[\begin{smallmatrix} A & B \\ C & D \end{smallmatrix} \right]; \mathcal{X}, \mathcal{U}, \mathcal{Y} \right)$ and $\Sigma^1_{i/s/o} = \left(\left[\begin{smallmatrix} A_1 & B_1 \\ C_1 & D_1 \end{smallmatrix} \right]; \mathcal{X}_1, \mathcal{U}, \mathcal{Y} \right)$ with the same input and output spaces are *similar* if there exists a boundedly invertible operator $R \in \mathcal{B}(\mathcal{X}; \mathcal{X}_1)$ such that

$$\begin{bmatrix} A_1 & B_1 \\ C_1 & D_1 \end{bmatrix} = \begin{bmatrix} RAR^{-1} & RB \\ CR^{-1} & D \end{bmatrix}.$$

We shall apply the same similarity notion to driving variable and output nulling representations, too, interpreting them as input/state/output systems (as explained in Remark 5.4).

Proposition 7.10. *Let* $\Sigma = (V; \mathcal{X}, \mathcal{W})$ *and* $\Sigma_1 = (V_1; \mathcal{X}_1, \mathcal{W})$ *be two state/signal systems with the same signal space* $\mathcal{W}$, *and let* R *be a boundedly invertible operator in* $\mathcal{B}(\mathcal{X}_1; \mathcal{X})$. *Then the following conditions are equivalent.*

1) Σ *and* Σ_1 *are similar with similarity operator* R.

2) $V_1 = \begin{bmatrix} R & 0 & 0 \\ 0 & R & 0 \\ 0 & 0 & 1_{\mathcal{W}} \end{bmatrix} V$.

3) Σ *and* Σ_1 *have driving variable representations* $\Sigma_{dv/s/s}$ *and* $\Sigma^1_{dv/s/s}$, *respectively, which are similar with similarity operator* R.

4) *To each driving variable representation* $\Sigma_{dv/s/s}$ *of* Σ *there is a (unique) driving variable representation* $\Sigma^1_{dv/s/s}$ *of* Σ_1 *such that these representations are similar with similarity operator* R.

5) Σ *and* Σ_1 *have output nulling representations* $\Sigma_{s/s/on}$ *and* $\Sigma^1_{s/s/on}$, *respectively, which are similar with similarity operator* R.

6) *To each output nulling representation* $\Sigma_{s/s/on}$ *of* Σ *there is a (unique) output nulling representation* $\Sigma^1_{s/s/on}$ *of* Σ_1 *such that these representations are similar with similarity operator* R.

7) *There exists some decomposition* $\mathcal{W} = \mathcal{Y} \dotplus \mathcal{U}$ *of* $\mathcal{W}$ *which is admissible both for* Σ *and for* Σ_1, *and the corresponding input/state/output representations* $\Sigma_{i/s/o}$ *and* $\Sigma^1_{i/s/o}$ *are similar with similarity operator* R.

8) *The systems* Σ *and* Σ_1 *have the same set of admissible decompositions* $\mathcal{W} = \mathcal{Y} \dotplus \mathcal{U}$ *of* $\mathcal{W}$, *and for every such decomposition the corresponding input/state/output representations* $\Sigma_{i/s/o}$ *and* $\Sigma^1_{i/s/o}$ *are similar with similarity operator* R.

We leave the easy proof to the reader.

Various partial converses to the statement that two similar systems are externally equivalent is also valid. Some additional conditions are always needed. One such condition is that both the systems are controllable and observable. In this case they need not actually be similar but only pseudo-similar. Two state/signal systems $\Sigma = (V; \mathcal{X}, \mathcal{W})$ and $\Sigma_1 = (V_1; \mathcal{X}_1, \mathcal{W})$ are called *pseudo-similar* if there exists an injective densely defined closed linear operator $R: \mathcal{X} \to \mathcal{X}_1$ with dense range such that the following conditions hold:

> If $(x(\cdot), w(\cdot))$ is a trajectory of Σ on $\mathbb{Z}^+$ with $x(0) \in \mathcal{D}(R)$, then $x(n) \in \mathcal{D}(R)$ for all $n \in \mathbb{Z}^+$ and $(Rx(\cdot), w(\cdot))$ is a trajectory of Σ_1 on $\mathbb{Z}^+$, and conversely, if $(x_1(\cdot), w(\cdot))$ is a trajectory of Σ_1 on $\mathbb{Z}^+$ with $x_1(0) \in \mathcal{R}(R)$, then $x_1(n) \in \mathcal{R}(R)$ for all $n \in \mathbb{Z}^+$ and $(R^{-1}x_1(\cdot), w(\cdot))$ is a trajectory of Σ on $\mathbb{Z}^+$.

Proposition 7.11. *Two controllable and observable state/signal systems $\Sigma = (V; \mathcal{X}, \mathcal{W})$ and $\Sigma_1 = (V_1; \mathcal{X}_1, \mathcal{W})$ with the same signal space $\mathcal{W}$ are externally equivalent if and only if they are pseudo-similar.*

Proof. In one direction the assertion is obvious: if Σ and Σ_1 are pseudo-similar, then they induce the same behavior (take $x(0) = 0$ and $x_1(0) = 0$).

Conversely, suppose that Σ and Σ and are controllable and observable state/signal systems which are externally equivalent. Then they have the same set of admissible input/output decompositions of the signal space $\mathcal{W}$. Let $\mathcal{W} = \mathcal{Y} \dotplus \mathcal{U}$ be such a decomposition, and denote the corresponding input/state/output representations of Σ and Σ_1 by $\Sigma_{i/s/o} = \left(\left[\begin{smallmatrix} A & B \\ C & D \end{smallmatrix} \right]; \mathcal{X}, \mathcal{U}, \mathcal{Y} \right)$ and $\Sigma^1_{i/s/o} = \left(\left[\begin{smallmatrix} A_1 & B_1 \\ C_1 & D_1 \end{smallmatrix} \right]; \mathcal{X}_1, \mathcal{U}, \mathcal{Y} \right)$, respectively. Then both $\Sigma_{i/s/o}$ and $\Sigma^1_{i/s/o}$ are controllable and observable, and also externally equivalent. This means that their input/output transfer functions coincide a neighborhood of zero. By [Aro79, Proposition 6], these two systems are pseudo-similar in the following sense: there exists an injective densely defined closed linear operator $R: \mathcal{X} \to \mathcal{X}_1$ with dense range such that

$$
\mathcal{R}(B) \subset \mathcal{D}(R), \quad A\mathcal{D}(R) \subset \mathcal{D}(R), \quad A_1\mathcal{R}(R) \subset \mathcal{R}(R),
$$
$$
A_1 R = RA|_{\mathcal{D}(R)}, \quad B_1 = RB, \quad C_1 R = C|_{\mathcal{D}(R)}, \quad D_1 = D. \tag{7.5}
$$

If $(x(\cdot), w(\cdot))$ and $(x_1(\cdot), w(\cdot))$ are externally generated trajectories of Σ and Σ_1, respectively, with $x(0) \in \mathcal{D}(R)$, $x_1(0) \in \mathcal{R}(R)$, and $x_1(0) = Rx(0)$, then for all $n \in \mathbb{Z}^+$,

$$
x(n) = A^n x(0) + \sum_{k=0}^{n-1} A^k Bu(n - k - 1),
$$
$$
x_1(n) = A_1^n Rx(0)(0) + \sum_{k=0}^{n-1} A_1^k B_1 u(n - k - 1),
$$

(7.6)

where $u(n) = P_{\mathcal{U}}^{\mathcal{Y}} w(n)$. This combined with (7.5) gives $x_1(n) = Rx(n)$ for all $n \in \mathbb{Z}^+$. Thus, Σ and Σ_1 are pseudo-similar. $\qquad\square$

8. Dilations of state/signal systems

In the classical finite-dimensional input/state/output systems theory a system is called *minimal* if the dimension of its state space is minimal among all systems with the same transfer function. By a classical result due to Kalman, such a finite-dimensional input/state/output system is minimal if and only if it is controllable and observable. We can reformulate this result is the state/signal setting as follows: a state/signal system with a finite-dimensional state space has a state space with minimal dimension among all externally equivalent systems if and only if it is controllable and observable.

In the case where the state space is infinite-dimensional the requirement that its state space should have minimal dimension becomes obscure (all infinite-dimensional separable Hilbert spaces has the same dimension). It is therefore necessary to define minimality in terms of some other property. One natural solution is to study *dilations* and *compressions* of systems. In the finite-dimensional case the minimality of the dimension of the state space is equivalent to the statement that the system cannot be compressed into a "smaller" system, and this characterization has a natural infinite-dimensional analogue. The notions of dilations and compressions of operators and of input/state/output systems have attracted a great deal of attention and it plays an important role in many works, see, e.g., [Aro79], [SF70], and [LP67] for Hilbert space versions, and [BGK79] and [Sta05] for Banach space versions.

Definition 8.1. The state/signal system $\widetilde{\Sigma} = (\widetilde{V}; \widetilde{\mathcal{X}}, \mathcal{W})$ is a *dilation along $\mathcal{Z}$* of the state/signal system $\Sigma = (V; \mathcal{X}, \mathcal{W})$, or equivalently, the state/signal system Σ is a *compression along $\mathcal{Z}$ onto $\mathcal{X}$* of the state/signal system $\widetilde{\Sigma}$, if the following conditions hold:

1) $\widetilde{\mathcal{X}} = \mathcal{X} \dotplus \mathcal{Z}$,
2) If $(\tilde{x}(\cdot), w(\cdot))$ is a trajectory of $\widetilde{\Sigma}$ on $\mathbb{Z}^+$ with $\tilde{x}(0) \in \mathcal{X}$, then $(P_{\mathcal{X}}^{\mathcal{Z}} \tilde{x}(\cdot), w(\cdot))$ is a trajectory of Σ on $\mathbb{Z}^+$.
3) There is at least one decomposition $\mathcal{W} = \mathcal{Y} \dotplus \mathcal{U}$ of $\mathcal{W}$ which is admissible for both $\widetilde{\Sigma}$ and Σ.

Note that, whereas the compressed system is determined uniquely by the dilated system and by the decomposition $\widetilde{\mathcal{X}} = \mathcal{X} \dotplus \mathcal{Z}$, the converse is clearly not true.

Lemma 8.2. *Let the state/signal system* $\widetilde{\Sigma} = (\widetilde{V}; \widetilde{\mathcal{X}}, \mathcal{W})$ *be a dilation along $\mathcal{Z}$ of* $\Sigma = (V; \mathcal{X}, \mathcal{W})$. *Then the following claims hold.*

1) *To each trajectory* $(x(\cdot), w(\cdot))$ *of* Σ *on* $\mathbb{Z}^+$ *there is a unique trajectory* $(\tilde{x}(\cdot), \tilde{w}(\cdot))$ *of* $\widetilde{\Sigma}$ *on* $\mathbb{Z}^+$ *satisfying* $\tilde{x}(0) = x(0)$ *and* $\tilde{w}(\cdot) = w(\cdot)$. *This trajectory has the additional property that* $x(\cdot) = P_{\mathcal{X}}^{\mathcal{Z}} \tilde{x}(\cdot)$.
2) $\widetilde{\Sigma}$ *and* Σ *are externally equivalent. In particular, they have the same admissible input/output decompositions of the signal space, and the input/output transfer functions and the input-to-output maps of the corresponding input/state/output representations of* $\widetilde{\Sigma}$ *and* Σ *coincide.*

Proof. Let $\mathcal{W} = \mathcal{Y} \dot{+} \mathcal{U}$ be a decomposition which is admissible both for $\widetilde{\Sigma}$ and for Σ, and denote the corresponding input/state/output representations of $\widetilde{\Sigma}$ and Σ by $\widetilde{\Sigma}_{i/s/o}$ and $\Sigma_{i/s/o}$, respectively. Let $(x(\cdot), w(\cdot))$ be a trajectory of Σ on $\mathbb{Z}^+$. Define $u(\cdot) = P_{\mathcal{U}}^{\mathcal{Y}} w(\cdot)$ and $y(\cdot) = P_{\mathcal{Y}}^{\mathcal{U}} w(\cdot)$. Then $(x(\cdot), u(\cdot), y(\cdot))$ is a trajectory of $\Sigma_{i/s/o}$, and $\widetilde{\Sigma}_{i/s/o}$ has a unique trajectory $(\tilde{x}(\cdot), u(\cdot), \tilde{y}(\cdot))$ on $\mathbb{Z}^+$ satisfying $\tilde{x}(0) = x(0)$. Define $\tilde{w}(\cdot) = \tilde{y}(\cdot) \dot{+} u(\cdot)$. Then $(\tilde{x}(\cdot), \tilde{w}(\cdot))$ is a trajectory of $\widetilde{\Sigma}$ on $\mathbb{Z}^+$. According to property 2) in Definition 8.1, $(P_{\mathcal{X}}^{\mathcal{Z}} \tilde{x}(\cdot), \tilde{w}(\cdot))$ must be a trajectory of Σ, and hence, if we define $\tilde{y}(\cdot) = P_{\mathcal{Y}}^{\mathcal{U}} \tilde{w}(\cdot)$, then $(P_{\mathcal{X}}^{\mathcal{Z}} \tilde{x}(\cdot), u(\cdot), \tilde{y}(\cdot))$ is a trajectory of $\Sigma_{i/s/o}$. But a trajectory of $\Sigma_{i/s/o}$ is determined uniquely by its initial state and input data, and therefore we must have $x(\cdot) = P_{\mathcal{X}}^{\mathcal{Z}} \tilde{x}(\cdot)$ and $\tilde{y}(\cdot) = y(\cdot)$. This proves assertion 1). Assertion 2) follows immediately from property 2) in Definition 8.1 together with assertion 1) . $\qquad\square$

Observability and controllability are preserved under compressions (but not under dilations).

Lemma 8.3. *Let the state/signal system $\widetilde{\Sigma} = (\widetilde{V}; \widetilde{\mathcal{X}}, \mathcal{W})$ be a dilation along $\mathcal{Z}$ of $\Sigma = (V; \mathcal{X}, \mathcal{W})$. Let $\widetilde{\mathfrak{R}}$ and $\mathfrak{R}$ be the reachable subspaces and let $\widetilde{\mathfrak{U}}$ and $\mathfrak{U}$ be the unobservable subspaces of $\widetilde{\Sigma}$ and Σ, respectively. Then $\mathfrak{U} = \widetilde{\mathfrak{U}} \cap \mathcal{X}$ and $\mathfrak{R}$ is the closure of $P_{\mathcal{X}}^{\mathcal{Z}} \widetilde{\mathfrak{R}}$. In particular, if $\widetilde{\Sigma}$ is controllable or observable, then Σ is controllable or observable, respectively.*

We leave the easy proof to the reader.

In order to be able to study the relationship between the two systems $\widetilde{\Sigma}$ and Σ in Definition 8.1 in more detail we need the following two invariance notions.[6]

Definition 8.4. Let $\Sigma = (V; \mathcal{X}, \mathcal{W})$ be a state/signal system.

1) A closed subspace $\mathcal{Z}$ of $\mathcal{X}$ is *outgoing invariant* for Σ if to each $x_0 \in \mathcal{Z}$ there is a (unique) trajectory $(x(\cdot), 0)$ of Σ on $\mathbb{Z}^+$ with $x(0) = x_0$ satisfying $x(n) \in \mathcal{Z}$ for all $n \in \mathbb{Z}^+$.

2) A closed subspace $\mathcal{Z}$ of $\mathcal{X}$ is *strongly invariant* for Σ if every trajectory $(x(\cdot), w(\cdot))$ of Σ on $\mathbb{Z}^+$ with $x(0) \in \mathcal{Z}$ satisfies $x(n) \in \mathcal{Z}$ for all $n \in \mathbb{Z}^+$.

These invariance properties can also be described in terms of the generating subspace V as follows.

Lemma 8.5. *Let $\Sigma = (V; \mathcal{X}, \mathcal{W})$ be a state/signal system, and let $\mathcal{Z}$ be a closed subspace of $\mathcal{X}$.*

1) *$\mathcal{Z}$ is outgoing invariant for Σ if and only if the following condition holds:*

$$\text{To each } x \in \mathcal{Z} \text{ there is a (unique) } z \in \mathcal{Z} \text{ such that } \begin{bmatrix} z \\ x \\ 0 \end{bmatrix} \in V. \qquad (8.1)$$

2) *$\mathcal{Z}$ is strongly invariant for Σ if and only if it the following implication is true:*

$$\text{If } \begin{bmatrix} z \\ x \\ w \end{bmatrix} \in V \text{ and } x \in \mathcal{Z}, \text{ then } z \in \mathcal{Z}. \qquad (8.2)$$

---

[6] The connections between these notions and the unobservable and reachable subspaces are explained in Lemma 8.6 below.

Proof. Proof of 1): The necessity of (8.1) for outgoing invariance is immediate (the solution $(x(\cdot), 0)$ mentioned in part 1) of Definition 8.4 satisfies $\begin{bmatrix} x(1) \\ x(0) \\ 0 \end{bmatrix} \in V.$)

Conversely, suppose that (8.1) holds. Let $x_0 \in \mathcal{Z}$. Then (8.1) with x replaced by x_0 gives the existence of $x(1) \in \mathcal{Z}$ such that $\begin{bmatrix} x(1) \\ x_0 \\ 0 \end{bmatrix} \in V$. Applying (8.1) once more with x replaced by $x(1)$ we get the existence of $x(2) \in \mathcal{Z}$ such that $\begin{bmatrix} x(2) \\ x(1) \\ 0 \end{bmatrix} \in V$. Continuing in the same way we get a sequence $x(\cdot)$ such that $x(0) = x_0$ and $(x(\cdot), 0)$ is a trajectory of Σ on $\mathbb{Z}^+$. According to Definition 8.4, $\mathcal{Z}$ is outgoing invariant.

Proof of 2): To see that (8.2) is necessary for $\mathcal{Z}$ to be strongly invariant we argue as follows. By part 1) of Proposition 2.2, the condition $\begin{bmatrix} z_0 \\ x_0 \\ w_0 \end{bmatrix} \in V$ implies that there exists a trajectory $(x(\cdot), w(\cdot))$ of Σ on $\mathbb{Z}^+$ with $x(0) = x_0$, $w(0) = w_0$, and $x(1) = z_0$. If, furthermore, $x_0 \in \mathcal{Z}$, then the strong invariance of $\mathcal{Z}$ implies that $x(n) \in \mathcal{Z}$ for all $n \in \mathbb{Z}^+$. In particular, $z_0 = x(1) \in \mathcal{Z}$.

The proof of the converse part is similar to the proof of the converse part of assertion 1), and it is left to the reader. $\qquad\square$

The two main examples of outgoing invariant and strongly invariant subspaces are the following:

Lemma 8.6. *Let $\Sigma = (V; \mathcal{X}, \mathcal{W})$ be a state/signal system.*

1) *The unobservable subspace is the maximal outgoing invariant subspace for Σ, i.e., it is outgoing invariant, and it contains every other outgoing invariant subspace.*
2) *The reachable subspace is the minimal closed strongly invariant subspace for Σ, i.e., it is strongly invariant, and it is contained in every other closed strongly invariant subspace.*

We leave the easy proof to the reader.

The following theorem is the main result of this section.

Theorem 8.7. *Let $\widetilde{\Sigma} = (\widetilde{V}; \widetilde{\mathcal{X}}, \mathcal{W})$ and $\Sigma = (V; \mathcal{X}, \mathcal{W})$ be two state/signal systems with $\widetilde{\mathcal{X}} = \mathcal{X} \dotplus \mathcal{Z}$ (and with the same signal space). Then $\widetilde{\Sigma}$ is a dilation along $\mathcal{Z}$ of Σ if and only if the following conditions hold:*

1) *V is given by*

$$V = \left\{ \begin{bmatrix} P_{\mathcal{X}}^{\mathcal{Z}} \tilde{z} \\ x \\ w \end{bmatrix} \;\middle|\; x \in \mathcal{X} \text{ and } \begin{bmatrix} \tilde{z} \\ x \\ w \end{bmatrix} \in \widetilde{V} \right\}. \tag{8.3}$$

2) *$\mathcal{Z}$ has a decomposition $\mathcal{Z} = \mathcal{Z}_o \dotplus \mathcal{Z}_i$ where $\mathcal{Z}_o$ is outgoing invariant for $\widetilde{\Sigma}$ and $\mathcal{Z}_o \dotplus \mathcal{X}$ is strongly invariant for $\widetilde{\Sigma}$.*

One possible choice of the subspaces $\mathcal{Z}_o$ and $\mathcal{Z}_i$ in 2) is to take $\mathcal{Z}_o = \mathcal{Z}_o^{\max}$ and to take $\mathcal{Z}_i$ to be an arbitrary direct complement of $\mathcal{Z}_o^{\max}$ in $\mathcal{Z}$, where

$$\mathcal{Z}_o^{\max} = \left\{ \tilde{x}_0 \in \tilde{X} \;\middle|\; \begin{array}{l} \text{there exists a trajectory } (\tilde{x}(\cdot), 0) \text{ of } \tilde{\Sigma} \text{ on } \mathbb{Z}^+ \text{ with} \\ \tilde{x}(0) = \tilde{x}_0 \text{ satisfying } P_{\mathcal{X}}^{\mathcal{Z}} \tilde{x}(n) = 0 \text{ for all } n \in \mathbb{Z}^+ \end{array} \right\}. \tag{8.4}$$

The subspace $\mathcal{Z}_o^{\max}$ is maximal in the sense that it contains every other space $\mathcal{Z}_o$ that can be used in the decomposition in 2).

We shall call $\mathcal{Z}_o$ an *outgoing subspace* and $\mathcal{Z}_i$ an *incoming subspace* of $\tilde{\Sigma}$.[7]

Proof. We begin by proving necessity of 1) and 2), assuming that $\tilde{\Sigma}$ is a dilation of Σ, and begin with condition 1). Let $\begin{bmatrix} \tilde{z}_0 \\ x_0 \\ w_0 \end{bmatrix} \in \tilde{V}$ with $x_0 \in \mathcal{X}$. By Proposition 2.2, $\tilde{\Sigma}$ has a trajectory $(\tilde{x}(\cdot), \tilde{w}(\cdot))$ on $\mathbb{Z}^+$ with $\tilde{x}(1) = \tilde{z}_0$, $\tilde{x}(0) = x_0$, and $w(0) = w_0$. By condition 2) in Definition 8.1, $(x(\cdot), \tilde{w}(\cdot))$ with $x(\cdot) = P_{\mathcal{X}}^{\mathcal{Z}} \tilde{x}(\cdot)$ is a trajectory of Σ. In particular, $\begin{bmatrix} x(1) \\ x(0) \\ w_0 \end{bmatrix} = \begin{bmatrix} P_{\mathcal{X}}^{\mathcal{Z}} \tilde{z}_0 \\ x_0 \\ w_0 \end{bmatrix} \in V$. This shows that the right-hand side of (8.3) is contained in V. The opposite inclusion follows from a similar argument which replaces condition 2) in Definition 8.1 by part 1) of Lemma 8.2.

To prove the existence of a decomposition of the type described in part 2) we define $\mathcal{Z}_o = \mathcal{Z}_o^{\max}$ by (8.4). It is easy to see that $\mathcal{Z}_o^{\max}$ is a closed subspace of $\mathcal{X}$, and it is contained in $\mathcal{Z}$ since $P_{\mathcal{X}}^{\mathcal{Z}} \mathcal{Z}_o^{\max} = 0$. Let $\mathcal{Z}_i$ be an arbitrary direct complement of $\mathcal{Z}_o^{\max}$ in $\mathcal{Z}$. We claim that this decomposition of $\mathcal{Z}$ has the two properties mentioned in 2).

It is easy to see from Definition 8.4 that $\mathcal{Z}_o^{\max}$ is outgoing invariant for $\tilde{\Sigma}$, so it remains to show that $\mathcal{Z}_o^{\max} \dotplus \mathcal{X}$ is strongly invariant for $\tilde{\Sigma}$. Let $(\tilde{x}(\cdot), w(\cdot))$ be a trajectory of $\tilde{\Sigma}$ on $\mathbb{Z}^+$ with $\tilde{x}(0) = z_0 + x_0$, where $z_0 \in \mathcal{Z}_o^{\max}$ and $x_0 \in \mathcal{X}$. Since $\mathcal{Z}_o^{\max}$ is outgoing invariant, there is a trajectory $(\tilde{x}_1(\cdot), 0)$ of $\tilde{\Sigma}$ on $\mathbb{Z}^+$ with $\tilde{x}_1(0) = z_0$ satisfying $\tilde{x}_1(n) \in \mathcal{Z}_o^{\max}$ for all $n \in \mathbb{Z}^+$. Define $\tilde{x}_2(\cdot) = \tilde{x}(\cdot) - \tilde{x}_1(\cdot)$. Then $(\tilde{x}_2(\cdot), w(\cdot))$ is a trajectory of $\tilde{\Sigma}$ on $\mathbb{Z}^+$ with $\tilde{x}_2(0) = x_0 \in \mathcal{X}$. Define $x(\cdot) = P_{\mathcal{X}}^{\mathcal{Z}} \tilde{x}_2(\cdot)$. By Condition 2) in Definition 8.1, $(x(\cdot), w(\cdot))$ is a trajectory of Σ on $\mathbb{Z}^+$. In particular, it is also a trajectory on $[1, \infty)$. By assertion 2) of Lemma 8.2, applied to the time interval $[1, \infty)$, there is a trajectory $(\tilde{x}_3(\cdot), w(\cdot))$ of $\tilde{\Sigma}$ on $[1, \infty)$ satisfying $\tilde{x}_3(1) = x(1)$ and $P_{\mathcal{X}}^{\mathcal{Z}} \tilde{x}_3(n) = x(n)$ for all $n \in [1, \infty)$. Define $\tilde{x}_4(\cdot) = \tilde{x}_2(\cdot) - \tilde{x}_3(\cdot)$. Then $(\tilde{x}_4(\cdot), 0)$ is a trajectory of $\tilde{\Sigma}$ on $[1, \infty)$, and it satisfies $P_{\mathcal{X}}^{\mathcal{Y}} \tilde{x}_4(n) = P_{\mathcal{X}}^{\mathcal{Y}} \tilde{x}_2(n) - P_{\mathcal{X}}^{\mathcal{Y}} \tilde{x}_3(n) = x(n) - x(n) = 0$ for all $n \in [1, \infty)$. It follows from (8.4) (after we have shifted the trajectory $(\tilde{x}(\cdot), 0)$ one step to the left) that $\tilde{x}_4(0) \in \mathcal{Z}_o^{\max}$. Thus, $\tilde{x}(1) = \tilde{x}_1(1) + \tilde{x}_3(1) + \tilde{x}_4(1)$ where $\tilde{x}_1(1) \in \mathcal{Z}_o^{\max}$, $\tilde{x}_3(1) = x(1) \in \mathcal{X}$, and $\tilde{x}_4(1) \in \mathcal{Z}_o^{\max}$, so $\tilde{x}(1) \in \mathcal{Z}_o^{\max} \dotplus \mathcal{X}$. This proves that the implication (8.2) holds with $\mathcal{Z}$ replaced by $\mathcal{Z}_o^{\max} \dotplus \mathcal{X}$. By Lemma 8.5, $\mathcal{Z}_o^{\max} \dotplus \mathcal{X}$ is strongly invariant.

To prove the maximality of $\mathcal{Z}_o^{\max}$ it suffices to observe that if $\mathcal{Z}_o$ is outgoing invariant, then for each $z_0 \in \mathcal{Z}_o$ there is a trajectory $(\tilde{x}(\cdot), 0)$ of $\tilde{\Sigma}$ on $\mathbb{Z}^+$ with

---

[7]The reason for these names will be explained elsewhere.

$\tilde{x}(0) = z_0$ satisfying $\tilde{x}(n) \in \mathcal{Z}_o \subset \mathcal{Z}$ for all $n \in \mathbb{Z}^+$, and hence $P_{\mathcal{X}}^{\mathcal{Z}}\tilde{x}(n) = 0$ for all $n \in \mathbb{Z}^+$. This implies that $z_0 \in \mathcal{Z}_o^{\max}$.

For the converse proof we assume that 1) and 2) hold. It follows from (8.3) that the two systems $\widetilde{\Sigma}$ and Σ have the same canonical input space $\mathcal{U}_0$, so condition 3) of Definition 8.1 is satisfied.

Our proof of the fact that also condition 2) of Definition 8.1 holds is based on the following implication:

$$\text{If } \begin{bmatrix} \tilde{z} \\ \tilde{x} \\ w \end{bmatrix} \in \widetilde{V} \text{ and } \tilde{x} \in \mathcal{Z}_o \dotplus \mathcal{X}, \text{ then } \begin{bmatrix} P_{\mathcal{X}}^{\mathcal{Z}} \tilde{z} \\ P_{\mathcal{X}}^{\mathcal{Z}} \tilde{x} \\ w \end{bmatrix} \in V. \tag{8.5}$$

The proof of (8.5) goes as follows. Let $\begin{bmatrix} \tilde{z} \\ \tilde{x} \\ w \end{bmatrix} \in \widetilde{V}$ with $\tilde{x} = z_0 + x_0$, where $z_0 \in \mathcal{Z}_o$ and $x_0 \in \mathcal{X}$. Since $\mathcal{Z}_o$ is outgoing invariant, there is some $z_1 \in \mathcal{Z}_o$ such that $\begin{bmatrix} z_1 \\ z_0 \\ 0 \end{bmatrix} \in \widetilde{V}$ (see Lemma 8.5). Since $\widetilde{V}$ is a subspace also $\begin{bmatrix} \tilde{z}-z_1 \\ x_0 \\ w \end{bmatrix} \in \widetilde{V}$. We can now apply (8.3) to conclude that $\begin{bmatrix} P_{\mathcal{X}}^{\mathcal{Z}}(\tilde{z}-z_1) \\ x_0 \\ w \end{bmatrix} \in V$. But $P_{\mathcal{X}}^{\mathcal{Z}}(\tilde{z} - z_1) = P_{\mathcal{X}}^{\mathcal{Z}}\tilde{z}$ since $z_1 \in \mathcal{Z}_o \subset \mathcal{Z}$ and $x_0 = P_{\mathcal{X}}^{\mathcal{Z}}\tilde{x}$ since $\tilde{x} - x_0 = z_0 \in \mathcal{Z}_o \subset \mathcal{Z}$. Thus, we conclude that $\begin{bmatrix} P_{\mathcal{X}}^{\mathcal{Z}} \tilde{z} \\ P_{\mathcal{X}}^{\mathcal{Z}} \tilde{x} \\ w \end{bmatrix} \in V$. This proves (8.5).

Let $(\tilde{x}(\cdot), w(\cdot))$ be a trajectory of $\widetilde{\Sigma}$ on $\mathbb{Z}^+$ with $\tilde{x}(0) \in \mathcal{X}$. Because of the strong invariance of $\mathcal{Z}_o \dotplus \mathcal{X}$, this implies that $\tilde{x}(n) \in \mathcal{Z}_o \dotplus \mathcal{X}$ for all $n \in \mathbb{Z}^+$. Define $x(\cdot) = P_{\mathcal{X}}^{\mathcal{Z}}\tilde{x}(\cdot)$. Then it follows from (8.5) that $(x(\cdot), w(\cdot))$ is a trajectory of Σ on $\mathbb{Z}^+$. Thus, condition 2) in Definition 8.1 holds, and we conclude that $\widetilde{\Sigma}$ is a dilation of Σ. $\qquad\square$

Let us record the following fact which we observed in the preceding proof.

Corollary 8.8. *Let the state/signal system $\widetilde{\Sigma} = (\widetilde{V}; \widetilde{\mathcal{X}}, \mathcal{W})$ be a dilation along $\mathcal{Z}$ of $\Sigma = (V; \mathcal{X}, \mathcal{W})$, and let $\widetilde{\mathcal{X}} = \mathcal{Z}_o \dotplus \mathcal{X} \dotplus \mathcal{Z}_i$ be the decomposition of $\widetilde{\mathcal{X}}$ given in Theorem 8.7. Denote $\mathcal{Z}_o \dotplus \mathcal{X}$ by $\mathcal{X}_o$. Then V is given by*

$$V = \left\{ \begin{bmatrix} P_{\mathcal{X}}^{\mathcal{Z}} \tilde{z} \\ P_{\mathcal{X}}^{\mathcal{Z}} \tilde{x} \\ w \end{bmatrix} \,\middle|\, \tilde{x} \in \mathcal{X}_o \text{ and } \begin{bmatrix} \tilde{z} \\ \tilde{x} \\ w \end{bmatrix} \in \widetilde{V} \right\}. \tag{8.6}$$

This follows from (8.3) and (8.5).

Corollary 8.9. *Let $\widetilde{\Sigma} = (\widetilde{V}; \widetilde{\mathcal{X}}, \mathcal{W})$ be a state/signal system. Assume that $\widetilde{\mathcal{X}} = \mathcal{X} \dotplus \mathcal{Z}$, and define V by (8.3). Then $\Sigma = (V; \mathcal{X}, \mathcal{W})$ is a state/signal node. It is a compression along $\mathcal{Z}$ onto $\mathcal{X}$ of $\widetilde{\Sigma}$ if and only if $\mathcal{Z}$ can be decomposed into $\mathcal{Z} = \mathcal{Z}_o \dotplus \mathcal{Z}_i$ in such a way that $\mathcal{Z}_o$ is outgoing invariant for $\widetilde{\Sigma}$ and $\mathcal{Z}_o \dotplus \mathcal{X}$ is strongly invariant for $\widetilde{\Sigma}$.*

Proof. If V is given by (8.3), then V clearly has properties (i) and (iii) in Definition 2.1. That it also has properties (i) and (iv) follows from Lemma 2.4, because if we denote the operator in part 3) of Lemma 2.3 corresponding to $\widetilde{V}$ and V by $\widetilde{F}$ and F, respectively, then $F = P_{\mathcal{X}}^{\mathcal{Z}}\widetilde{F}$ with $\mathcal{D}(F) = \mathcal{D}(\widetilde{F})$. Thus Σ is a state/signal node. The remaining claims follow from Theorem 8.7. $\qquad\square$

Remark 8.10. It is possible to reformulate condition 2) in Theorem 8.7 by focusing on the subspace $\mathcal{X}_o := \mathcal{Z}_o \dotplus \mathcal{X}$ instead of focusing on $\mathcal{Z}_o$. We claim that condition 2) in Theorem 8.7 is equivalent to the following condition:

    2') $\widetilde{\mathcal{X}}$ has a decomposition $\widetilde{\mathcal{X}} = \mathcal{X}_o \dotplus \mathcal{Z}_i$, where $\mathcal{Z}_i \subset \mathcal{Z}$, $\mathcal{X} \subset \mathcal{X}_o$, $\mathcal{X}_o$ is strongly invariant for $\widetilde{\Sigma}$, and $\mathcal{X}_o \cap \mathcal{Z}$ is outgoing invariant for $\widetilde{\Sigma}$.

Clearly, 2') follows from from 2) if we take $\mathcal{X}_o = \mathcal{Z}_o \dotplus \mathcal{X}$. It is almost as easy to derive 2) from 2'), with $\mathcal{Z}_o = \mathcal{X}_o \cap \mathcal{Z}$; the only slightly nontrivial part is to show that $\widetilde{\mathcal{X}} = \mathcal{Z}_o \dotplus \mathcal{X} \dotplus \mathcal{Z}_i$, or equivalently, that $\mathcal{X}_o = (\mathcal{X}_o \cap \mathcal{Z}) \dotplus \mathcal{X}$. However, this follows from the assumptions that $\widetilde{\mathcal{X}} = \mathcal{X} \dotplus \mathcal{Z} = \mathcal{X}_o \dotplus \mathcal{Z}_i$ where $\mathcal{Z}_i \subset \mathcal{Z}$ and $\mathcal{X} \subset \mathcal{X}_o$, which implies that $P_{\mathcal{Z}}^{\mathcal{X}} - P_{\mathcal{Z}_i}^{\mathcal{X}_o}$ is a projection with kernel $\mathcal{X} \dotplus \mathcal{Z}_i$ and range $\mathcal{X}_o \cap \mathcal{Z}$ (we leave the proof of this to the reader). The same replacement of 2) by 2') can be carried out in Corollary 8.9, too. The final conclusion of Theorem 8.7 says that if $\mathcal{Z}_o$ is an arbitrary subspace of $\mathcal{Z}$ satisfying the properties listed in 2), then $\mathcal{Z}_o \subset \mathcal{Z}_o^{\max}$. This result implies that the subspace $\mathcal{X}_o^{\max} := \mathcal{Z}_o^{\max} \dotplus \mathcal{X}$ has an analogous maximality property: if $\mathcal{X}_o$ is an arbitrary subspace of $\widetilde{\mathcal{X}}$ satisfying the properties listed in 2'), then $\mathcal{X}_o \subset \mathcal{X}_o^{\max}$. A similar argument shows that all the subspaces $\mathcal{Z}_o$ in 2) and all the subspaces $\mathcal{X}_o$ in 2') satisfy $\mathcal{Z}_o^{\min} \subset \mathcal{Z}_o$ and $\mathcal{X}_o^{\min} \subset \mathcal{X}_o$, where $\mathcal{Z}_o^{\min}$ and $\mathcal{X}_o^{\min}$ are defined in Theorem 8.11 below.

Theorem 8.11. *Among all the decompositions* $\widetilde{\mathcal{X}} = \mathcal{Z}_o \dotplus \mathcal{X} \dotplus \mathcal{Z}_i$ *in Theorem* 8.7 *there is one for which the outgoing subspace* $\mathcal{Z}_o$ *is the smallest possible, i.e., there is an outgoing invariant subspace* $\mathcal{Z}_o = \mathcal{Z}_o^{\min}$ *which can be used in this decomposition, and which is contained in every outgoing subspace* $\mathcal{Z}_o$ *for every other choice of decomposition. The subspace* $\mathcal{Z}_o^{\min}$ *can be constructed as follows: Let* $\mathcal{X}_o^{\min}$ *be the closure in* $\widetilde{\mathcal{X}}$ *of all the possible values of the state components* $\widetilde{x}(\cdot)$ *of all trajectories* $(\widetilde{x}(\cdot), w(\cdot))$ *of* $\widetilde{\Sigma}$ *on* $\mathbb{Z}^+$ *satisfying* $\widetilde{x}(0) \in \mathcal{X}$, *and define* $\mathcal{Z}_o^{min} = \mathcal{X}_o^{\min} \cap \mathcal{Z}$.

Proof. Define $\mathcal{X}_o^{\min}$ and $\mathcal{Z}_o^{\min}$ as described in Theorem 8.11, and let $\mathcal{Z}_i$ be an arbitrary direct complement to $\mathcal{Z}_o^{\min}$ in $\mathcal{Z}$. Then $\widetilde{\mathcal{X}} = \mathcal{X} \dotplus \mathcal{Z} = \mathcal{X} \dotplus \mathcal{Z}_o^{\min} \dotplus \mathcal{Z}_i$. We have both $\mathcal{X} \subset \mathcal{X}_o^{\min}$ and $\mathcal{Z}_o^{\min} \subset \mathcal{X}_o^{\min}$, so $\mathcal{Z}_o^{\min} \dotplus \mathcal{X} \subset \mathcal{X}_o^{\min}$. To see that we actually have $\mathcal{Z}_o^{\min} \dotplus \mathcal{X} = \mathcal{X}_o^{\min}$ it suffices to show that $\mathcal{X}_o^{\min} \cap \mathcal{Z}_i = \{0\}$ (since $\mathcal{X}_o^{\min} \subset \widetilde{\mathcal{X}} = (\mathcal{Z}_o^{\min} \dotplus \mathcal{X}) \dotplus \mathcal{Z}_i$). But this is true because

$$\mathcal{X}_o^{\min} \cap \mathcal{Z}_i = (\mathcal{X}_o^{\min} \cap \mathcal{Z}) \cap \mathcal{Z}_i = \mathcal{Z}_o^{\min} \cap \mathcal{Z}_i = \{0\}.$$

Thus $\mathcal{X}_o^{\min} = \mathcal{Z}_o^{\min} \dotplus \mathcal{X}$ and $\widetilde{\mathcal{X}} = \mathcal{X}_o^{\min} \dotplus \mathcal{Z}_i = \mathcal{Z}_o^{\min} \dotplus \mathcal{X} \dotplus \mathcal{Z}_i$.

    It is easy to see that $\mathcal{X}_o^{\min}$ is the smallest (closed) strongly invariant subspace of $\widetilde{\mathcal{X}}$ which contains $\mathcal{X}$. In particular, for each decomposition $\widetilde{\mathcal{X}} = \mathcal{X}_o \dotplus \mathcal{Z}_i$ satisfying condition 2') in Remark 8.10 we must have $\mathcal{X}_o^{\min} \subset \mathcal{X}_o$. As we saw in Remark 8.10, this implies that if $\mathcal{Z}_o$ is an arbitrary subspace which satisfies the conditions listed in 2) of Theorem 8.7, then $\mathcal{Z}_o^{\min} \subset \mathcal{Z}_o$. This proves the claim about the minimality of $\mathcal{Z}_o^{\min}$ (and of $\mathcal{X}_o^{\min}$). It only remains to show that $\mathcal{Z}_o^{\min}$ is outgoing invariant for $\widetilde{\Sigma}$. To do this we argue as follows.

    Choose some arbitrary decomposition $\widetilde{\mathcal{X}} = \mathcal{Z}_o \dotplus \mathcal{X} \dotplus \mathcal{Z}_i$ of the type given in Theorem 8.7, and define $\mathcal{X}_o := \mathcal{Z}_o \dotplus \mathcal{X}$. Since $\mathcal{X}_o^{\min}$ is the smallest (closed) strongly invariant subspace of $\widetilde{\mathcal{X}}$ which contains $\mathcal{X}$ we must have $\mathcal{X} \subset \mathcal{X}_o^{\min} \subset \mathcal{X}_o$. It follows from (8.3) and (8.6) that (8.6) also holds if we replace $\mathcal{X}_o$ by $\mathcal{X}_o^{\min}$. Take some arbitrary $z_0 \in \mathcal{Z}_0^{\min} = \mathcal{X}_o^{\min} \cap \mathcal{Z} \subset \mathcal{X}_o \cap \mathcal{Z} = \mathcal{Z}_o$. Then there exists some trajectory $(\tilde{x}(\cdot), w(\cdot))$ of $\widetilde{\Sigma}$ on $\mathbb{Z}^+$ with $\tilde{x}(0) = z_0$. By (8.6) with $\mathcal{X}_o$ replaced by $\mathcal{X}_o^{\min}$, if we define $x(\cdot) = P_{\mathcal{X}}^{\mathcal{Z}}\tilde{x}(\cdot)$, then $(x(\cdot), w(\cdot))$ is a trajectory of Σ. Observe that $x(0) = 0$. By part 1) Lemma 8.2, there exists a (unique) trajectory $(\tilde{x}_1(\cdot), w(\cdot))$ of $\widetilde{\Sigma}$ on $\mathbb{Z}^+$ with $P_{\mathcal{X}}^{\mathcal{Z}}\tilde{x}_1(\cdot) = x(\cdot)$ (in particular, $\tilde{x}_1(0) = 0$). Define $\tilde{x}_2(\cdot) = \tilde{x}(\cdot) - \tilde{x}_1(\cdot)$. Then $(\tilde{x}_2(\cdot), 0)$ is a trajectory of $\widetilde{\Sigma}$ with $\tilde{x}_2(0) = z_0$ and $P_{\mathcal{X}}^{\mathcal{Z}}\tilde{x}_2(\cdot) = x(\cdot) - x(\cdot) = 0$. Thus $\tilde{x}_2(n) \subset \mathcal{Z}$ for all $n \in \mathbb{Z}^+$. But on the other hand, by the strong invariance of $\mathcal{X}_o^{\min}$, $\tilde{x}_2(n) \subset \mathcal{X}_o^{\min}$ for all $n \in \mathbb{Z}^+$. Thus, $\tilde{x}_2(n) \subset \mathcal{Z} \cap \mathcal{X}_o^{\min} = \mathcal{Z}_o^{\min}$ for all $n \in \mathbb{Z}^+$ and, as we recall, $\tilde{x}_2(0) = z_0$. This proves that $\mathcal{Z}_o^{\min}$ is outgoing invariant. $\qquad\square$

It is often useful to split a compression or dilation into the product of two successive dilations or compression.

Lemma 8.12. *Let $\widehat{\Sigma} = (\widehat{V}; \widehat{\mathcal{X}}, \mathcal{W})$ be a compression of $\widetilde{\Sigma} = (\widetilde{V}; \widetilde{\mathcal{X}}, \mathcal{W})$ along $\widehat{\mathcal{Z}}$ onto $\widehat{\mathcal{X}}$, and let $\Sigma = (V; \mathcal{X}, \mathcal{W})$ be a compression of $\widehat{\Sigma}$ along $\mathcal{Z}$ onto $\mathcal{X}$. Then $\Sigma = (V; \mathcal{X}, \mathcal{W})$ is a compression of $\widetilde{\Sigma}$ along $\widehat{\mathcal{Z}} \dotplus \mathcal{Z}$ onto $\mathcal{X}$, and $P_{\mathcal{X}}^{\widehat{\mathcal{Z}}+\mathcal{Z}} = P_{\mathcal{X}}^{\mathcal{Z}} P_{\widehat{\mathcal{X}}}^{\widehat{\mathcal{Z}}}$.*

The easy proof is left to the reader.

Two particularly simple types of dilations are those where one of the two subspaces $\mathcal{Z}_o$ and $\mathcal{Z}_i$ in Theorem 8.7 can be taken to be zero.

Definition 8.13. *The state/signal system $\widetilde{\Sigma} = (\widetilde{V}; \widetilde{\mathcal{X}}, \mathcal{W})$ is an* outgoing dilation along $\mathcal{Z}$ *of the state/signal system $\Sigma = (V; \mathcal{X}, \mathcal{W})$, or equivalently, the state/signal system Σ is an* outgoing compression along $\mathcal{Z}$ onto $\mathcal{X}$ *of the state/signal system $\widetilde{\Sigma}$, if the following conditions hold:*

1) $\widetilde{\mathcal{X}} = \mathcal{X} \dotplus \mathcal{Z}$,
2) *If $(\tilde{x}(\cdot), w(\cdot))$ is a trajectory of $\widetilde{\Sigma}$ on $\mathbb{Z}^+$, then $(P_{\mathcal{X}}^{\mathcal{Z}}\tilde{x}(\cdot), w(\cdot))$ is a trajectory of Σ on $\mathbb{Z}^+$.*
3) *There is at least one decomposition $\mathcal{W} = \mathcal{Y} \dotplus \mathcal{U}$ of $\mathcal{W}$ which is admissible for both $\widetilde{\Sigma}$ and Σ.*

Clearly, every outgoing dilation is also a dilation.

Lemma 8.14. *Let $\widetilde{\Sigma} = (\widetilde{V}; \widetilde{\mathcal{X}}, \mathcal{W})$ and $\Sigma = (V; \mathcal{X}, \mathcal{W})$ be two state/signal systems with $\widetilde{\mathcal{X}} = \mathcal{X} \dotplus \mathcal{Z}$ (and with the same signal space). Then the following conditions are equivalent.*

1) *$\widetilde{\Sigma}$ is an outgoing dilation along $\mathcal{Z}$ of Σ,*
2) *V is given by*

$$
V = \begin{bmatrix} P_{\mathcal{X}}^{\mathcal{Z}} & 0 & 0 \\ 0 & P_{\mathcal{X}}^{\mathcal{Z}} & 0 \\ 0 & 0 & 1_{\mathcal{X}} \end{bmatrix} \widetilde{V} = \left\{ \begin{bmatrix} P_{\mathcal{X}}^{\mathcal{Z}}\tilde{z} \\ P_{\mathcal{X}}^{\mathcal{Z}}\tilde{x} \\ w \end{bmatrix} \; \middle| \; \begin{bmatrix} \tilde{z} \\ \tilde{x} \\ w \end{bmatrix} \in \widetilde{V} \right\}. \tag{8.7}
$$

3) *(8.3) holds and $\mathcal{Z}$ is outgoing invariant for $\widetilde{\Sigma}$.*

Proof. The proof of the fact that 1) implies 2) is essentially the same as the proof of the necessity of (8.3) in Theorem 8.7, and the proof of the converse implication is a simplified version of the sufficiency part of the proof of the same theorem. That 1) and 2) together imply 3) is a simplified version of the final paragraph of the proof of Theorem 8.11 (replace $\mathcal{Z}_o^{\min}$ by $\mathcal{Z}$, replace $\mathcal{X}_o^{\min}$ by $\widetilde{\mathcal{X}}$, and use the facts that $\widetilde{\Sigma}$ is a dilation of Σ and that (8.6) now holds with $\mathcal{X}_o$ replaced by $\widetilde{\mathcal{X}}$). Finally, that 3) implies 2) follows from Corollary 8.8. $\square$

Definition 8.15. The state/signal system $\widetilde{\Sigma} = (\widetilde{V}; \widetilde{\mathcal{X}}, \mathcal{W})$ is an *incoming dilation along $\mathcal{Z}$* of the state/signal system $\Sigma = (V; \mathcal{X}, \mathcal{W})$, or equivalently, the state/signal system Σ is an *incoming compression along $\mathcal{Z}$ onto $\mathcal{X}$* of the state/signal system $\widetilde{\Sigma}$, if the following conditions hold:

1) $\widetilde{\mathcal{X}} = \mathcal{X} \dotplus \mathcal{Z}$,
2) If $(\tilde{x}(\cdot), w(\cdot))$ is a trajectory of $\widetilde{\Sigma}$ on $\mathbb{Z}^+$ with $\tilde{x}(0) \in \mathcal{X}$, then $\tilde{x}(n) \in \mathcal{X}$ for all $n \in \mathbb{Z}^+$ and $(x(\cdot), w(\cdot))$ is a trajectory of Σ on $\mathbb{Z}^+$.
3) There is at least one decomposition $\mathcal{W} = \mathcal{Y} \dotplus \mathcal{U}$ of $\mathcal{W}$ which is admissible for both $\widetilde{\Sigma}$ and Σ.

Lemma 8.16. *Let $\widetilde{\Sigma} = \left(\widetilde{V}; \widetilde{\mathcal{X}}, \mathcal{W}\right)$ and $\Sigma = (V; \mathcal{X}, \mathcal{W})$ be two state/signal systems with $\widetilde{\mathcal{X}} = \mathcal{X} \dotplus \mathcal{Z}$ (and with the same signal space). Then the following conditions are equivalent.*

1) $\widetilde{\Sigma}$ *is an incoming dilation along $\mathcal{Z}$ of Σ,*
2) V *is given by*

$$V = \left\{ \begin{bmatrix} \tilde{z} \\ x \\ w \end{bmatrix} \in \widetilde{V} \,\middle|\, x \in \mathcal{X} \right\}. \tag{8.8}$$

3) (8.3) *holds and $\mathcal{X}$ is strongly invariant for $\widetilde{\Sigma}$.*

This proof is similar to the proof of Lemma 8.14 and it is left to the reader.

Definition 8.17. A state/signal system is *minimal* if it is not a (nontrivial) dilation of any other state/signal system (along any direction).

Theorem 8.18. *An state/signal system is minimal if and only if it is controllable and observable.*

Proof. Let $\widetilde{\Sigma}$ be state/signal system, and let Σ be a compression of $\widetilde{\Sigma}$. If $\widetilde{\Sigma}$ is observable, then the outgoing subspace $\mathcal{Z}_o$ in the decomposition in Theorem 8.7 is trivial (since it is part of the unobservable subspace), and if $\widetilde{\Sigma}$ is controllable, then the incoming subspace $\mathcal{Z}_i$ in the decomposition in Theorem 8.7 is trivial (since $\mathcal{Z}_o \dotplus \mathcal{X}$ contains the reachable subspace). Thus, if $\widetilde{\Sigma}$ is both controllable and observable, then it does not have any nontrivial dilation.

The converse claim follows from Theorem 8.19 below (which shows that every non-observable or non-controllable system has a nontrivial compression). $\square$

Theorem 8.19. $\Sigma = (V; \mathcal{X}, \mathcal{W})$ *be a state/signal system. Denote the reachable subspace of Σ by $\mathfrak{R}$ and the unobservable subspace of Σ by $\mathfrak{U}$.*

1) *Let $\mathfrak{O}$ be a direct complement to $\mathfrak{U}$ in $\mathcal{X}$, define $\mathcal{X}_\circ := \overline{P_{\mathfrak{O}}^{\mathfrak{U}}\mathfrak{R}}$, and let $\mathfrak{O}_i$ be a direct complement to $\mathcal{X}_\circ$ in $\mathfrak{O}$. Define $V_\circ$ by*

$$V_\circ := \left\{ \begin{bmatrix} P_{\mathcal{X}_\circ}^{\mathfrak{U}+\mathfrak{O}_i}z \\ x \\ w \end{bmatrix} \,\middle|\, x \in \mathcal{X}_\circ, \begin{bmatrix} z \\ x \\ w \end{bmatrix} \in V \right\}. \tag{8.9}$$

Then $\Sigma_\circ = (V_\circ; \mathcal{X}_\circ, \mathcal{W})$ is a minimal state/signal systems which is a compression of Σ along $\mathfrak{U} \dotplus \mathfrak{O}_i$. Here $\mathfrak{U}$ is outgoing invariant for Σ and $\mathfrak{U} \dotplus \mathcal{X}_\circ$ is strongly invariant for Σ, so that we can take $\mathcal{Z}_o = \mathfrak{U}$ and $\mathcal{Z}_i = \mathfrak{O}_i$ in the decomposition in Theorem 8.7.

2) *Let $\mathfrak{Q}$ be a direct complement to $\mathfrak{R}$ in $\mathcal{X}$, define $\mathfrak{R}_o = \mathfrak{R} \cap \mathfrak{U}$, and let $\mathcal{X}_\bullet$ be a direct complement to $\mathfrak{R}_o$ in $\mathfrak{R}$. Define*

$$V_\bullet := \left\{ \begin{bmatrix} P_{\mathcal{X}_\bullet}^{\mathfrak{R}_o+\mathfrak{Q}}z \\ x \\ w \end{bmatrix} \,\middle|\, x \in \mathcal{X}_\bullet, \begin{bmatrix} z \\ x \\ w \end{bmatrix} \in V \right\}. \tag{8.10}$$

Then $\Sigma_\bullet = (V_\bullet; \mathcal{X}_\bullet, \mathcal{W})$ is a minimal state/signal systems which is a compression of Σ along $\mathfrak{R}_o \dotplus \mathfrak{Q}$ onto $\mathcal{X}_\bullet$. Here $\mathfrak{R}_o$ is outgoing invariant for Σ and $\mathfrak{R}_o \dotplus \mathcal{X}_\bullet$ is strongly invariant for Σ, so that we can take $\mathcal{Z}_o = \mathfrak{R}_o$ and $\mathcal{Z}_i = \mathfrak{Q}$ in the decomposition in Theorem 8.7.

Proof. Proof of 1). We begin by performing an outgoing compression of Σ along $\mathfrak{U}$ onto $\mathfrak{O}$, i.e., we define

$$V_\circ^1 := \left\{ \begin{bmatrix} P_{\mathfrak{O}}^{\mathfrak{U}}z \\ x \\ w \end{bmatrix} \,\middle|\, x \in \mathfrak{O}, \begin{bmatrix} z \\ x \\ w \end{bmatrix} \in V \right\}.$$

According to Lemma 8.6, $\mathfrak{U}$ is outgoing invariant for Σ, so by Corollary 8.9, $\Sigma_\circ^1 := (V_\circ^1; \mathfrak{O}, \mathcal{W})$ is a compression of Σ along $\mathfrak{U}$. Moreover, it follows from Lemma 8.3 that $\Sigma_\circ^1$ is observable.

We continue by performing an incoming compression of $\Sigma_\circ^1$ along $\mathfrak{O}_i$ onto its reachable subspace, which according to Lemma 8.3 is equal to $\mathcal{X}_\circ$. Thus, we define

$$V_\circ := \left\{ \begin{bmatrix} P_{\mathcal{X}_\circ}^{\mathfrak{O}_i}z \\ x \\ w \end{bmatrix} \,\middle|\, x \in \mathcal{X}_\circ, \begin{bmatrix} z \\ x \\ w \end{bmatrix} \in V_\circ^1 \right\}.$$

The subspace $\mathcal{X}_\circ$ is strongly invariant for $\Sigma_\circ^1$ (see Lemma 8.6), so by Corollary 8.9, $\Sigma_\circ := (V_\circ; \mathcal{X}_\circ, \mathcal{W})$ is a compression of $\Sigma_\circ^1$ along $\mathfrak{O}_i$. By Lemma 8.12, this system is the same one which we defined in Part 1), and by Lemma 8.3, $\Sigma_\circ$ is both controllable and observable.

It remains to show that $\mathfrak{U} \dotplus \mathcal{X}_\circ$ is strongly invariant for Σ. However, this follows from the fact that the maximal outgoing subspace $\mathcal{Z}_o^{\max}$ defined in (8.4) always is contained in the unobservable subspace $\mathfrak{U}$, and in this particular case it coincides with $\mathfrak{U}$. Thus, $\mathfrak{U} \dotplus \mathcal{X}_\circ$ coincides with the space $\mathcal{Z}_o^{\max} \dotplus \mathcal{X}_\circ$, and it must therefore be strongly invariant.

Proof of 2). We begin by performing an incoming compression of Σ along $\mathfrak{Q}$ onto $\mathfrak{R}$, i.e., we define

$$V_\bullet^1 := \left\{ \begin{bmatrix} P_{\mathfrak{R}}^{\mathfrak{Q}}z \\ x \\ w \end{bmatrix} \,\middle|\, x \in \mathfrak{R}, \begin{bmatrix} z \\ x \\ w \end{bmatrix} \in V \right\}.$$

According to Lemma 8.6, $\mathfrak{R}$ is strongly invariant for Σ, so by Corollary 8.9, $\Sigma^1_\bullet :=$ $(V^1_\bullet; \mathfrak{R}, \mathcal{W})$ is a compression of Σ along $\mathfrak{Q}$. Moreover, it follows from Lemma 8.3 that $\Sigma^1_\bullet$ is controllable.

We continue by performing an outgoing compression of $\Sigma^1_\bullet$ along its unobservable subspace, which according to Lemma 8.3 is equal to $\mathfrak{R}_o$. That is, we define

$$V_\bullet := \left\{ \begin{bmatrix} P^{\mathfrak{R}_o}_{\mathcal{X}_\bullet} z \\ x \\ w \end{bmatrix} \;\middle|\; x \in \mathcal{X}_\bullet, \; \begin{bmatrix} z \\ x \\ w \end{bmatrix} \in V^1_\bullet \right\}.$$

The subspace $\mathfrak{R}_o$ is outgoing invariant for $\Sigma^1_\bullet$ (see Lemma 8.6), so by Corollary 8.9, $\Sigma_\bullet := (V_\bullet; \mathcal{X}_\bullet, \mathcal{W})$ is a compression of $\Sigma^1_\bullet$ along $\mathfrak{R}_o$. By Lemma 8.12, this system is the same one which we defined in Part 1), and by Lemma 8.3, $\Sigma_\bullet$ is both controllable and observable. We already observed above that $\mathfrak{R}_o$ is outgoing invariant and that $\mathfrak{R}_o \dotplus \mathcal{X}_\bullet = \mathfrak{R}$ is strongly invariant for Σ. $\qquad\square$

Theorem 8.20. *Every realizable signal behavior has a minimal state/signal realization (i.e., the behavior has a state/signal realization which is minimal).*

This follows from Theorem 8.19 (since a compressed system is externally equivalent to the original system).

Up to now we have not used any specific representation of a state/signal system in our study of dilations and compressions. For completeness we interpret some of our results in terms of driving variable, output nulling, and input/state/output representations. We begin with the following description of the crucial formula (8.3) in Theorem 8.7.

Lemma 8.21. *Let* $\widetilde{\Sigma} = (\widetilde{V}; \widetilde{\mathcal{X}}, \mathcal{W})$ *and* $\Sigma = (V; \mathcal{X}, \mathcal{W})$ *be two state/signal systems with* $\widetilde{\mathcal{X}} = \mathcal{X} \dotplus \mathcal{Z}$ *(and with the same signal space).*

1) *The following conditions are equivalent:*
   (a) *V is given by (8.3).*
   (b) *If* $\left(\begin{bmatrix} \widetilde{A}' & \widetilde{B}' \\ \widetilde{C}' & \widetilde{D}' \end{bmatrix}; \widetilde{\mathcal{X}}, \widetilde{\mathcal{L}}, \mathcal{W} \right)$ *is a driving variable representation of* $\widetilde{\Sigma}$*, then* $\left(\begin{bmatrix} P^{\mathcal{Z}}_{\mathcal{X}} \widetilde{A}'|_{\mathcal{X}} & P^{\mathcal{Z}}_{\mathcal{X}} \widetilde{B}' \\ \widetilde{C}'|_{\mathcal{X}} & \widetilde{D}' \end{bmatrix}; \mathcal{X}, \widetilde{\mathcal{L}}, \mathcal{W} \right)$ *is a driving variable representation of* Σ*.*
   (c) *If* $\left(\begin{bmatrix} \widetilde{A}'' & \widetilde{B}'' \\ \widetilde{C}'' & \widetilde{D}'' \end{bmatrix}; \widetilde{\mathcal{X}}, \mathcal{W}, \widetilde{\mathcal{K}} \right)$ *is an output nulling representation of* $\widetilde{\Sigma}$*, then* $\left(\begin{bmatrix} P^{\mathcal{Z}}_{\mathcal{X}} \widetilde{A}''|_{\mathcal{X}} & P^{\mathcal{Z}}_{\mathcal{X}} \widetilde{B}'' \\ \widetilde{C}''|_{\mathcal{X}} & \widetilde{D}'' \end{bmatrix}; \mathcal{X}, \mathcal{W}, \widetilde{\mathcal{K}} \right)$ *is an output nulling representation of* Σ*.*
   (d) *If* $\left(\begin{bmatrix} \widetilde{A} & \widetilde{B} \\ \widetilde{C} & \widetilde{D} \end{bmatrix}; \widetilde{\mathcal{X}}, \mathcal{U}, \mathcal{Y} \right)$ *is an input/state/output representation of* $\widetilde{\Sigma}$ *corresponding to some admissible input/output decomposition* $\mathcal{W} = \mathcal{Y} \dotplus \mathcal{U}$*, then* $\left(\begin{bmatrix} P^{\mathcal{Z}}_{\mathcal{X}} \widetilde{A}|_{\mathcal{X}} & P^{\mathcal{Z}}_{\mathcal{X}} \widetilde{B} \\ \widetilde{C}|_{\mathcal{X}} & \widetilde{D} \end{bmatrix}; \mathcal{X}, \mathcal{U}, \mathcal{Y} \right)$ *is an input/state/output representation of* Σ *corresponding to the same admissible decomposition of* $\mathcal{W}$*.*
2) *Assume that the equivalent conditions (a)–(d) above hold. Then every driving variable representation of* Σ *is of the form described in (b), every output nulling representation of* Σ *is of the form described in (c), and every input/output representation of* Σ *is of the form described in (d).*

Proof. The equivalence of (a)–(d) follows from (3.3), (4.3), (5.2), and (8.3).

That *every* input/state/output representations of V must be of the type given in (d) follows from the uniqueness of such a representation (see Theorem 5.1). The proof of the claim that all possible output nulling representations of V are of the type (c) is similar to the proof of the claim that all possible driving variable representations of V are of the type (b), so let us only prove the latter claim.

Let $\left(\left[\begin{smallmatrix} A' & B' \\ C' & D' \end{smallmatrix}\right]; \mathcal{X}, \mathcal{L}, \mathcal{W}\right)$ be an arbitrary driving variable representation of Σ, and let $\left(\left[\begin{smallmatrix} \widetilde{A}' & \widetilde{B}' \\ \widetilde{C}' & \widetilde{D}' \end{smallmatrix}\right]; \widetilde{\mathcal{X}}, \widetilde{\mathcal{L}}, \mathcal{W}\right)$ be the driving variable representation of $\widetilde{\Sigma}$ mentioned in part (b). Then by Theorem 6.1, there exist operators $K' \in \mathcal{B}(\mathcal{X}; \widetilde{\mathcal{L}})$ and $M' \in \mathcal{B}(\mathcal{L}; \widetilde{\mathcal{L}})$, with M' boundedly invertible, such that

$$\begin{bmatrix} A' & B' \\ C' & D' \end{bmatrix} = \begin{bmatrix} P_{\mathcal{X}}^{\mathcal{Z}} \widetilde{A}'|_{\mathcal{X}} + P_{\mathcal{X}}^{\mathcal{Z}} \widetilde{B}' K_1' & P_{\mathcal{X}}^{\mathcal{Z}} \widetilde{B}' M' \\ \widetilde{C}'|_{\mathcal{X}} + \widetilde{D}' K_1' & \widetilde{D}' M' \end{bmatrix}.$$

Define $\widetilde{K}' = K' P_{\mathcal{X}}^{\mathcal{Z}}$. Then

$$\begin{bmatrix} A' & B' \\ C' & D' \end{bmatrix} = \begin{bmatrix} P_{\mathcal{X}}^{\mathcal{Z}} (\widetilde{A}' + \widetilde{B}' \widetilde{K}')|_{\mathcal{X}} & P_{\mathcal{X}}^{\mathcal{Z}} \widetilde{B}' M' \\ (\widetilde{C}' + \widetilde{D}' \widetilde{K}')|_{\mathcal{X}} & \widetilde{D}' M' \end{bmatrix}.$$

By Theorem 6.1, $\left(\left[\begin{smallmatrix} \widetilde{A}' + \widetilde{B}' \widetilde{K}' & \widetilde{B}' M' \\ \widetilde{C}' + \widetilde{D}' \widetilde{K}' & \widetilde{D}' M' \end{smallmatrix}\right]; \widetilde{\mathcal{X}}, \widetilde{\mathcal{L}}, \mathcal{W}\right)$ is a driving variable representation of $\widetilde{\Sigma}$, and hence $\left(\left[\begin{smallmatrix} A' & B' \\ C' & D' \end{smallmatrix}\right]; \mathcal{X}, \mathcal{L}, \mathcal{W}\right)$ is of the type (b). $\qquad\square$

Definition 8.1 is very closely related to the following definition of a dilation of a input/state/output system.

Definition 8.22. We say that the input/state/output system

$$\widetilde{\Sigma}_{i/s/o} = \left(\begin{bmatrix} \widetilde{A} & \widetilde{B} \\ \widetilde{C} & \widetilde{D} \end{bmatrix}; \widetilde{\mathcal{X}}, \mathcal{U}, \mathcal{Y}\right)$$

is a *dilation along* $\mathcal{Z}$ of the input/state/output system

$$\Sigma_{i/s/o} = \left([\begin{smallmatrix} A & B \\ C & D \end{smallmatrix}]; \mathcal{X}, \mathcal{U}, \mathcal{Y}\right),$$

or equivalently, that $\Sigma_{i/s/o}$ is a *compression along* $\mathcal{Z}$ onto $\mathcal{X}$ of $\widetilde{\Sigma}_{i/s/o}$, if $\widetilde{\mathcal{X}} = \mathcal{X} \dotplus \mathcal{Z}$ and the following condition holds: For each $x_0 \in \mathcal{X}$ and each input sequence $u(\cdot) \in \mathcal{U}^{\mathbb{Z}^+}$ the corresponding trajectories $(\widetilde{x}(\cdot), u(\cdot), \widetilde{y}(\cdot))$ and $(x(\cdot), u(\cdot), y(\cdot))$ of $\widetilde{\Sigma}_{i/s/o}$, respectively, $\Sigma_{i/s/o}$, with initial state $\widetilde{x}(0) = x(0) = x_0$, satisfy $x(\cdot) = P_{\mathcal{X}}^{\mathcal{Z}} \widetilde{x}(\cdot)$ and $\widetilde{y}(\cdot) = y(\cdot)$.

As usual, we shall call an input/state/output system *minimal* if it is not a (nontrivial) dilation of any other input/state/output system (along any direction).

Lemma 8.23. *Let* $\widetilde{\Sigma} = (\widetilde{V}; \widetilde{\mathcal{X}}, \mathcal{W})$ *and* $\Sigma = (V; \mathcal{X}, \mathcal{W})$ *be two state/signal systems with* $\widetilde{\mathcal{X}} = \mathcal{X} \dotplus \mathcal{Z}$ *(and with the same signal space* $\mathcal{W}$*).*

1) *Suppose that* $\widetilde{\Sigma}$ *and* Σ *have a common admissible input/output decomposition* $\mathcal{W} = \mathcal{Y} \dotplus \mathcal{U}$*. Denote the corresponding input/state/output representations by* $\widetilde{\Sigma}_{i/s/o}$*, respectively,* $\Sigma_{i/s/o}$*. If* $\widetilde{\Sigma}_{i/s/o}$ *is a dilation along* $\mathcal{Z}$ *of* $\Sigma_{i/s/o}$*, then* $\widetilde{\Sigma}$ *is a dilation along* $\mathcal{Z}$ *of* Σ*.*

2) *Conversely, if $\widetilde{\Sigma}$ is a dilation along $\mathcal{Z}$ of Σ, then the two systems have the same admissible input/output decompositions $\mathcal{W} = \mathcal{Y} \dotplus \mathcal{U}$, and if we denote the corresponding input/state/output representations by $\widetilde{\Sigma}_{i/s/o}$ and $\Sigma_{i/s/o}$, respectively, then $\widetilde{\Sigma}_{i/s/o}$ is a dilation along $\mathcal{Z}$ of $\Sigma_{i/s/o}$.*

Proof. Proof of 1). Let $(\widetilde{x}(\cdot), w(\cdot))$ be a trajectory of $\widetilde{\Sigma}$ on $\mathbb{Z}^+$ with $\widetilde{x}(0) \in \mathcal{X}$. Then $(\widetilde{x}(\cdot), u(\cdot), y(\cdot))$ with $u(\cdot) = P_{\mathcal{U}}^{\mathcal{Y}} w(\cdot)$ and $y(\cdot) = P_{\mathcal{Y}}^{\mathcal{U}} w(\cdot)$ is a trajectory of $\widetilde{\Sigma}_{i/s/o}$. By Definition 8.22, $(x(\cdot), u(\cdot), y(\cdot))$ with $x(\cdot) = P_{\mathcal{X}}^{\mathcal{Z}} \widetilde{x}(\cdot)$ is a trajectory of $\Sigma_{i/s/o}$, and hence $(P_{\mathcal{X}}^{\mathcal{Z}} \widetilde{x}(\cdot), w(\cdot))$ is a trajectory of Σ. Thus, $\widetilde{\Sigma}$ is a dilation along $\mathcal{Z}$ of Σ.

Proof of 2). That the two systems have the same admissible input/output decompositions follows from Lemmas 5.7 and 8.2. Let $\mathcal{W} = \mathcal{Y} \dotplus \mathcal{U}$ be a decomposition which is admissible both for $\widetilde{\Sigma}$ and for Σ. Let $(\widetilde{x}(\cdot), u(\cdot), y(\cdot))$ be a trajectory of $\widetilde{\Sigma}_{i/s/o}$ on $\mathbb{Z}^+$ with $\widetilde{x}(0) = x_0 \in \mathcal{X}$. Then $(\widetilde{x}(\cdot), w(\cdot))$ with $w(\cdot) = y(\cdot) + u(\cdot)$ is a trajectory of $\widetilde{\Sigma}$, and by Definition 8.1, $(P_{\mathcal{X}}^{\mathcal{Y}} \widetilde{x}(\cdot), w(\cdot))$ is a trajectory of Σ. Hence $(P_{\mathcal{X}}^{\mathcal{Y}} \widetilde{x}(\cdot), u(\cdot), y(\cdot))$ is a trajectory of $\Sigma_{i/s/o}$. More precisely, it is the unique trajectory of Σ with the initial state x_0 and the input data $u(\cdot)$. Thus, $\widetilde{\Sigma}_{i/s/o}$ is a dilation along $\mathcal{Z}$ of $\Sigma_{i/s/o}$. $\qquad\square$

Theorem 8.24. *Let $\widetilde{\Sigma}_{i/s/o} = \left(\begin{bmatrix} \widetilde{A} & \widetilde{B} \\ \widetilde{C} & \widetilde{D} \end{bmatrix}; \widetilde{\mathcal{X}}, \mathcal{U}, \mathcal{Y} \right)$ and $\Sigma_{i/s/o} = \left(\begin{bmatrix} A & B \\ C & D \end{bmatrix}; \mathcal{X}, \mathcal{U}, \mathcal{Y} \right)$ be two input/state/output systems with $\widetilde{\mathcal{X}} = \mathcal{X} \dotplus \mathcal{Z}$ (and with the same input and output spaces). Then $\Sigma_{i/s/o}$ is a compression along $\mathcal{Z}$ onto $\mathcal{X}$ of $\widetilde{\Sigma}_{i/s/o}$ if and only if $\mathcal{Z}$ can be decomposed into $\mathcal{Z} = \mathcal{Z}_o \dotplus \mathcal{Z}_i$ such that the decomposition of $\begin{bmatrix} \widetilde{A} & \widetilde{B} \\ \widetilde{C} & \widetilde{D} \end{bmatrix}$ with respect to the decomposition $\widetilde{\mathcal{X}} = \mathcal{Z}_o \dotplus \mathcal{X} \dotplus \mathcal{Z}_i$ has the following form (where $*$ stands for an irrelevant block)*

$$
\begin{bmatrix} \widetilde{A} & \widetilde{B} \\ \widetilde{C} & \widetilde{D} \end{bmatrix} =
\left[\begin{array}{ccc|c} * & * & * & * \\ 0 & A & * & B \\ 0 & 0 & * & 0 \\ \hline 0 & C & * & D \end{array} \right].
\tag{8.11}
$$

This is a non-orthogonal version of [Aro79, Proposition 4]. For completeness we include a short proof based on Theorem 8.7.

Proof of Theorem 8.24. Let $\widetilde{\Sigma}$ and Σ be the state/signal systems induced by $\widetilde{\Sigma}_{i/s/o}$ and $\Sigma_{i/s/o}$, respectively.

If $\begin{bmatrix} \widetilde{A} & \widetilde{B} \\ \widetilde{C} & \widetilde{D} \end{bmatrix}$ is of the form (8.11), then it is easy to see that $\mathcal{Z}_o$ is outgoing invariant and $\mathcal{Z}_o \dotplus \mathcal{X}$ is strongly invariant for $\widetilde{\Sigma}$. Moreover, it follows from Lemma 8.21 that (8.3) holds. Thus, by Theorem 8.7, $\widetilde{\Sigma}$ is a dilation along $\mathcal{Z}$ of Σ, and consequently, by Lemma 8.23, $\widetilde{\Sigma}_{i/s/o}$ is a dilation along $\mathcal{Z}$ of $\Sigma_{i/s/o}$.

Conversely, suppose that $\widetilde{\Sigma}_{i/s/o}$ is a dilation along $\mathcal{Z}$ of $\Sigma_{i/s/o}$. Then, by Lemma 8.23, $\widetilde{\Sigma}$ is a dilation along $\mathcal{Z}$ of Σ. Let $\widetilde{\mathcal{X}} = \mathcal{Z}_o \dotplus \mathcal{X} \dotplus \mathcal{Z}_i$ be the decomposition in Theorem 8.7. Then it is easy to see that the fact that $\mathcal{Z}_o$ is outgoing invariant and $\mathcal{Z}_o \dotplus \mathcal{X}$ is strongly invariant imposes the structure (8.11) on $\begin{bmatrix} \widetilde{A} & \widetilde{B} \\ \widetilde{C} & \widetilde{D} \end{bmatrix}$. That the entries in positions $(2,2)$, $(2,4)$, $(4,2)$, and $(4,4)$ are A, B, C, and D follows from (8.3) and Lemma 8.21. $\qquad\square$

It is not difficult to see that the decomposition (8.11) of $\begin{bmatrix} \widetilde{A} & \widetilde{B} \\ \widetilde{C} & \widetilde{D} \end{bmatrix}$ with respect to the decomposition $\widetilde{\mathcal{X}} = \mathcal{Z}_o \dotplus \mathcal{X} \dotplus \mathcal{Z}_i$ is valid if and only if (we denote $\mathcal{Z}_o \dotplus \mathcal{Z}_i$ by $\mathcal{Z}$)

$$\mathcal{R}(\widetilde{B}) \subset \mathcal{Z}_o \dotplus \mathcal{X}, \quad \mathcal{Z}_o \subset \mathcal{N}(\widetilde{C}),$$
$$\mathcal{R}(\widetilde{A}|_{\mathcal{Z}_o}) \subset \mathcal{Z}_o, \quad \mathcal{R}(\widetilde{A}|_{\mathcal{Z}_o \dotplus \mathcal{X}}) \subset \mathcal{Z}_o \dotplus \mathcal{X}, \tag{8.12}$$
$$A = P^{\mathcal{Z}}_{\mathcal{X}} \widetilde{A}|_{\mathcal{X}}, \; B = P^{\mathcal{Z}}_{\mathcal{X}} \widetilde{B}, \; C = \widetilde{C}|_{\mathcal{X}}, \; D = \widetilde{D}.$$

Thus, in particular, $\widetilde{A} \in \mathcal{B}(\widetilde{\mathcal{X}})$ is an dilation of $A \in \mathcal{B}(\mathcal{X})$, i.e.,

$$A^n = P^{\mathcal{Z}}_{\mathcal{X}} \widetilde{A}|^n_{\mathcal{X}}, \quad n \in \mathbb{Z}^+. \tag{8.13}$$

Orthogonal dilations (i.e., dilations where $\mathcal{X}$ and $\mathcal{Z}$ are orthogonal) play an essential role in the Nagy–Foiaş theory of harmonic analysis for operators in Hilbert space (see [SF70]) which is intimately connected with the Lax–Phillips scattering theory (see [LP67] and [AA70]).

Theorem 8.25. *An input/state/output system is minimal if and only if it is controllable and observable. Moreover, an input/state/output system Σ which is not minimal can be compressed into a minimal system (i.e., there is a minimal input/ state/output system which is an compression of Σ).*

This is a non-orthogonal version of [Aro79, Propositions 3 and 4, p. 151]. It is easy to deduce this theorem from Theorems 8.18 and 8.19 in the same way as we derived Theorem 8.24 from Theorem 8.7. We leave the details to the reader.

Theorem 8.26. *Let Σ be a state/signal system. Then the following conditions are equivalent:*

1) Σ *is minimal.*
2) Σ *is controllable and observable.*
3) Σ *has a minimal input/state/output representation.*
4) Σ *has a controllable driving variable representation and an observable output nulling representation.*
5) *Every input/state/output representation of Σ is minimal.*
6) *Every driving variable representation of Σ is controllable, and every output nulling representation of Σ is observable.*

Proof. This follows from Propositions 3.5, 4.5, and 5.5, and Theorems 8.18 and 8.25. $\qquad\square$

Lemma 8.27. *Let $\widetilde{\Sigma}_{i/s/o} = \left(\begin{bmatrix} \widetilde{A} & \widetilde{B} \\ \widetilde{C} & \widetilde{D} \end{bmatrix} ; \widetilde{\mathcal{X}}, \mathcal{U}, \mathcal{Y} \right)$ and $\Sigma_{i/s/o} = \left(\begin{bmatrix} A & B \\ C & D \end{bmatrix} ; \mathcal{X}, \mathcal{U}, \mathcal{Y} \right)$ be two input/state/output systems with $\widetilde{\mathcal{X}} = \mathcal{X} \dotplus \mathcal{Z}$. Denote the four block transfer functions of $\widetilde{\Sigma}_{i/s/o}$ and $\Sigma_{i/s/o}$ by $\begin{bmatrix} \widetilde{\mathfrak{A}}(z) & \widetilde{\mathfrak{B}}(z) \\ \widetilde{\mathfrak{C}}(z) & \widetilde{\mathfrak{D}}(z) \end{bmatrix}$ and $\begin{bmatrix} \mathfrak{A}(z) & \mathfrak{B}(z) \\ \mathfrak{C}(z) & \mathfrak{D}(z) \end{bmatrix}$, respectively. Then the following conditions are equivalent:*

1) $\widetilde{\Sigma}_{i/s/o}$ is a dilation along $\mathcal{Z}$ of $\Sigma_{i/s/o}$.

2) For all $n \in \mathbb{Z}^+$,

$$A^n = P_{\mathcal{X}}^{\mathcal{Z}} \widetilde{A}^n|_{\mathcal{X}}, \qquad A^n B = P_{\mathcal{X}}^{\mathcal{Z}} \widetilde{A}^n \widetilde{B},$$
$$CA^n = \widetilde{C} \widetilde{A}^n|_{\mathcal{X}}, \qquad CA^n B = \widetilde{C} \widetilde{A}^n \widetilde{B}, \qquad D = \widetilde{D}. \tag{8.14}$$

3) For all z in some neighborhood at zero,

$$\begin{bmatrix} \mathfrak{A}(z) & \mathfrak{B}(z) \\ \mathfrak{C}(z) & \mathfrak{D}(z) \end{bmatrix} = \begin{bmatrix} P_{\mathcal{X}}^{\mathcal{Z}} \widetilde{\mathfrak{A}}(z)|_{\mathcal{X}} & P_{\mathcal{X}}^{\mathcal{Z}} \widetilde{\mathfrak{B}}(z) \\ \mathfrak{C}(z)|_{\mathcal{X}} & \mathfrak{D}(z) \end{bmatrix}. \tag{8.15}$$

Proof. The equivalence of 1) and 2) follows from (6.7), and the equivalence of 2) and 3) follows from (6.6). $\qquad\square$

Theorem 8.28. *Let $\widetilde{\Sigma} = (\widetilde{V}; \widetilde{\mathcal{X}}, \mathcal{W})$ and $\Sigma = (V; \mathcal{X}, \mathcal{W})$ be two state/signal systems with $\widetilde{\mathcal{X}} = \mathcal{X} \dotplus \mathcal{Z}$ (and with the same signal space $\mathcal{W}$).*

1) *$\widetilde{\Sigma}$ is a dilation of Σ if and only if there exist driving variable representations $\widetilde{\Sigma}_{dv/s/s}$ and $\Sigma_{dv/s/s}$ of $\widetilde{\Sigma}$ and Σ, respectively, with the property that $\widetilde{\Sigma}_{dv/s/s}$ is a dilation along $\mathcal{Z}$ of $\Sigma_{dv/s/s}$ (in the input/state/output sense; in particular they have the same driving variable space).*

2) *If $\widetilde{\Sigma}$ is a dilation of Σ, then to every driving variable representation $\Sigma_{dv/s/s}$ of Σ there exists at least one driving variable representation $\widetilde{\Sigma}_{dv/s/s}$ of $\widetilde{\Sigma}$ such that $\widetilde{\Sigma}_{dv/s/s}$ is a dilation along $\mathcal{Z}$ of $\Sigma_{dv/s/s}$ (in the input/state/output sense).*

Proof. Assertion 1) follows from Remark 5.2 and Lemma 8.23.

To prove assertion 2) we take an arbitrary driving-variable representation $\Sigma_{dv/s/s} = \left(\begin{bmatrix} A' & B' \\ C' & D' \end{bmatrix}; \mathcal{X}, \mathcal{L}, \mathcal{W}\right)$ of Σ. Let $\widetilde{\Sigma}_{dv/s/s} = \left(\begin{bmatrix} \widetilde{A}' & \widetilde{B}' \\ \widetilde{C}' & \widetilde{D}' \end{bmatrix}; \widetilde{\mathcal{X}}, \widetilde{\mathcal{L}}, \mathcal{W}\right)$ be the driving variable representation of $\widetilde{\Sigma}$ mentioned in part 1). Then by Theorem 6.1, there exist operators $K' \in \mathcal{B}(\mathcal{X}; \widetilde{\mathcal{L}})$ and $M' \in \mathcal{B}(\mathcal{L}; \widetilde{\mathcal{L}})$, with M' boundedly invertible, such that

$$\begin{bmatrix} \mathfrak{A}'(z) & \mathfrak{B}'(z) \\ \mathfrak{C}'(z) & \mathfrak{D}'(z) \end{bmatrix} = \begin{bmatrix} P_{\mathcal{X}}^{\mathcal{Z}} \widetilde{\mathfrak{A}}'(z) & P_{\mathcal{X}}^{\mathcal{Z}} \widetilde{\mathfrak{B}}'(z) \\ \mathfrak{C}'(z) & \mathfrak{D}'(z) \end{bmatrix}$$
$$\times \begin{bmatrix} 1_{\mathcal{X}} & 0 \\ -K' P_{\mathcal{X}}^{\mathcal{Z}} \widetilde{\mathfrak{A}}'(z) & 1_{\mathcal{L}} - K' P_{\mathcal{X}}^{\mathcal{Z}} \widetilde{\mathfrak{B}}'(z) \end{bmatrix}^{-1} \begin{bmatrix} 1_{\mathcal{X}} & 0 \\ 0 & M' \end{bmatrix},$$

Define $\widetilde{K}' = K' P_{\mathcal{X}}^{\mathcal{Z}}$. Then the right-hand side is the compression along $\mathcal{Z}$ of the function

$$\begin{bmatrix} \widetilde{\mathfrak{A}}'(z) & \widetilde{\mathfrak{B}}'(z) \\ \widetilde{\mathfrak{C}}'(z) & \widetilde{\mathfrak{D}}'(z) \end{bmatrix} \begin{bmatrix} 1_{\widetilde{\mathcal{X}}} & 0 \\ -\widetilde{K}' \widetilde{\mathfrak{A}}'(z) & 1_{\mathcal{L}} - \widetilde{K}' \widetilde{\mathfrak{B}}'(z) \end{bmatrix}^{-1} \begin{bmatrix} 1_{\widetilde{\mathcal{X}}} & 0 \\ 0 & M' \end{bmatrix},$$

which according to Theorem 6.1 is the four-block transfer function of the driving variable representation $\begin{bmatrix} \widetilde{A}' & \widetilde{B}' \\ \widetilde{C}' & \widetilde{D}' \end{bmatrix} \begin{bmatrix} 1_{\widetilde{\mathcal{X}}} & 0 \\ \widetilde{K}' & M' \end{bmatrix}$ of $\widetilde{\Sigma}$. By Lemma 8.27, $\widetilde{\Sigma}_{dv/s/s}$ is a dilation along $\mathcal{Z}$ of $\Sigma_{dv/s/s}$. $\qquad\square$

Theorem 8.29. *Let* $\widetilde{\Sigma} = (\widetilde{V}; \widetilde{\mathcal{X}}, \mathcal{W})$ *and* $\Sigma = (V; \mathcal{X}, \mathcal{W})$ *be two state/signal systems with* $\widetilde{\mathcal{X}} = \mathcal{X} \dotplus \mathcal{Z}$ *(and with the same signal space* $\mathcal{W}$*).*

1) *$\widetilde{\Sigma}$ is a dilation of Σ if and only if there exist output nulling representations $\widetilde{\Sigma}_{s/s/on}$ and $\Sigma_{s/s/on}$ of $\widetilde{\Sigma}$ and Σ, respectively, with the property that $\widetilde{\Sigma}_{s/s/on}$ is a dilation along $\mathcal{Z}$ of $\Sigma_{s/s/on}$ (in the input/state/output sense; in particular they have the same error space).*

2) *If $\widetilde{\Sigma}$ is a dilation of Σ, then to every output nulling representation $\Sigma_{s/s/on}$ of Σ there exists at least one output nulling representation $\widetilde{\Sigma}_{s/s/on}$ of $\widetilde{\Sigma}$ such that $\widetilde{\Sigma}_{s/s/on}$ is a dilation along $\mathcal{Z}$ of $\Sigma_{s/s/on}$ (in the input/state/output sense).*

The proof of this theorem is similar to the proof of Theorem 8.28, and we leave it to the reader.

9. Stability

Below we shall introduce and study different stability notions for state/signal systems. These are related to the stability of different representations of the system. In this connection we interpret each representation as an input/state/output system, and apply the following notion of stability.

Definition 9.1. A input/state/output system is

1) *stable*, if the following implication holds for all its trajectories $(x(\cdot), u(\cdot), y(\cdot))$:

$$u(\cdot) \in \ell^2(\mathbb{Z}^+; \mathcal{U}) \Rightarrow x(\cdot) \in \ell^\infty(\mathbb{Z}^+; \mathcal{X}) \text{ and } y(\cdot) \in \ell^2(\mathbb{Z}^+; \mathcal{Y}). \tag{9.1}$$

2) *strongly stable*, if the following implication holds for all its trajectories $(x(\cdot), u(\cdot), y(\cdot))$:

$$u(\cdot) \in \ell^2(\mathbb{Z}^+; \mathcal{U}) \Rightarrow \lim_{n\to\infty} x(n) = 0 \text{ and } y(\cdot) \in \ell^2(\mathbb{Z}^+; \mathcal{Y}). \tag{9.2}$$

3) *power stable*, if there exists a constant $r > 1$ such that the following implication holds for all its trajectories $(x(\cdot), u(\cdot), y(\cdot))$:

$$u(\cdot) = 0 \Rightarrow \lim_{n\to\infty} r^n \|x(n)\| = 0. \tag{9.3}$$

It is clear that (9.2) implies (9.1).

Lemma 9.2. *An input/state/output system $\Sigma_{i/s/o} = \left(\left[\begin{smallmatrix} A & B \\ C & D \end{smallmatrix}\right]; \mathcal{X}, \mathcal{U}, \mathcal{Y}\right)$ with the four block transfer function $\left[\begin{smallmatrix} \mathfrak{A} & \mathfrak{B} \\ \mathfrak{C} & \mathfrak{D} \end{smallmatrix}\right]$ is stable if and only if the following four conditions hold:*

1) *There is a constant $C > 0$ such that $\|A^n\| \le C$ for all $n \in \mathbb{Z}^+$.*
2) *$\mathfrak{B}(\bar{z})^* x \in H^2(\mathbb{D}; \mathcal{U})$ for all $x \in \mathcal{X}$.*
3) *$\mathfrak{C}(z) x \in H^2(\mathbb{D}; \mathcal{Y})$ for all $x \in \mathcal{X}$.*
4) *$\mathfrak{D} \in H^\infty(\mathbb{D}; \mathcal{U}, \mathcal{Y})$.*

This lemma is undoubtedly known, but we have not been able to find an explicit statement in the literature. (A continuous time version of this lemma can easily be derived from [Sta05].) For completeness we therefore include a short proof.

168   D.Z. Arov and O.J. Staffans

Proof. Clearly, $\Sigma_{i/s/o}$ is stable if and only if the four input-state-output maps listed in (6.8) have the following properties:

1') $\breve{\mathfrak{A}}$ maps $\mathcal{X}$ into $\ell^\infty(\mathbb{Z}^+;\mathcal{X})$;
2') $\breve{\mathfrak{B}}$ maps $\ell^2(\mathbb{Z}^+;\mathcal{U})$ into $\ell^\infty(\mathbb{Z}^+;\mathcal{X})$;
3') $\breve{\mathfrak{C}}$ maps $\mathcal{X}$ into $\ell^2(\mathbb{Z}^+;\mathcal{X})$;
4') $\breve{\mathfrak{D}}$ maps $\ell^2(\mathbb{Z}^+;\mathcal{U})$ into $\ell^2(\mathbb{Z}^+;\mathcal{Y})$.

We claim that each one of these conditions is equivalent to the corresponding condition listed in the statement Lemma 9.2. It is easy to see that all of these operators are always *closed* as operators between the indicated spaces, so by the closed graph theorem, 1')–4') are equivalent to the corresponding statements where we require each of these maps to be bounded, i.e.,

1'') $\breve{\mathfrak{A}} \in \mathcal{B}(\mathcal{X};\ell^\infty(\mathbb{Z}^+;\mathcal{X}))$;
2'') $\breve{\mathfrak{B}} \in \mathcal{B}(\ell^2(\mathbb{Z}^+;\mathcal{U});\ell^\infty(\mathbb{Z}^+;\mathcal{X}))$;
3'') $\breve{\mathfrak{C}} \in \mathcal{B}(\mathcal{X};\ell^2(\mathbb{Z}^+;\mathcal{X}))$;
4'') $\breve{\mathfrak{D}} \in \mathcal{B}(\ell^2(\mathbb{Z}^+;\mathcal{U});\ell^2(\mathbb{Z}^+;\mathcal{Y}))$.

Obviously, 1) is equivalent to 1''). Condition 1) implies that $\mathbb{D} \subset \rho(A)$, and hence all the transfer functions listed in 2)–4) are defined and analytic on $\mathbb{D}$. That 3) is equivalent to 3'') follows from the fact that the Z-transform is a bounded linear map from $\ell^2(\mathbb{Z}^+;\mathcal{U})$ onto $H^2(\mathbb{D};\mathcal{Y})$ with a bounded inverse. The equivalence of 4) and 4'') is well known: a causal convolution operator $\breve{\mathfrak{D}}$ maps $\ell^2(\mathbb{Z}^+;\mathcal{U})$ into $\ell^2(\mathbb{Z}^+;\mathcal{Y})$ if and only if its symbol $\mathfrak{D}$ belongs to $H^\infty(\mathbb{D};\mathcal{U},\mathcal{Y})$.

The equivalence of 2) and 2'') remains to be established. It is easy to see that 2'') is equivalent to the following condition:

2''') the sequence $\{\widetilde{\mathfrak{B}}_n\}_{n\in\mathbb{Z}^+}$ of operators defined by $\widetilde{\mathfrak{B}}_n u = \sum_{k=0}^n A^k Bu(-k-1)$ is uniformly bounded in $\mathcal{B}(\ell^2(\mathbb{Z}^-;\mathcal{U});\mathcal{X})$.

Assume that 2''') holds. Then, for each $u \in \ell^2(\mathbb{Z}^-;\mathcal{U})$, the sequence $\widetilde{\mathfrak{B}}_n u$ is a Cauchy sequence in $\mathcal{X}$ (since the norm in $\ell^2(\mathbb{Z}^-;\mathcal{U})$ of the sequence $\{u(k)\}_{k<m}$ tends to zero as $m \to -\infty$). Denote the limit by $\widetilde{\mathfrak{B}}$. Then $\widetilde{\mathfrak{B}}u = \sum_{k=0}^\infty A^k Bu(-k-1)$ and $\widetilde{\mathfrak{B}} \in \mathcal{B}(\ell^2(\mathbb{Z}^-;\mathcal{U});\mathcal{X})$. By duality, $\widetilde{\mathfrak{B}}^* \in \mathcal{B}(\mathcal{X};\ell^2(\mathbb{Z}^-;\mathcal{U}))$. This is equivalent to the statement that the operator $x \mapsto B^*(A^*)^n x$, $n \in \mathbb{Z}^+$, maps $\mathcal{X}$ into $\ell^2(\mathbb{Z}^+;\mathcal{Y})$, which equivalent to 2) (in the same way as 3) is equivalent to 3')). Thus, 2''') $\Rightarrow$ 2). Conversely, if 2) holds, then the operator that we denoted by $\widetilde{\mathfrak{B}}^*$ above is bounded, hence so is $\widetilde{\mathfrak{B}}$, and this implies 2'''). $\qquad\square$

Lemma 9.3. *An input/state/output system $\Sigma_{i/s/o} = \left(\left[\begin{smallmatrix} A & B \\ C & D \end{smallmatrix}\right];\mathcal{X},\mathcal{U},\mathcal{Y}\right)$ is strongly stable if and only if it is stable and A is strongly stable, i.e., $\lim_{n\to\infty} A^n x = 0$ for all $x \in \mathcal{X}$.*

Also this lemma must be known, but we have not found an explicit proof in the literature (a proof of the well-posed continuous time version of this lemma is given in [Sta05], and the discrete time proof is the same). For the convenience of the reader we therefore again include a short proof.

Proof. It is easy to see that if $\Sigma_{i/s/o}$ is strongly stable then it is stable, and $\lim_{n\to\infty} A^n x = 0$ for all $x \in \mathcal{X}$. Let us therefore only prove the converse part.

Let $(x(\cdot), u(\cdot), y(\cdot))$ be a trajectory of $\Sigma_{i/s/o}$ on $\mathbb{Z}^+$ with $u \in \ell^2(\mathbb{Z}^+; \mathcal{U})$. Fix $\epsilon > 0$. Choose m large enough so that $\sum_{k=m}^{\infty} \|u(k)\|^2 \le \epsilon^2$. Then we have for all $n \ge m$,

$$x(n) = A^{n-m} x(m) + \sum_{k=0}^{n-m-1} A^{n-k-1} Bu(m+k)$$

Here $A^{n-m} x(m) \to 0$ as $n \to \infty$ (because of the strong stability of A), and the norm of the second term is at most $C\epsilon$, where C is the norm of the mapping $\check{\mathfrak{B}} \in \mathcal{B}(\ell^2(\mathbb{Z}^+; \mathcal{U}); \ell^\infty(\mathbb{Z}^+; \mathcal{X}))$. Since ϵ was arbitrary, this implies that $x(k) \to 0$ as $k \to \infty$. $\qquad\square$

Remark 9.4. As is well known, conditions 2) and 3) in Lemma 9.2 imply that the sums

$$\mathcal{C} := \sum_{n \in \mathbb{Z}^+} A^n BB^* (A^*)^n, \tag{9.4}$$

$$\mathcal{O} := \sum_{n \in \mathbb{Z}^+} (A^*)^n C^* C A^n, \tag{9.5}$$

converge monotonically in the strong sense to nonnegative operators $\mathcal{O} \in \mathcal{B}(\mathcal{X})$ and $\mathcal{C} \in \mathcal{B}(\mathcal{X})$, respectively. These are called the infinite time controllability, respectively, observability Gramians of the system. They are the minimal nonnegative solutions of the Stein equations

$$H - AHA^* = BB^*, \tag{9.6}$$

$$G - A^*GA = C^*C, \tag{9.7}$$

respectively. If A is strongly stable, then the nonnegative solution H of (9.6) is unique (hence $H = \mathcal{C}$), and if A^* is strongly stable (i.e., $(A^*)^n x \to 0$ for all $x \in \mathcal{X}$), then the nonnegative solution G of (9.7) is unique (hence $G = \mathcal{O}$).

Lemma 9.5. *Let* $\Sigma_{i/s/o} = \left(\left[\begin{smallmatrix} A & B \\ C & D \end{smallmatrix}\right]; \mathcal{X}, \mathcal{U}, \mathcal{Y}\right)$ *be an input/state/output system. Then the following conditions are equivalent:*

1) $\Sigma_{i/s/o}$ *is power stable;*
2) $\overline{\mathbb{D}} := \{z \in \mathbb{C} \mid |z| \le 1\} \subset \Lambda_A;$
3) *There exists constants* $q < 1$ *and* $C > 0$ *such that* $\|A^n\| \le Cq^n$.

Proof. Clearly 2) and 3) are equivalent. It is also clear that 3) implies 1). For the converse implication we observe that condition 1) says that there is some $r > 1$ such that $\lim_{n\to\infty} r^n A^n x = 0$ for all $x \in \mathcal{X}$. By the uniform boundedness principle, $\sup_{n \in \mathbb{Z}^+} r^n \|A^n\| < \infty$. This implies 3) with $\gamma = 1/r$. $\qquad\square$

Lemma 9.6. *Every power stable input/state/output system is strongly stable.*

Proof. This follows from Lemmas 9.2, 9.3, and 9.5. $\qquad\square$

Thus, power stability implies strong stability, which further implies stability.

We call a driving variable or output nulling representation of a state/signal system stable, or strongly stable, or power stable, if it has this property when it is interpreted as an input/state/output system, as explained in Remark 5.4.

Definition 9.7. A state/signal system is

1) *stabilizable* (or strongly stabilizable, or power stabilizable) if it has a stable (or strongly stable, or power stable, respectively) driving variable representation.
2) *detectable* (or strongly detectable, or power detectable) if it has a stable (or strongly stable, or power stable, respectively) output nulling representation.
3) *LFT-stabilizable*[8] (or strongly LFT-stabilizable, or power LFT-stabilizable), if it has a stable (or strongly stable, or power stable, respectively) input/state/output representation.

Next we shall show that the above notions are closely connected to the corresponding (better known) notions for input/state/output systems.[9]

Definition 9.8. An input/state/output system $\Sigma_{i/s/o} = \left(\left[\begin{smallmatrix} A & B \\ C & D \end{smallmatrix} \right]; \mathcal{X}, \mathcal{U}, \mathcal{Y} \right)$ is

1) *stabilizable* (or strongly stabilizable, or power stabilizable) if there exists an operator $L \in \mathcal{B}(\mathcal{X}; \mathcal{U})$, called a *state feedback operator*, such that the new input/state/output system with input $\ell(\cdot)$ and output $w(\cdot) = \left[\begin{smallmatrix} y(\cdot) \\ u(\cdot) \end{smallmatrix} \right]$, described by the system of equations

$$\begin{aligned} x(n+1) &= Ax(n) + Bu(n), \\ y(n) &= Cx(n) + Du(n), \\ u(n) &= Lx(n) + \ell(n), \qquad z \in \mathbb{Z}^+, \end{aligned} \tag{9.8}$$

is stable (or strongly stable, or power stable, respectively).
2) *detectable* (or strongly detectable, or power detectable) if there exists an operator $H \in \mathcal{B}(\mathcal{Y}; \mathcal{X})$, called an *output injection operator*, such that the new input/state/output system with input $w(\cdot) = \left[\begin{smallmatrix} e(\cdot) \\ u(\cdot) \end{smallmatrix} \right]$ and output $y(\cdot)$, described by the system of equations

$$\begin{aligned} x(n+1) &= Ax(n) + Hy(n) + Bu(n), \\ y(n) &= Cx(n) + e(n) + Du(n), \qquad z \in \mathbb{Z}^+, \end{aligned} \tag{9.9}$$

is stable (or strongly stable, or power stable, respectively).
3) *output feedback stabilizable* (or strongly output feedback stabilizable, or power output feedback stabilizable) if there exists an operator $K \in \mathcal{B}(\mathcal{Y}; \mathcal{U})$, called a *output feedback operator*, such that $1_{\mathcal{U}} - KD$ has a bounded inverse and the

---

[8]LFT stands for Linear Fractional Transformation.
[9]A number of slightly different ways of presenting these notions do exist. We have chosen to present a version which makes the connection to the state/signal theory as simple as possible. This is a discrete time analogue of the approach used in [Sta05, Chapter 7].

new input/state/output system with input $\ell(\cdot)$ and output $y(\cdot)$, described by the (implicit) system of equations (where $u(n)$ should be eliminated)

$$
\begin{aligned}
x(n+1) &= Ax(n) + Bu(n), \\
y(n) &= Cx(n) + Du(n), \\
u(n) &= Ky(n) + \ell(n), \qquad z \in \mathbb{Z}^+,
\end{aligned}
\tag{9.10}
$$

is stable (or strongly stable, or power stable, respectively).

4) *LFT-stabilizable* (or strongly LFT-stabilizable, or power LFT-stabilizable), if there exists Hilbert spaces $\widetilde{\mathcal{Y}}$ and $\widetilde{\mathcal{U}}$ and an operator $\Psi = \left[\begin{smallmatrix} \Psi_{11} & \Psi_{12} \\ \Psi_{21} & \Psi_{22} \end{smallmatrix}\right] \in \mathcal{B}(\left[\begin{smallmatrix} \mathcal{Y} \\ \mathcal{U} \end{smallmatrix}\right] ; \left[\begin{smallmatrix} \widetilde{\mathcal{Y}} \\ \widetilde{\mathcal{U}} \end{smallmatrix}\right])$, called an *LFT-feedback operator*, such that both Ψ itself and $\Psi_{21}D + \Psi_{22}$ have bounded inverses, and such that the new input/state/output system with input $u_1(\cdot)$ and output $y_1(\cdot)$, described by the (implicit) system of equations (where $u(n)$ and $y(n)$ should be eliminated)

$$
\begin{aligned}
x(n+1) &= Ax(n) + Bu(n), \\
y(n) &= Cx(n) + Du(n), \\
y_1(n) &= \Psi_{11}y(n) + \Psi_{12}u(n), \\
u_1(n) &= \Psi_{21}y(n) + \Psi_{22}u(n), \qquad z \in \mathbb{Z}^+,
\end{aligned}
\tag{9.11}
$$

is stable (or strongly stable, or power stable, respectively).

More explicitly, the resulting input/state/output systems have the following structure. If we denote the system in part 1) by $\Sigma^L = \left(\left[\begin{smallmatrix} A^L & B^L \\ C^L & D^L \end{smallmatrix}\right] ; \mathcal{X}, \mathcal{U}, \left[\begin{smallmatrix} \mathcal{Y} \\ \mathcal{U} \end{smallmatrix}\right]\right)$, then

$$
\begin{bmatrix} A^L & B^L \\ C^L & D^L \end{bmatrix} = \left[\begin{array}{c|c} A+BL & B \\ \hline C+DL & D \\ L & 1_{\mathcal{U}} \end{array} \right].
\tag{9.12}
$$

If we denote the system in part 2) by $\Sigma^H = \left(\left[\begin{smallmatrix} A^H & B^H \\ C^H & D^H \end{smallmatrix}\right] ; \mathcal{X}, \left[\begin{smallmatrix} \mathcal{Y} \\ \mathcal{U} \end{smallmatrix}\right], \mathcal{Y}\right)$, then

$$
\begin{bmatrix} A^H & B^H \\ C^H & D^H \end{bmatrix} = \left[\begin{array}{c|cc} A+HC & H & B+HD \\ \hline C & 1_{\mathcal{Y}} & D \end{array} \right].
\tag{9.13}
$$

If we denote the system in part 3) by $\Sigma^K = \left(\left[\begin{smallmatrix} A^K & B^K \\ C^K & D^K \end{smallmatrix}\right] ; \mathcal{X}, \mathcal{U}, \mathcal{Y}\right)$, then

$$
\begin{aligned}
\begin{bmatrix} A^K & B^K \\ C^K & D^K \end{bmatrix} &= \begin{bmatrix} A+BK\left(1_{\mathcal{Y}}-DK\right)^{-1}C & B\left(1_{\mathcal{U}}-KD\right)^{-1} \\ \left(1_{\mathcal{Y}}-DK\right)^{-1}C & D\left(1_{\mathcal{U}}-KD\right)^{-1} \end{bmatrix} \\
&= \begin{bmatrix} A & B \\ C & D \end{bmatrix} \begin{bmatrix} 1_{\mathcal{K}} & 0 \\ -KC & 1_{\mathcal{U}}-KD \end{bmatrix}^{-1} \\
&= \begin{bmatrix} 1_{\mathcal{Y}} & -BK \\ 0 & 1_{\mathcal{Y}}-DK \end{bmatrix}^{-1} \begin{bmatrix} A & B \\ C & D \end{bmatrix}.
\end{aligned}
\tag{9.14}
$$

If we denote the system in part 4) by $\Sigma^{\Psi} = \left(\begin{bmatrix} A^{\Psi} & B^{\Psi} \\ C^{\Psi} & D^{\Psi} \end{bmatrix} ; \mathcal{X}, \widetilde{\mathcal{U}}, \widetilde{\mathcal{Y}} \right)$, then (see also Lemma 10.1)

$$\begin{bmatrix} A^{\Psi} & B^{\Psi} \\ C^{\Psi} & D^{\Psi} \end{bmatrix} = \begin{bmatrix} A & B \\ \Psi_{11}C & \Psi_{11}D + \Psi_{12} \end{bmatrix} \begin{bmatrix} 1_{\mathcal{X}} & 0 \\ \Psi_{21}C & \Psi_{21}D + \Psi_{22} \end{bmatrix}^{-1}, \tag{9.15}$$

or equivalently,

$$\begin{aligned}
A^{\Psi} &= A - B(\Psi_{21}D + \Psi_{22})^{-1}\Psi_{21}C, \\
B^{\Psi} &= B(\Psi_{21}D + \Psi_{22})^{-1}, \\
C^{\Psi} &= \Psi_{11}C - (\Psi_{11}D + \Psi_{12})(\Psi_{21}D + \Psi_{22})^{-1}\Psi_{21}C, \\
D^{\Psi} &= (\Psi_{11}D + \Psi_{12})(\Psi_{21}D + \Psi_{22})^{-1}.
\end{aligned} \tag{9.16}$$

When we apply Definition 9.8 to various systems it is often more convenient to use the following equivalent characterization:

Lemma 9.9. *Let $\Sigma_{i/s/o} = \left(\begin{bmatrix} A & B \\ C & D \end{bmatrix} ; \mathcal{X}, \mathcal{U}, \mathcal{Y} \right)$ be an input/state/output system.*

1) *The system $\Sigma^{L} = \left(\begin{bmatrix} A^{L} & B^{L} \\ C^{L} & D^{L} \end{bmatrix} ; \mathcal{X}, \mathcal{U}, \begin{bmatrix} \mathcal{Y} \\ \mathcal{U} \end{bmatrix} \right)$ whose coefficient matrix is given by (9.12) is stable (or strongly stable, or power stable) if and only if the system $\left(\left[\begin{array}{c|c} A^{L} & B \\ \hline \begin{smallmatrix} C \\ L \end{smallmatrix} & \begin{smallmatrix} 0 \\ 0 \end{smallmatrix} \end{array} \right] ; \mathcal{X}, \mathcal{U}, \begin{bmatrix} \mathcal{Y} \\ \mathcal{U} \end{bmatrix} \right)$ has the same property.*

2) *The system $\Sigma^{H} = \left(\begin{bmatrix} A^{H} & B^{H} \\ C^{H} & D^{H} \end{bmatrix} ; \mathcal{X}, \begin{bmatrix} \mathcal{Y} \\ \mathcal{U} \end{bmatrix}, \mathcal{Y} \right)$ whose coefficient matrix is given by (9.13) is stable (or strongly stable, or power stable) if and only if the system $\left(\left[\begin{array}{c|c} A^{H} & H\ B \\ \hline C & 0\ 0 \end{array} \right] ; \mathcal{X}, \begin{bmatrix} \mathcal{Y} \\ \mathcal{U} \end{bmatrix}, \mathcal{Y} \right)$ has the same property.*

3) *The system $\Sigma^{K} = \left(\begin{bmatrix} A^{K} & B^{K} \\ C^{K} & D^{K} \end{bmatrix} ; \mathcal{X}, \mathcal{U}, \mathcal{Y} \right)$ whose coefficient matrix is given by (9.14) is stable (or strongly stable, or power stable) if and only if the system $\left(\begin{bmatrix} A^{K} & B \\ C & 0 \end{bmatrix} ; \mathcal{X}, \mathcal{U}, \mathcal{Y} \right)$ has the same property.*

4) *The system $\Sigma^{\Psi} = \left(\begin{bmatrix} A^{\Psi} & B^{\Psi} \\ C^{\Psi} & D^{\Psi} \end{bmatrix} ; \mathcal{X}, \widetilde{\mathcal{U}}, \widetilde{\mathcal{Y}} \right)$ whose coefficient matrix is given by (9.15) is stable (or strongly stable, or power stable) if and only if the system $\left(\begin{bmatrix} A^{\Psi} & B \\ C & 0 \end{bmatrix} ; \mathcal{X}, \mathcal{U}, \mathcal{Y} \right)$ has the same property.*

Proof. Proof of 1): The latter system differs from Σ^{L} only in the sense that we have subtracted a multiple of the first input from the second input and modified the feedthrough term, and this does not affect stability.

Proof of 2): The latter system differs from Σ^{H} only in the sense that we have subtracted a multiple of the second output from the first output and modified the feedthrough term, and this does not affect stability.

Proof of 3): The latter system differs from Σ^{K} only in the sense that we have multiplied both the input and the output by bounded invertible operators and modified the feedthrough term, and this does not affect stability.

Proof of 4): The latter system differs from Σ^{Ψ} only in the sense that we have multiplied both the input and the output by bounded invertible operators and modified the feedthrough term, and this does not affect stability. Indeed, the operator multiplying C to the left is invertible, because of the invertibility of Ψ and the following Schur factorization:

$$\begin{bmatrix} \Psi_{11} & \Psi_{12} \\ \Psi_{21} & \Psi_{22} \end{bmatrix} \begin{bmatrix} 1_{\mathcal{Y}} & D \\ 0 & 1_{\mathcal{U}} \end{bmatrix} \begin{bmatrix} 1_{\mathcal{Y}} & 0 \\ -(\Psi_{21}D + \Psi_{22})^{-1}\Psi_{21} & 1_{\mathcal{U}} \end{bmatrix}$$
$$= \begin{bmatrix} \Psi_{11} - (\Psi_{11}D + \Psi_{12})(\Psi_{21}D + \Psi_{22})^{-1}\Psi_{21} & \Psi_{11}D + \Psi_{12} \\ 0 & \Psi_{21}D + \Psi_{22} \end{bmatrix}. \quad \square$$

Lemma 9.10. *Let $\Sigma_{i/s/o} = ([\begin{smallmatrix} A & B \\ C & D \end{smallmatrix}]; \mathcal{X}, \mathcal{U}, \mathcal{Y})$ be an input/state/output system.*

1) *If $\Sigma_{i/s/o}$ is output feedback stabilizable (or strongly output feedback stabilizable, or power output feedback stabilizable), then it is also LFT-stabilizable (or strongly LFT-stabilizable, or power LFT-stabilizable, respectively).*

2) *If $\Sigma_{i/s/o}$ is LFT-stabilizable (or strongly LFT-stabilizable) with an LFT-feedback operator $\Psi = [\begin{smallmatrix} \Psi_{11} & \Psi_{12} \\ \Psi_{21} & \Psi_{22} \end{smallmatrix}]$ where Ψ_{22} has a bounded inverse, then it is also output feedback stabilizable (or strongly output feedback stabilizable, respectively).*

3) *If $\Sigma_{i/s/o}$ is LFT-stabilizable (or strongly LFT-stabilizable) and $D = 0$, then it is also output feedback stabilizable (or strongly output feedback stabilizable, respectively).*

4) *If $\Sigma_{i/s/o}$ is power LFT-stabilizable then it is also power output feedback stabilizable.*

5) *If $\Sigma_{i/s/o}$ is LFT-stabilizable, then it is both stabilizable and detectable.*

Proof. Proof of 1) : Take $\widetilde{\mathcal{Y}} = \mathcal{Y}$, $\widetilde{\mathcal{U}} = \mathcal{U}$, and $\Psi = [\begin{smallmatrix} 1_{\mathcal{Y}} & 0 \\ -K & 1_{\mathcal{U}} \end{smallmatrix}]$.

Proof of 2): Use parts 3)–4) of Lemma 9.9, and take $K = -\Psi_{22}^{-1}\Psi_{21}$.

Proof of 3): This follows from 2), since the assumption that $D = 0$ implies that $\Psi_{22} = \Psi_{21}D + \Psi_{22}$ has a bounded inverse.

Proof of 4): The claim 2) remains valid also in the power stabilizable case, with the same proof. However, in the power stabilizable case the spectral radius of the operator A^{Ψ} lies strictly inside the unit disk. This implies that the set of all LFT-feedbacks Ψ which power stabilize $\Sigma_{i/s/o}$ is open. Therefore, if it is nonempty, it must contain some element Ψ for which Ψ_{22} is invertible. Thus, by the power stable version of part 2), $\Sigma_{i/s/o}$ is power output feedback stabilizable.

Proof of 5): Use parts 1), 2) and 4) of Lemma 9.9, and take $L = -(\Psi_{21}D + \Psi_{22})^{-1}\Psi_{21}C$ and $H = -B(\Psi_{21}D + \Psi_{22})^{-1}\Psi_{21}$. Also note that L is a left multiple of C and that H is a right multiple of B, which implies that in this case Σ^L is stable if and only if $([\begin{smallmatrix} A^L & B \\ C & 0 \end{smallmatrix}]; \mathcal{X}, \mathcal{U}, \mathcal{Y})$ is stable, and Σ^H is stable if and only if $([\begin{smallmatrix} A^H & B \\ C & 0 \end{smallmatrix}]; \mathcal{X}, \mathcal{U}, \mathcal{Y})$ is stable. $\quad \square$

Theorem 9.11. *Let $\Sigma = (V; \mathcal{X}, \mathcal{W})$ be a state/signal node.*

1) *The following conditions are equivalent.*
   (a) *Σ is stabilizable (or strongly stabilizable, or power stabilizable);*
   (b) *Σ has a stabilizable (or strongly stabilizable, or power stabilizable) input/state/output representation;*

    (c) *every input/state/output representation of Σ is stabilizable (or strongly stabilizable, or power stabilizable).*

2) *The following conditions are equivalent.*
    (a) *Σ is detectable (or strongly detectable, or power detectable);*
    (b) *Σ has a detectable (or strongly detectable, or power detectable) input/state/output representation;*
    (c) *every input/state/output representation of Σ is detectable (or strongly detectable, or power detectable).*

3) *The following conditions are equivalent.*
    (a) *Σ is LFT-stabilizable (or strongly LFT-stabilizable, or power LFT-stabilizable);*
    (b) *Σ has a LFT-stabilizable (or strongly LFT-stabilizable, or power LFT-stabilizable) input/state/output representation;*
    (c) *every input/state/output representation of Σ is LFT-stabilizable (or strongly LFT-stabilizable, or power LFT-stabilizable).*

Proof. The proofs of the strongly stable and power stable versions of this theorem are identical to the proofs of the basic version, so below we shall only prove the basic "stable" version.

Proof of 1): We prove this by showing that (b) $\Rightarrow$ (a) $\Rightarrow$ (c) (the implication (c) $\Rightarrow$ (b) is trivial).

Let $\Sigma_{i/s/o} = \left(\left[\begin{smallmatrix} A & B \\ C & D \end{smallmatrix}\right]; \mathcal{X}, \mathcal{U}, \mathcal{Y}\right)$ be a stabilizable input/state/output representation of Σ, and let $L \in \mathcal{B}(\mathcal{Y}; \mathcal{U})$ be a stabilizing state feedback operator. Then the system Σ^L whose coefficient matrix is given by (9.12) is stable. This system has an obvious interpretation as a driving variable representation of Σ (with driving variable space $\mathcal{U}$). Thus, according to Definition 9.7, Σ is stabilizable.

Conversely, suppose that Σ is stabilizable (in the sense of Definition 9.7). Let $\Sigma_{dv/s/s} = \left(\left[\begin{smallmatrix} A' & B' \\ C' & D' \end{smallmatrix}\right]; \mathcal{X}, \mathcal{L}, \mathcal{W}\right)$ be a stable driving variable representation of Σ, and let $\Sigma_{i/s/o} = \left(\left[\begin{smallmatrix} A & B \\ C & D \end{smallmatrix}\right]; \mathcal{X}, \mathcal{U}, \mathcal{Y}\right)$ be an arbitrary input/state/output representation of Σ. We can alternatively interpret this representation, too, as a driving variable representation as explained in Remark 5.2. Split C' and D' into $C' = \left[\begin{smallmatrix} C'_1 \\ C'_2 \end{smallmatrix}\right]$ and $D' = \left[\begin{smallmatrix} D'_1 \\ D'_2 \end{smallmatrix}\right]$ in accordance with the splitting $\mathcal{W} = \mathcal{Y} \dotplus \mathcal{U}$. Then, by Theorem 3.3, there exist operators $L \in \mathcal{B}(\mathcal{X}; \mathcal{U})$ and $M' \in \mathcal{B}(\mathcal{L}; \mathcal{U})$, with M' boundedly invertible, such that

$$\begin{bmatrix} A' & B' \\ C'_1 & D'_1 \\ C'_2 & D'_2 \end{bmatrix} = \begin{bmatrix} A & B \\ C & D \\ 0 & 1_{\mathcal{U}} \end{bmatrix} \begin{bmatrix} 1 & 0 \\ L & M' \end{bmatrix} = \begin{bmatrix} A + BL & BM' \\ C + DL & DM' \\ L & M' \end{bmatrix}. \qquad (9.17)$$

This coefficient matrix is identical to the one in (9.12) apart from the fact that the input variable has been multiplied by the invertible operator M'. This means that L is a stabilizing state feedback operator for $\Sigma_{i/s/o}$.

The proof of Part 2) is similar to the proof of Part 1), and it is left to the reader (this time we interpret the input/state/output representation as an output nulling representation as explained in Remark 5.2).

Proof of 3): The implication (a) $\Rightarrow$ (c) follows from Theorem 5.11 (take Ψ to be the operator Θ defined in (1.6)), and the implication (c) $\Rightarrow$ (b) is trivial. Thus, it remains to prove the implication (b) $\Rightarrow$ (a).

Let $\Sigma_{i/s/o} = \left(\left[\begin{smallmatrix} A & B \\ C & D \end{smallmatrix}\right]; \mathcal{X}, \mathcal{U}, \mathcal{Y}\right)$ be an input/state/output representation of Σ with a LFT-stabilizing feedback operator $\Psi \in \mathcal{B}\left(\left[\begin{smallmatrix} \mathcal{Y} \\ \mathcal{U} \end{smallmatrix}\right]; \left[\begin{smallmatrix} \tilde{\mathcal{Y}} \\ \tilde{\mathcal{U}} \end{smallmatrix}\right]\right)$, and let $\Sigma^{\Psi} = \left(\left[\begin{smallmatrix} A^{\Psi} & B^{\Psi} \\ C^{\Psi} & D^{\Psi} \end{smallmatrix}\right]; \mathcal{X}, \tilde{\mathcal{U}}, \tilde{\mathcal{Y}}\right)$ be the stable input/state/output system whose coefficient matrix is given by (9.15). We claim that there exists an admissible input/output decomposition $\mathcal{W} = \mathcal{Y}_1 \dotplus \mathcal{U}_1$ of $\mathcal{W}$ such that the corresponding input/state/output representation is stable. The proof of this claim is by direct construction.

We begin by interpreting Ψ as an operator $\Psi = \left[\begin{smallmatrix} \Psi_1 \\ \Psi_2 \end{smallmatrix}\right] \in \mathcal{B}(\mathcal{W}; \left[\begin{smallmatrix} \tilde{\mathcal{Y}} \\ \tilde{\mathcal{U}} \end{smallmatrix}\right])$, where $\Psi_1 = \Psi_{11} P_{\mathcal{Y}}^{\mathcal{U}} + \Psi_{12} P_{\mathcal{U}}^{\mathcal{Y}}$ and $\Psi_2 = \Psi_{21} P_{\mathcal{Y}}^{\mathcal{U}} + \Psi_{22} P_{\mathcal{U}}^{\mathcal{Y}}$. The bounded inverse of this operator belongs to $\mathcal{B}(\left[\begin{smallmatrix} \tilde{\mathcal{Y}} \\ \tilde{\mathcal{U}} \end{smallmatrix}\right]; \mathcal{W})$, and it can be decomposed into $\tilde{\Psi} := \Psi^{-1} = \left[\begin{smallmatrix} \tilde{\Psi}_1 & \tilde{\Psi}_2 \end{smallmatrix}\right]$. Define

$$\mathcal{Y}_1 = \mathcal{N}\left(\left[\tilde{\Psi}_2\right]\right), \quad \mathcal{U}_1 = \mathcal{N}\left(\left[\tilde{\Psi}_1\right]\right).$$

Define $P \in \mathcal{B}(\mathcal{W})$ and $Q \in \mathcal{B}(\mathcal{W})$ by

$$P := \tilde{\Psi} \begin{bmatrix} \Psi_1 \\ 0 \end{bmatrix}, \quad Q := \tilde{\Psi} \begin{bmatrix} 0 \\ \Psi_2 \end{bmatrix}.$$

Clearly $P + Q = 1_{\mathcal{W}}$. For all $w \in \mathcal{Y}_1$ we have $Qw = 0$, hence $Pw = w$, and for all $w \in \mathcal{U}_1$ we have $Pw = 0$, hence $Qw = w$. This implies that P and Q are complementary projections in $\mathcal{W}$, with $\mathcal{R}(P) = \mathcal{N}(Q) = \mathcal{Y}_1$ and $\mathcal{N}(P) = \mathcal{R}(Q) = \mathcal{U}_1$, i.e., $P = P_{\mathcal{Y}_1}^{\mathcal{U}_1}$ and $Q = P_{\mathcal{U}_1}^{\mathcal{Y}_1}$. In particular, this implies that $\mathcal{W} = \mathcal{Y}_1 \dotplus \mathcal{U}_1$. Furthermore, Ψ_1 maps $\mathcal{Y}_1$ one-to-one onto $\tilde{\mathcal{Y}}$ with the bounded inverse $\tilde{\Psi}_1$, and Ψ_1 maps $\mathcal{U}_1$ one-to-one onto $\tilde{\mathcal{U}}$ with the bounded inverse $\tilde{\Psi}_2$.

Let $\Phi := \begin{bmatrix} P_{\mathcal{Y}_1}^{\mathcal{U}_1}|_{\mathcal{Y}} & P_{\mathcal{Y}_1}^{\mathcal{U}_1}|_{\mathcal{U}} \\ P_{\mathcal{U}_1}^{\mathcal{Y}_1}|_{\mathcal{Y}} & P_{\mathcal{U}_1}^{\mathcal{Y}_1}|_{\mathcal{U}} \end{bmatrix}$. This is the same operator that we find in (1.6), corresponding to the two decompositions $\mathcal{W} = \mathcal{Y} \dotplus \mathcal{U} = \mathcal{Y}_1 \dotplus \mathcal{U}_1$, and it is explicitly given by

$$\Phi = \begin{bmatrix} \tilde{\Psi}_1 \Psi_{11} & \tilde{\Psi}_1 \Psi_{12} \\ \tilde{\Psi}_2 \Psi_{21} & \tilde{\Psi}_2 \Psi_{22} \end{bmatrix}.$$

In particular, $\Phi_{12} D + \Phi_{22} = \tilde{\Psi}_2(\Psi_{12} D + \Psi_{22})$ is invertible, and by Theorem 5.11, the decomposition $\mathcal{W} = \mathcal{Y}_1 \dotplus \mathcal{U}_1$ is admissible. Let us denote the corresponding input/state/output system by $\Sigma^1_{i/s/o} = \left(\left[\begin{smallmatrix} A_1 & B_1 \\ C_1 & D_1 \end{smallmatrix}\right]; \mathcal{X}, \mathcal{U}_1, \mathcal{Y}_1\right)$. This system is obtained from Σ^{Ψ} by multiplying the input by $\tilde{\Psi}_2^{-1}$ and the output by $\tilde{\Psi}_1$. Thus, $\Sigma^1_{i/s/o}$ is stable.

$\square$

10. Appendix

Lemma 10.1. *Let $A \in \mathcal{B}(\mathcal{X}; \mathcal{Z})$ and $B \in \mathcal{B}(\mathcal{Z}; \mathcal{X})$.*

1) *$1_{\mathcal{X}} - BA$ has a bounded inverse if and only if $1_{\mathcal{Z}} - AB$ has a bounded inverse.*
2) *If $1_{\mathcal{X}} - BA$ has a bounded inverse, then*

$$
\begin{aligned}
(1_{\mathcal{Z}} - AB)^{-1} &= 1_{\mathcal{X}} + A(1_{\mathcal{X}} - BA)^{-1}B, \\
B(1_{\mathcal{Z}} - AB)^{-1} &= (1_{\mathcal{X}} - BA)^{-1}B.
\end{aligned}
\tag{10.1}
$$

For a proof see, e.g., [Sta05, Appendix A4].

Acknowlegment

Damir Z. Arov thanks Åbo Akademi for its hospitality and the Academy of Finland for its financial support during his visits to Åbo in 2003–2005. He also gratefully acknowledges the partial financial support by the joint grant UM1-2567-OD-03 from the U.S. Civilian Research and Development Foundation (CRDF) and the Ukrainian Government. Olof J. Staffans gratefully acknowledges the financial support by grant 203991 from the Academy of Finland.

References

[AA70] Vadim M. Adamyan and Damir Z. Arov, *On unitary couplings of semiunitary operators*, Eleven Papers in Analysis (Providence, R.I.), American Mathematical Society Translations, vol. 95, American Mathematical Society, 1970, pp. 75–129.

[Aro74] Damir Z. Arov, *Scattering theory with dissipation of energy*, Dokl. Akad. Nauk SSSR. **216** (1974), 713–716, Translated in Soviet Math. Dokl. 15 (1974), 848–854.

[Aro79] ______, *Passive linear stationary dynamic systems*, Sibir. Mat. Zh. **20** (1979), 211–228, Translated in Sib. Math. J. 20 (1979), 149-162.

[Bel68] Vitold Belevitch, *Classical network theory*, Holden-Day, San Francisco, Calif.-Cambridge-Amsterdam, 1968.

[BGK79] Harm Bart, Israel Gohberg, and Marinus A. Kaashoek, *Minimal factorization of matrix and operator functions*, Operator Theory: Advances and Applications, vol. 1, Birkhäuser-Verlag, Basel Boston Berlin, 1979.

[BS05] Joseph A. Ball and Olof J. Staffans, *Conservative state-space realizations of dissipative system behaviors*, To appear in Integral Equations Operator Theory (2005), 63 pp.

[Fuh74] Paul A. Fuhrmann, *On realization of linear systems and applications to some questions of stability*, Math. Systems Theory **8** (1974), 132–140.

[Hel74] J. William Helton, *Discrete time systems, operator models, and scattering theory*, J. Funct. Anal. **16** (1974), 15–38.

[LP67] Peter D. Lax and Ralph S. Phillips, *Scattering theory*, Academic Press, New York, 1967.

[PW98]   Jan Willem Polderman and Jan C. Willems, *Introduction to mathematical systems theory: A behavioral approach*, Springer-Verlag, New York, 1998.

[SF70]   Béla Sz.-Nagy and Ciprian Foiaş, *Harmonic analysis of operators on Hilbert space*, North-Holland, Amsterdam London, 1970.

[Sta05] Olof J. Staffans, *Well-posed linear systems*, Cambridge University Press, Cambridge and New York, 2005.

[WT02]   Jan C. Willems and Harry L. Trentelman, *Synthesis of dissipative systems using quadratic differential forms: Part I*, IEEE Trans. Autom. Control **47** (2002), 53–69.

Damir Z. Arov
Division of Mathematical Analysis
Institute of Physics and Mathematics
South-Ukrainian Pedagogical University
65020 Odessa, Ukraine

Olof J. Staffans
Åbo Akademi University
Department of Mathematics
FIN-20500 Åbo, Finland
URL: `http://www.abo.fi/~staffans/`

Operator Theory:
Advances and Applications, Vol. 161, 179–223
© 2005 Birkhäuser Verlag Basel/Switzerland

Conservative Structured Noncommutative Multidimensional Linear Systems

Joseph A. Ball, Gilbert Groenewald and Tanit Malakorn

Abstract. We introduce a class of conservative structured multidimensional linear systems with evolution along a free semigroup. The system matrix for such a system is unitary and the associated transfer function is a formal power series in noncommuting indeterminates. A formal power series $T(z_1, \ldots, z_d)$ in the noncommuting indeterminates $z_1, \ldots, z_d$ arising in this way satisfies a noncommutative von Neumann inequality, i.e., substitution of a d-tuple of noncommuting operators $\delta = (\delta_1, \ldots, \delta_d)$ on a fixed separable Hilbert space which is contractive in the appropriate sense yields a contraction operator $T(\delta) = T(\delta_1, \ldots, \delta_d)$. We also obtain the converse realization theorem: any formal power series satisfying such a von Neumann inequality can be realized as the transfer function of such a conservative structured multidimensional linear system.

Mathematics Subject Classification (2000). Primary 47A56; Secondary 13F25, 47A60, 93B28.

Keywords. Formal power series, noncommuting indeterminates, energy balance, Hahn-Banach separation argument, noncommutative Schur-Agler class.

Contents

The first author was partially supported by US National Science Foundation under Grant Number DMS-9987636; The second author is supported by the National Research Foundation of South Africa under Grant Number 2053733; The third author was supported by a grant from Naresuan University, Thailand.

1. Introduction

This paper concerns extensions of the classical theory of conservative discrete-time linear systems to the setting of conservative structured multidimensional linear systems with evolution along a finitely generated free semigroup (words in a finite set of letters). By way of introduction we first review the relevant points of the classical theory.

By a (classical) conservative discrete-time input/state/output (i/s/o) linear system, we mean a system of equations of the form

$$\Sigma = \Sigma(U): \quad \begin{cases} x(n+1) &=& Ax(n) + Bu(n) \\ y(n) &=& Cx(n) + Du(n) \end{cases} \tag{1.1}$$

such that the so-called *connection matrix* or *colligation*

$$U = \begin{bmatrix} A & B \\ C & D \end{bmatrix} : \begin{bmatrix} \mathcal{H} \\ \mathcal{U} \end{bmatrix} \to \begin{bmatrix} \mathcal{H} \\ \mathcal{Y} \end{bmatrix} \tag{1.2}$$

is unitary. Here we assume that $x(n)$ takes values in the *state space* $\mathcal{H}$, $u(n)$ takes values in the *input space* $\mathcal{U}$ and $y(n)$ takes values in the output space $\mathcal{Y}$ where $\mathcal{H}$, $\mathcal{U}$ and $\mathcal{Y}$ are all assumed to be Hilbert spaces. The unitary property of the colligation U leads to the energy balance relation

$$\|x(n+1)\|^2 - \|x(n)\|^2 = \|u(n)\|^2 - \|y(n)\|^2. \tag{1.3}$$

Summing over all n with $0 \leq n \leq N$ leads to

$$\|x(N+1)\|^2 - \|x(0)\|^2 = \sum_{n=0}^{N} \left[\|u(n)\|^2 - \|y(n)\|^2 \right].$$

In particular, if we assume that $x(0) = 0$ and let $N \to \infty$ we get

$$\sum_{n=0}^{\infty} \|y(n)\|^2 \leq \sum_{n=0}^{\infty} \|u(n)\|^2. \tag{1.4}$$

Application of the Z-transform

$$\{x(n)\}_{n \in \mathbb{Z}_+} \mapsto \widehat{x}(z) := \sum_{n \in \mathbb{Z}_+} x(n) z^n$$

to the system equations (1.1) leads to the frequency-domain formulas

$$\widehat{x}(z) = (I - zA)^{-1} x(0) + z(I - zA)^{-1} B\widehat{u}(z) \tag{1.5}$$

$$\widehat{y}(z) = C(I - zA)^{-1} x(0) + T_\Sigma(z) \cdot \widehat{u}(z) \tag{1.6}$$

where

$$T_\Sigma(z) = D + zC(I - zA)^{-1} B. \tag{1.7}$$

In particular, if we assume $x(0) = 0$ we get the input-output relation

$$\widehat{y}(z) = T_\Sigma(z) \cdot \widehat{u}(z).$$

From (1.4) and the Plancherel theorem we then see that

$$\|T_\Sigma \cdot \widehat{u}\|_{H^2(\mathbb{D}, \mathcal{Y})} \leq \|\widehat{u}\|_{H^2(\mathbb{D}, \mathcal{U})}$$

for all $\widehat{u} \in H^2(\mathbb{D}, \mathcal{U})$ (the *Hardy space* of $\mathcal{U}$-valued functions $\widehat{u}(z) = \sum_{n=0}^{\infty} u(n) z^n$ on the unit disk $\mathbb{D}$ with norm-square summable Taylor coefficients ($\|u\|_{H^2(\mathbb{D}, \mathcal{U})}^2 = \sum_{n=0}^{\infty} \|u(n)\|^2 < \infty$). As a result it follows that T_Σ is in the operator-valued *Schur class* $\mathcal{S}(\mathcal{U}, \mathcal{Y})$ consisting of functions $S(z) = \sum_{n=0}^{\infty} S_n z^n$ analytic on $\mathbb{D}$ with values equal to contraction operators from $\mathcal{U}$ into $\mathcal{Y}$. Conversely, it is well known that any Schur class function $S(z) = \sum_{n=0}^{\infty} S_n z^n \in \mathcal{S}(\mathcal{U}, \mathcal{Y})$ can be realized as the transfer function of a conservative linear system, i.e., any $S \in \mathcal{S}(\mathcal{U}, \mathcal{Y})$ can be written in the form $S(z) = D + zC(I - zA)^{-1}B$ for some unitary colligation $U = \left[\begin{smallmatrix} A & B \\ C & D \end{smallmatrix}\right] : \left[\begin{smallmatrix} \mathcal{H} \\ \mathcal{U} \end{smallmatrix}\right] \to \left[\begin{smallmatrix} \mathcal{H} \\ \mathcal{Y} \end{smallmatrix}\right]$. Moreover any such S satisfies a *von Neumann inequality*: *if $\mathcal{K}$ is another Hilbert space and $T \in \mathcal{L}(\mathcal{K})$ has $\|T\| < 1$, then $\|S(T)\| \le 1$ where $S(T) \in \mathcal{L}(\mathcal{U} \otimes \mathcal{K}, \mathcal{Y} \otimes \mathcal{K})$ is given by $S(T) = \sum_{n=0}^{\infty} S_n \otimes T^n$. The following theorem is a convenient summary of the various equivalent characterizations of the operator-valued Schur class $\mathcal{S}(\mathcal{U}, \mathcal{Y})$.

Theorem 1.1. *Let $z \mapsto S(z)$ be a $\mathcal{L}(\mathcal{U}, \mathcal{Y})$-valued function defined on the unit disk $\mathbb{D}$. Then the following conditions are equivalent:*

1. *$S \in \mathcal{S}(\mathcal{U}, \mathcal{Y})$, i.e., S is analytic on $\mathbb{D}$ and $\|S(z)\| \le 1$ for all $z \in \mathbb{D}$.*
2. *S is analytic on $\mathbb{D}$ and $\|S(T)\| \le 1$ for any operator T on some Hilbert space $\mathcal{K}$ with $\|T\| < 1$.*
3. *There exists a Hilbert space $\mathcal{H}'$ and an operator-valued function $z \mapsto H(z) \in \mathcal{L}(\mathcal{H}', \mathcal{Y})$ so that*

$$\frac{I - S(z)S(w)^*}{1 - z\overline{w}} = H(z)H(w)^* \quad \text{for all} \quad z, w \in \mathbb{D}.$$

4. *$S(z)$ can be realized as the transfer function of a conservative discrete-time i/s/o linear system, i.e., there is a unitary colligation U of the form (1.2) so that*

$$S(z) = D + zC(I_\mathcal{H} - zA)^{-1}B.$$

For more information on the Schur class and its applications in both operator theory and engineering, we refer the reader to [42, 25, 26, 46, 38, 47]. Recent work has generalized these ideas to multivariable settings in several ways. We mention [1, 2, 17, 15, 16] for extensions to the polydisk $\mathbb{D}^n \subset \mathbb{C}^n$ setting, [33, 52, 10, 30, 3, 8, 18, 45, 37, 4] for extensions to the unit ball $\mathbb{B}^n \subset \mathbb{C}^n$ (where additional refinements concerning Nevanlinna-Pick-type interpolation and lifting theorems are also explored), and the recent work [56, 7, 6, 12] which suggests how a unification of these two settings can be achieved.

In the present paper we generalize these ideas to other types of conservative structured multidimensional linear systems. This paper can be considered as a sequel to our paper [13] where we introduced and studied a general class of systems called *structured noncommutative multidimensional linear systems* (SN-MLSs). These systems have evolution along a free semigroup rather than along an integer lattice as is usually taken in work in multidimensional linear system theory, and the transfer function is a formal power series in noncommuting indeterminates rather than an analytic function of several complex variables. In [13] it is assumed that the input space, state space and output space were all finite-dimensional linear spaces, and analogues of the standard results in finite-dimensional linear system theory (such as controllability, observability, Kalman decomposition, state space similarity theorem, Hankel operators and realization theory) were developed. Here we use the same notion of SNMLS as introduced in [13] but take the input space, state space and output space all to be Hilbert spaces and introduce a notion of *conservative* SNMLS for which the system and its adjoint satisfy an *energy balance* relation. The main result is Theorem 5.3 which can be viewed as a far-reaching generalization of Theorem 1.1. In this generalization, the unit disk is replaced by a tuple of (not necessarily commuting) operators $\delta = (\delta_1, \ldots, \delta_d)$ on some Hilbert space $\mathcal{K}$ in an appropriate noncommutative Cartan domain ($\| \sum_{i=1}^d I_i \otimes \delta_i \| < 1$ for an appropriate collection of $n \times m$ matrices $I_1, \ldots, I_d$), and analytic operator-valued functions $z \mapsto T(z) = \sum_{n=0}^\infty T_n z^n$ on the unit disk are replaced by formal power series

$$T(z) = \sum_{w \in \mathcal{F}_d} T_w z^w \tag{1.8}$$

in a set of noncommuting formal indeterminates $z = (z_1, \ldots, z_d)$, where the coefficients T_w are operators from $\mathcal{U}$ to $\mathcal{Y}$. Here $\mathcal{F}_d$ is the free semigroup generated by the set of letters $\{1, \ldots, d\}$; thus elements of $\mathcal{F}_d$ are words w of the form $w = i_N \cdots i_1$ where $i_k \in \{1, \ldots, d\}$ for each $k = 1, \ldots, N$. We also consider the empty word $\emptyset$ as an element of $\mathcal{F}_d$ which serves as the unit element for $\mathcal{F}_d$: $\emptyset \cdot w = w \cdot \emptyset = w$ for all $w \in \mathcal{F}_d$. Given a formal power series $T(z)$ as in (1.8) and an operator-tuple $\delta = (\delta_1, \ldots, \delta_d)$ we may define $T(\delta) \in \mathcal{L}(\mathcal{U} \otimes \mathcal{K}, \mathcal{Y} \otimes \mathcal{K})$ by

$$T(\delta) = \sum_{w \in \mathcal{F}_d} T_w \otimes \delta^w \tag{1.9}$$

whenever the series converges, where

$$\delta^w = \delta_{i_N} \cdots \delta_{i_1} \in \mathcal{L}(\mathcal{K}) \text{ if } w = i_N \cdots i_1 \text{ and } \delta_{i_k} \in \mathcal{L}(\mathcal{K}) \text{ for } k = 1, \ldots, N.$$

Theorem 5.3 characterizes formal power series $T(z)$ for which a noncommutative von Neumann inequality $\|T(\delta)\| \leq 1$ holds for all operator tuples $\delta = (\delta_1, \ldots, \delta_d)$ in a suitable noncommutative Cartan domain $\| \sum_{i=1}^d I_i \otimes \delta_i \| < 1$ in terms analogous to those in Theorem 1.1. One can view the result as a noncommutative analogue of the recent work of Ambrozie-Timotin [7], Ball-Bolotnikov [12] and Ambrozie-Eschmeier [6] on extensions of the so-called Schur-Agler class to more general domains in $\mathbb{C}^d$. In this more general setting there is no analogue of condition (1) in Theorem 1.1. In the classical case, the implication (2) $\implies$ (3) follows in a rather straightforward way as a consequence of the fact that the Schur class can be identified with the space of contractive multipliers on the Hardy space over the unit disk. This type of argument applies in our setting only in special cases (see Remark 5.11); the general case requires a rather involved separation argument of Hahn-Banach type first used in this context by Agler for the (commutative) polydisk setting (see [1]). The analogue of implication (3) $\implies$ (4) in Theorem 1.1 follows the now standard "lurking isometry" argument which now has appeared in many contexts (see [11] for a survey), while the implication (4) $\implies$ (1) is elementary but in our setting requires some care (see Theorem 4.2).

This functional calculus of formal power series considered as functions of noncommuting operator-tuples has been used in the context of robust control and the theory of structured singular values (μ-analysis) – see [22, 23, 24, 44, 57]; we explore these connections further in our paper [14].

We mention that results on formal power series (including polynomials in noncommuting indeterminates) closely related to our Theorem 5.3 below have appeared in the recent work of Helton, McCullough and Putinar [39, 40, 41] on representations of polynomials in noncommuting indeterminates as sums of squares as well as in related work of Kalyuzhnyĭ-Verbovetzkiĭ and Vinnikov [43]. This work has motivation from somewhat different connections with system theory. We indicate more precise connections between this work and our Theorem 5.3 in Remark 5.15 below.

In a different direction, the paper of Alpay and Kalyuzhnyĭ-Verbovetzkiĭ [5] introduces the notion of a rational, inner formal power series and develops a realization theory for these (see Remarks 5.2 and 5.5 below).

The paper is organized as follows. Following the present Introduction, in Section 2 we review the needed material from [13] on structured noncommutative multidimensional linear systems (SNMLSs). In Section 3 we define the adjoint of a SNMLS (having all signal spaces equal to Hilbert spaces). This gives the natural setting for the definition of a conservative SNMLS in Section 4. Section 5 contains the main Theorem 5.3 on the identification of the structured noncommutative Schur-Agler class with the set of formal power series capable of being realized as the transfer function of a conservative SNMLS.

2. Structured noncommutative multidimensional linear systems: basic definitions and properties

We present an infinite-dimensional Hilbert-space version of the structured noncommutative multidimensional linear systems (SNMLS) introduced in [13]. As in graph theory, a graph G consists of a set of vertices $V = V(G)$ and edges $E = E(G)$ connecting vertices. We assume throughout that the sets V and E are both finite, i.e., that G is a *finite graph*. We are interested only in what we call *admissible graphs*, i.e., *a bipartite graph such that each connected component is a complete bipartite graph*. This means simply that:

1. the set of vertices V has a disjoint partitioning $V = S \dot\cup R$ into the set of *source* vertices S and *range* vertices R,
2. S and R in turn have disjoint partitionings $S = \dot\cup_{k=1}^{K} S_k$ and $R = \dot\cup_{k=1}^{K} R_k$ into nonempty subsets $S_1, \ldots, S_K$ and $R_1, \ldots, R_K$ such that, for each $s_k \in S_k$ and $r_k \in R_k$ (with the same value of k) there is a unique edge $e = e_{s_k, r_k}$ connecting s_k to r_k ($\mathbf{s}(e) = s_k$, $\mathbf{r}(e) = r_k$), and
3. *every* edge of G is of this form.

If v is a vertex of G (so either $v \in S$ or $v \in R$) we denote by $[v]$ the path-connected component p (i.e., the complete bipartite graph $p = G_k$ with set of source vertices equal to S_k and set of range vertices equal to R_k for some $k = 1, \ldots, K$) containing v. Thus, given two distinct vertices $v_1, v_2 \in S \cup R$, there is a path of G connecting v_1 to v_2 if and only if $[v_1] = [v_2]$ and this path has length 2 if both v_1 and v_2 are either in S or in R and has length 1 otherwise. In case $s \in S$ and $r \in R$ are such that $[s] = [r]$, we shall use the notation $e_{s,r}$ for the unique edge having s as source vertex and r as range vertex:

$$e_{s,r} \in E \text{ determined by } \mathbf{s}(e_{s,r}) = s, \ \mathbf{r}(e_{s,r}) = r. \tag{2.1}$$

Note that $e_{s,r}$ is well defined only for $s \in S$ and $r \in R$ with $[s] = [r]$.

We define a *structured noncommutative multidimensional linear system* (SN-MLS) to be a collection $\Sigma = (G, \ \mathcal{H}, \ U)$ where G is an *admissible graph*, $\mathcal{H} = \{\mathcal{H}_p \colon p \in P\}$ is a collection of (separable) Hilbert spaces (called auxiliary state spaces) indexed by the path-connected components P of the graph G, and where U is a *connection matrix* (sometimes also called *colligation*) of the form

$$U = \begin{bmatrix} A & B \\ C & D \end{bmatrix} = \begin{bmatrix} [A_{r,s}] & [B_r] \\ [C_s] & D \end{bmatrix} \colon \begin{bmatrix} \oplus_{s \in S} \mathcal{H}_{[s]} \\ \mathcal{U} \end{bmatrix} \to \begin{bmatrix} \oplus_{r \in R} \mathcal{H}_{[r]} \\ \mathcal{Y} \end{bmatrix} \tag{2.2}$$

where $\mathcal{U}$ and $\mathcal{Y}$ are additional (separable) Hilbert spaces (to be interpreted as the *input space* and the *output space* respectively). This definition differs from that in [13] in that here we take the auxiliary state spaces $\mathcal{H}_p$, the input space $\mathcal{U}$ and the output space $\mathcal{Y}$ to be separable (possibly infinite-dimensional) Hilbert spaces rather than finite-dimensional linear spaces.

With any SNMLS we associate an input/state/output linear system with evolution along a free semigroup as follows. We denote by $\mathcal{F}_E$ the free semigroup generated by the edge set E. An element of $\mathcal{F}_E$ is then a word w of the form $w = e_N \cdots e_1$ where each e_k is an edge of G for $k = 1, \ldots, N$. We denote the empty word (consisting of no letters) by $\emptyset$. The semigroup operation is concatenation: if $w = e_N \cdots e_1$ and $w' = e'_{N'} \cdots e'_1$, then ww' is defined to be

$$ww' = e_N \cdots e_1 e'_{N'} \cdots e'_1.$$

Note that the empty word $\emptyset$ acts as the identity element for this semigroup. On occasion we shall have use of the notation we^{-1} for a word $w \in \mathcal{F}_E$ and an edge $e \in E$; by this notation we mean

$$we^{-1} = \begin{cases} w' & \text{if } w = w'e, \\ \text{undefined} & \text{otherwise.} \end{cases} \tag{2.3}$$

with a similar convention for $e^{-1}w$.

If $\Sigma = (G, \mathcal{H}, U)$ is an SNMLS, we associate the system equations (with evolution along $\mathcal{F}_E$)

$$\Sigma \colon \begin{cases} x_{\mathbf{s}(e)}(ew) &= \ \Sigma_{s \in S} A_{\mathbf{r}(e),s} x_s(w) + B_{\mathbf{r}(e)} u(w) \\ x_{s'}(ew) &= \ 0 \text{ if } s' \neq \mathbf{s}(e) \\ y(w) &= \ \Sigma_{s \in S} C_s x_s(w) + D u(w). \end{cases} \tag{2.4}$$

Here the *state vector* $x(w)$ at position w (for $w \in \mathcal{F}_E$) has the form of a column vector

$$x(w) = \mathrm{col}_{s \in S}\, x_s(w)$$

with column entries indexed by the source vertices $s \in S$ and with column entry $x_s(w)$ taking values in the auxiliary state space $\mathcal{H}_{[s]}$ (and thus $x(w)$ takes values in the *state space* $\oplus_{s \in S} \mathcal{H}_{[s]}$), while $u(w) \in \mathcal{U}$ denotes the *input* at position w and $y(w) \in \mathcal{Y}$ denotes the *output* at position w. Just as in the classical case, if we specify an initial condition $x(\emptyset) \in \oplus_{s \in S} \mathcal{H}_{[s]}$ and feed in an input string $\{u(w)\}_{w \in \mathcal{F}_E}$, then equations (2.4) enables us to recursively compute $x(w)$ for all $w \in \mathcal{F}_E \setminus \{\emptyset\}$ and $y(w)$ for all $w \in \mathcal{F}_E$.

The solution of these recursions can be made more explicit as follows. Note first of all that a consequence of the system equations is that

$$x(ew) \in \mathcal{H}_{\mathbf{s}(e)} := \mathrm{col}_{s \in S}[\delta_{s,\mathbf{s}(e)} \mathcal{H}_{[\mathbf{s}(e)]}] \text{ for all } e \in E \text{ and } w \in \mathcal{F}_E$$

(where $\delta_{s,s'}$ is the Kronecker delta function). Given $x(\emptyset)$ and $\{u(w)\}_{w \in \mathcal{F}(E)}$, we can solve the system equations (2.4) or (2.7) uniquely for $\{x(w)\}_{w \in \mathcal{F}_E \setminus \{\emptyset\}}$ and $\{y(w)\}_{w \in \mathcal{F}_E}$ as follows:

$$x_{\mathbf{s}(e_N)}(e_N \cdots e_1) = \sum_{s \in S} A_{\mathbf{r}(e_N),\mathbf{s}(e_{N-1})} A_{\mathbf{r}(e_{N-1}),\mathbf{s}(e_{N-2})} \cdots A_{\mathbf{r}(e_1),s} x_s(\emptyset)$$

$$+ \sum_{r=1}^{N} A_{\mathbf{r}(e_N),\mathbf{s}(e_{N-1})} \cdots A_{\mathbf{r}(e_{r+1}),\mathbf{s}(e_r)} B_{\mathbf{r}(e_r)} u(e_{r-1} \cdots e_1) \qquad (2.5)$$

where we interpret $u(e_{r-1} \cdots e_1)$ to be $u(\emptyset)$ where $r = 1$, and $x_s(e_N e_{N-1} \cdots e_1) = 0$ if $s \neq \mathbf{s}(e_N)$. Also,

$$y(e_N \cdots e_1) = \sum_{s \in S} C_{\mathbf{s}(e_N)} A_{\mathbf{r}(e_N),\mathbf{s}(e_{N-1})} A_{\mathbf{r}(e_{N-1}),\mathbf{s}(e_{N-2})} \cdots A_{\mathbf{r}(e_1),s} x_s(\emptyset)$$

$$+ \sum_{r=1}^{N} C_{\mathbf{s}(e_N)} A_{\mathbf{r}(e_N),\mathbf{s}(e_{N-1})} \cdots A_{\mathbf{r}(e_{r+1}),\mathbf{s}(e_r)} B_{\mathbf{r}(e_r)} u(e_{r-1} \cdots e_1)$$

$$+ D u(e_N \cdots e_1). \qquad (2.6)$$

This formula must be interpreted appropriately for special cases. As examples, for the particular cases $r = 1$ and $r = N$ we have the interpretations

$$A_{\mathbf{r}(e_N),\mathbf{s}(e_{N-1})} \cdots A_{\mathbf{r}(e_{r+1}),\mathbf{s}(e_r)} B_{\mathbf{r}(e_r)} u(e_{r-1} \cdots e_1)|_{r=1}$$
$$= A_{\mathbf{r}(e_N),\mathbf{s}(e_{N-1})} \cdots A_{\mathbf{r}(e_2),\mathbf{s}(e_1)} B_{\mathbf{r}(e_1)} u(\emptyset),$$
$$A_{\mathbf{r}(e_N),\mathbf{s}(e_{N-1})} \cdots A_{\mathbf{r}(e_{r+1}),\mathbf{s}(e_r)} B_{\mathbf{r}(e_r)} u(e_{r-1} \cdots e_1)|_{r=N} = B_{\mathbf{r}(e_N)} u(e_{N-1} \cdots e_1).$$

The system equations (2.4) can be written more compactly in operator-theoretic form as

$$\Sigma: \begin{cases} x(ew) &= I_{\Sigma;e} A x(w) + I_{\Sigma;e} B u(w) \\ y(w) &= C x(w) + D u(w) \end{cases} \qquad (2.7)$$

186   J.A. Ball, G. Groenewald and T. Malakorn

where $I_{\Sigma;e}\colon \oplus_{r\in R}\mathcal{H}_{[r]} \to \oplus_{s\in S}\mathcal{H}_{[s]}$ is given via matrix entries

$$[I_{\Sigma;e}]_{s,r} = \begin{cases} I_{\mathcal{H}_{[\mathbf{s}(e)]}} = I_{\mathcal{H}_{[\mathbf{r}(e)]}} & \text{if } s = \mathbf{s}(e) \text{ and } r = \mathbf{r}(e), \\ 0 & \text{otherwise.} \end{cases} \tag{2.8}$$

A consequence of the system equations (2.8) is the identity

$$\begin{bmatrix} \operatorname{col}_{r\in R} x_{s_{[r]}}(e_{s_{[r]},r}w) \\ y(w) \end{bmatrix} = U \begin{bmatrix} \operatorname{col}_{s\in S} x_s(w) \\ u(w) \end{bmatrix} \tag{2.9}$$

for any choice of source-vertex cross-section $p \mapsto s_p$. Here we say that a map $p \mapsto s_p$ from the set of path-connected components P of G into the set of source vertices S of G is a *source-vertex cross-section* if, for each path-connected component $p \in P$, the path-connected component of G containing $s_p \in S$ is equal to p:

$$s_p \in S \text{ for each } p \in P \text{ and } [s_p] = p. \tag{2.10}$$

More precisely, the system of equations (2.9) is equivalent to (2.7) in the sense that *the function* $w \mapsto (u(w), x(w), y(w))$ *satisfies* (2.7) *if and only if the function* $w \mapsto (u(w), x(w), y(w))$ *satisfies* (2.9) *for every choice of source-vertex cross-section map* $p \mapsto s_p \in S$ (see (2.10)). From the fact that (2.9) holds for any choice of source-vertex cross-section $p \mapsto s_p$ we deduce that the state vector $w \mapsto x(w)$ of any system trajectory $w \mapsto (u(w), x(w), y(w))$ satisfies the compatibility condition

$$x_s(e_{s,r}w) \text{ is independent of } s \in [r] \text{ for each fixed } r \in R \text{ and } w \in \mathcal{F}_E, \tag{2.11}$$

as can also be seen directly from the system equations (2.4).

Note that $I_{\Sigma;e}$ is already determined by the first two pieces G and $\mathcal{H}$ of $\Sigma = (G, \mathcal{H}, U)$. On occasion we shall need these objects in situations where we have a graph G and a collection of Hilbert spaces $\mathcal{H} = \{\mathcal{H}_p\colon p \in P\}$ without the presence of any particular connection matrix U. In this situation we shall use the notation $I_{G,\mathcal{H};e}$ in place of $I_{\Sigma;e}$. We shall also have occasion to need the operator pencil $Z_\Sigma(z) = \sum_{e\in E} I_{\Sigma,e}z_e$, also written as $Z_{G,\mathcal{H}}(z) = \sum_{e\in E} I_{G,\mathcal{H};e}z_e$ when U is absent or suppressed.

Also just as in the classical case, it is convenient to introduce "frequency-domain" notation for explicit representation of system trajectories. For any linear space $\mathcal{H}$, we define the formal noncommutative Z-transform of a sequence of $\mathcal{H}$-valued functions as a formal power series in several noncommuting indeterminates $z = (z_e : e \in E)$ as follows:

$$\{h(w)\}_{w\in\mathcal{F}_E} \mapsto \widehat{h}(z) = \sum_{w\in\mathcal{F}_E} h(w)z^w, \tag{2.12}$$

where $z^\emptyset = 1$, $z^w = z_{e_N}z_{e_{N-1}}\cdots z_{e_1}$ if $w = e_N e_{N-1}\cdots e_1$. Thus

$$z^w \cdot z^{w'} = z^{ww'}, \qquad z^w \cdot z_e = z^{we} \text{ for } w, w' \in \mathcal{F}_E \text{ and } e \in E.$$

On occasion we shall have need of multiplication on the right or left by z_e^{-1}; we use the convention

$$z^w z_e^{-1} = \begin{cases} z^{we^{-1}} & \text{if } we^{-1} \in \mathcal{F}_E \text{ is defined;} \\ 0 & \text{if } we^{-1} \text{ is undefined.} \end{cases} \tag{2.13}$$

where we use the convention (2.3) for the meaning of we^{-1}. We use the obvious analogous convention to define $z_e^{-1} z^w$. As derived in [13], application of the formal noncommutative Z-transform to the system equations (2.4) and solving gives a frequency-domain formula for the state and output trajectory:

$$\widehat{x}(z) = (I - Z_\Sigma(z)A)^{-1} x(\emptyset) + (I - Z_\Sigma(z)A)^{-1} Z_\Sigma(z) B \widehat{u}(z)$$
$$\widehat{y}(z) = C(I - Z_\Sigma(z)A)^{-1} x(\emptyset) + T_\Sigma(z)\widehat{u}(z) \tag{2.14}$$

where we have set

$$Z_\Sigma(z) = \sum_{e \in E} I_{\Sigma;e} z_e \tag{2.15}$$

and where the formal power series given by

$$T_\Sigma(z) = D + C(I - Z_\Sigma(z)A)^{-1} Z_\Sigma(z) B \tag{2.16}$$

$$= T_\emptyset + \sum_{N=1}^{\infty} \sum_{e_1,\ldots,e_N \in E} C_{\mathbf{s}(e_N)} A_{\mathbf{r}(e_N),\mathbf{s}(e_{N-1})} \cdots A_{\mathbf{r}(e_2),\mathbf{s}(e_1)} B_{\mathbf{r}(e_1)} z_{e_N} z_{e_{N-1}} \cdots z_{e_2} z_{e_1} \tag{2.17}$$

is the *transfer function* of the SNMLS Σ.

As explained in [13], there are three particular examples worth special mention; we refer to these as (1) *noncommutative Fornasini-Marchesini systems*, (2) *noncommutative Givone-Roesser systems*, and (3) *noncommutative full-structured multidimensional linear systems*. These special cases are defined as follows.

Example 2.1. **Noncommutative Fornasini-Marchesini systems.** We let G^{FM} be the admissible graph with source-vertex set S^{FM} consisting of a single element $S^{FM} = \{1\}$ and with range-vertex set R^{FM} and set of edges E^{FM} both equal to the finite set $\{1,\ldots,d\}$, with edge j having source vertex 1 and range vertex j:

$$\mathbf{s}^{FM}(j) = 1 \text{ and } \mathbf{r}^{FM}(j) = j \text{ for } j = 1,\ldots,d.$$

Suppose now that $\Sigma = (G^{FM}, \mathcal{H}, U^{FM})$ is a SNMLS with structure graph G^{FM}. As G^{FM} has only one path-connected component ($P^{FM} = \{p_1\}$, the collection of Hilbert spaces $\mathcal{H} = \{\mathcal{H}_p : p \in P\}$ collapses to a single Hilbert space $\mathcal{H}$. The connection matrix U^{FM} then has the form

$$U^{FM} = \begin{bmatrix} A & B \\ C & D \end{bmatrix} = \begin{bmatrix} A_1 & B_1 \\ \vdots & \vdots \\ A_d & B_d \\ C & D \end{bmatrix} : \begin{bmatrix} \mathcal{H} \\ \mathcal{U} \end{bmatrix} \to \begin{bmatrix} \oplus_{j=1}^{d} \mathcal{H} \\ \mathcal{Y} \end{bmatrix}$$

and the system equations (2.4) have the form

$$\Sigma^{FM}: \left\{ \begin{array}{rcl} x(1w) &=& A_1 x(w) + B_1 u(w) \\ \vdots & & \vdots \\ x(dw) &=& A_d x(w) + B_d u(w) \\ y(w) &=& C x(w) + D u(w). \end{array} \right. \tag{2.18}$$

or in more compact form

$$\Sigma^{FM}: \left\{ \begin{array}{rcl} x(jw) &=& I_{\Sigma^{FM},j} A x(w) + I_{\Sigma^{FM},j} B u(w) \\ y(w) &=& C x(w) + D u(w) \end{array} \right. \tag{2.19}$$

where we set

$$I_{\Sigma^{FM},j} = \begin{bmatrix} 0 & \cdots & I_{\mathcal{H}} & \cdots & 0 \end{bmatrix} \quad \text{where } I_{\mathcal{H}} \text{ occurs in the } j\text{th column.}$$

The transfer function $T_{\Sigma^{FM}}(z)$ then has the form

$$\begin{aligned} T_{\Sigma^{FM}}(z) &= D + C(I - Z_{\Sigma^{FM}}(z)A)^{-1} Z_{\Sigma^{FM}}(z) B \\ &= D + C(I - z_1 A_1 - \cdots - z_d A_d)^{-1}(z_1 B_1 + \cdots + z_d B_d) \\ &= D + \sum_{v \in \mathcal{F}_d} \sum_{j=1}^{d} C A^v B_j z^v z_j \end{aligned} \tag{2.20}$$

where the structure matrix $Z_{\Sigma^{FM}}(z)$ is given by

$$Z_{\Sigma^{FM}}(z) = \sum_{j=1}^{d} I_{\Sigma^{FM},j} z_j = \begin{bmatrix} z_1 I_{\mathcal{H}} & \cdots & z_d I_{\mathcal{H}} \end{bmatrix}.$$

We consider these as noncommutative Fornasini-Marchesini systems (with evolution along the free semigroup generated by $\{1, \ldots, d\}$) to be a noncommutative analogue of the commutative multidimensional systems (with evolution along an integer lattice rather than a tree or free semigroup) introduced and studied by Fornasini and Marchesini (see, e.g., [34]).

Example 2.2. **Noncommutative Givone-Roesser systems.** We let G^{GR} be the graph with source-vertex set S^{GR}, range vertex set R^{GR} and set of edges E^{GR} all equal to the finite set $\{1, \ldots, d\}$ with edge j having source vertex j and range vertex j:

$$\mathbf{s}^{FM}(j) = j \text{ and } \mathbf{r}^{FM}(j) = j \text{ for } j = 1, \ldots, d.$$

Suppose now that $\Sigma = (G^{GR}, \mathcal{H}, U^{GR})$ is a SNMLS with structure graph G^{GR}. As G^{GR} has d path-connected components ($P^{GR} = \{p_1, \ldots, p_d\}$), the collection of Hilbert spaces $\mathcal{H}$ can be labeled as $\mathcal{H} = \{\mathcal{H}_j : j = 1, \ldots, d\}$. The connection matrix U^{GR} then has the form

$$U^{GR} = \begin{bmatrix} A & B \\ C & D \end{bmatrix} = \begin{bmatrix} A_{11} & \cdots & A_{1d} & B_1 \\ \vdots & & \vdots & \vdots \\ A_{d1} & \cdots & A_{dd} & B_d \\ C_1 & \cdots & C_d & D \end{bmatrix} : \begin{bmatrix} \oplus_{i=1}^{d} \mathcal{H}_i \\ \mathcal{U} \end{bmatrix} \rightarrow \begin{bmatrix} \oplus_{j=1}^{d} \mathcal{H}_j \\ \mathcal{Y} \end{bmatrix}$$

and the system equations (2.4) have the form

$$\Sigma^{GR}: \begin{cases} x_1(1w) &= A_{11}x_1(w) + \cdots + A_{1d}x_d(w) + B_1u(w) \\ \quad\vdots & \quad\vdots \\ x_d(dw) &= A_{d1}x_1(w) + \cdots + A_{dd}x_d(w) + B_du(w) \\ x_{i'}(iw) &= 0 \text{ if } i' \neq i, \\ y(w) &= C_1x_1(w) + \cdots + C_dx_d(w) + Du(w). \end{cases} \tag{2.21}$$

or in more compact form

$$\Sigma^{GR}: \begin{cases} x(jw) &= I_{\Sigma^{GR},j}Ax(w) + I_{\Sigma^{GR},j}Bu(w) \\ y(w) &= Cx(w) + Du(w) \end{cases} \tag{2.22}$$

where we set

$$I_{\Sigma^{GR},j} = \begin{bmatrix} 0 & & & & \\ & \ddots & & & \\ & & I_{\mathcal{H}_j} & & \\ & & & \ddots & \\ & & & & 0 \end{bmatrix}$$

where the nonzero entry occurs in the jth diagonal slot.

The transfer function $T_{\Sigma^{GR}}(z)$ then has the form

$$T_{\Sigma^{GR}}(z) = D + C(I - Z_{\Sigma^{GR}}(z)A)^{-1}Z_{\Sigma^{GR}}(z)B \tag{2.23}$$

$$= D + \begin{bmatrix} C_1 & \cdots & C_d \end{bmatrix} \left(\begin{bmatrix} I_{\mathcal{H}_1} & & \\ & \ddots & \\ & & I_{\mathcal{H}_d} \end{bmatrix} - \begin{bmatrix} z_1 A_{11} & \cdots & z_1 A_{1d} \\ \vdots & & \vdots \\ z_d A_{d1} & \cdots & z_d A_{dd} \end{bmatrix} \right)^{-1} \begin{bmatrix} z_1 B_1 \\ \vdots \\ z_d B_d \end{bmatrix}$$

$$= D + \sum_{N=1}^{\infty} \sum_{i_1,\ldots,i_N \in \{1,\ldots,d\}} C_{i_N} A_{i_N,i_{N-1}} A_{i_{N-1},i_{N-2}} \cdots A_{i_2,i_1} B_{i_1} z_{i_N} z_{i_{N-1}} \cdots z_{i_2} z_{i_1} \tag{2.24}$$

where the structure matrix $Z_{\Sigma^{GR}}(z)$ is given by

$$Z_{\Sigma^{GR}}(z) = \sum_{j=1}^{d} I_{\Sigma^{GR},j}z_j = \begin{bmatrix} z_1 I_{\mathcal{H}_1} & & \\ & \ddots & \\ & & z_d I_{\mathcal{H}_d} \end{bmatrix}.$$

We consider these noncommutative Givone-Roesser systems to be a noncommutative analogue of the commutative multidimensional systems introduced and studied by Givone and Roesser (see, e.g., [35, 36]).

Example 2.3. **Noncommutative full-structured multidimensional systems.** We take G^{full} to be the complete bipartite graph on source-vertex set $S^{\mathrm{full}} = \{1,\ldots,n\}$ and range-vertex set $R^{\mathrm{full}} = \{1,\ldots,m\}$. Thus we may label the edge set E^{full} as $E^{\mathrm{full}} = \{(i,j): i = 1,\ldots,n; j = 1,\ldots,m\}$ with

$$\mathbf{s}^{\mathrm{full}}(i,j) = i, \qquad \mathbf{r}^{\mathrm{full}}(i,j) = j.$$

We let $\mathcal{F}_{n,m}$ denote the free semigroup generated by the set

$$E^{\text{full}} = \{1,\ldots,n\} \times \{1,\ldots,m\}.$$

Thus elements of $\mathcal{F}_{n,m}$ are words w of the form $(i_N, j_N)(i_{N-1}, j_{N-1}) \cdots (i_1, j_1)$ where $i_k \in \{1,\ldots,n\}$ for all $k = 1,\ldots,N$ and $j_k \in \{1,\ldots,m\}$ for all $k = 1,\ldots,N$. Suppose that $\Sigma^{\text{full}} = (G^{\text{full}}, \mathcal{H}, U^{\text{full}})$ is a SNMLS with structure graph equal to G^{full}. As G^{full} has only one connected component in this case, the collection of Hilbert spaces $\mathcal{H} = \{\mathcal{H}_p : p \in P^{\text{full}}\}$ collapses to a single Hilbert space denoted as $\mathcal{H}$. The connection matrix U^{full} then has the form

$$U^{\text{full}} = \begin{bmatrix} A & B \\ C & D \end{bmatrix} = \begin{bmatrix} A_{11} & \cdots & A_{1n} & B_1 \\ \vdots & & \vdots & \vdots \\ A_{m1} & \cdots & A_{mn} & B_m \\ C_1 & \cdots & C_n & D \end{bmatrix} : \begin{bmatrix} \oplus_{i=1}^{n} \mathcal{H} \\ \mathcal{U} \end{bmatrix} \to \begin{bmatrix} \oplus_{j=1}^{m} \mathcal{H} \\ \mathcal{Y} \end{bmatrix}$$

and the associated system equations have the form

$$\Sigma^{\text{full}} : \begin{cases} x_1((1,j) \cdot w) & = A_{j1} x_1(w) + \cdots + A_{jn} x_n(w) + B_j u(w) \text{ for } j = 1,\ldots,m, \\ & \vdots \\ x_n((n,j) \cdot w) & = A_{j1} x_1(w) + \cdots + A_{jn} x_n(w) + B_j u(w) \text{ for } j = 1,\ldots,m, \\ x_{i'}((i,j) \cdot w) & = 0 \text{ if } i' \neq i, \\ y(w) & = C_1 x_1(w) + \cdots + C_n x_n(w) + D u(w). \end{cases} \tag{2.25}$$

Note that, as is consistent with (2.11), $x_i((i,j) \cdot w)$ is independent of i for each fixed $j \in \{1,\ldots,m\}$ and $w \in \mathcal{F}_{n,m}$. The transfer function $T_{\Sigma^{\text{full}}}$ then has the form

$$T_{\Sigma^{\text{full}}}(z) = D + C(I - Z_{\Sigma^{\text{full}}}(z)A)^{-1} Z_{\Sigma^{\text{full}}}(z) B \tag{2.26}$$

$$= D + \begin{bmatrix} C_1 & \cdots & C_n \end{bmatrix}$$

$$\cdot \left(\begin{bmatrix} I_{\mathcal{H}} & & \\ & \ddots & \\ & & I_{\mathcal{H}} \end{bmatrix} - \begin{bmatrix} \sum_{j=1}^{m} z_{1j} A_{j1} & \cdots & \sum_{j=1}^{m} z_{1j} A_{jn} \\ \vdots & & \vdots \\ \sum_{j=1}^{m} z_{nj} A_{j1} & \cdots & \sum_{j=1}^{m} z_{nj} A_{jn} \end{bmatrix} \right)^{-1} \begin{bmatrix} \sum_{j=1}^{m} z_{1j} B_j \\ \vdots \\ \sum_{j=1}^{m} z_{nj} B_j \end{bmatrix}$$

$$= D + \sum_{N=1}^{\infty} \sum_{i_1,\ldots,i_N \in \{1,\ldots,n\}} \sum_{j_1,\ldots,j_N \in \{1,\ldots,m\}} C_{i_N} A_{j_N, i_{N-1}} A_{j_{N-1}, i_{N-2}} \cdots A_{j_2, i_1} B_{i_1}$$

$$\cdot z_{i_N, j_N} z_{i_{N-1}, j_{N-1}} \cdots z_{i_2, j_2} z_{i_1, j_1} \tag{2.27}$$

and where $Z_{\Sigma^{\text{full}}}(z)$ is given by

$$Z_{\Sigma^{\text{full}}}(z) = \begin{bmatrix} z_{1,1} I_{\mathcal{H}} & \cdots & z_{1,m} I_{\mathcal{H}} \\ \vdots & & \vdots \\ z_{n,1} I_{\mathcal{H}} & \cdots & z_{n,m} I_{\mathcal{H}} \end{bmatrix}.$$

3. Adjoint systems

It turns out that the adjoint system for a SNMLS Σ has a somewhat different form. Let us say that the collection $\Sigma_* = (G, \mathcal{H}_*, U_*)$ is a *SNMLS of adjoint form* if

1. G is an admissible finite graph,
2. $\mathcal{H}_* = \{\mathcal{H}_{*p} : p \in P\}$ is a collection of Hilbert spaces (the *auxiliary state spaces for Σ_**) indexed by the set P of path-connected components of G, and
3. the *connection matrix U_* for Σ_** has the form

$$U_* = \begin{bmatrix} A_* & B_* \\ C_* & D_* \end{bmatrix} = \begin{bmatrix} [A_{*s,r}] & [B_{*s}] \\ [C_{*r}] & D_* \end{bmatrix} : \begin{bmatrix} \oplus_{r \in R} \mathcal{H}_{*[r]} \\ \mathcal{U}_* \end{bmatrix} \to \begin{bmatrix} \oplus \mathcal{H}_{*[s]} \\ \mathcal{Y}_* \end{bmatrix} \tag{3.1}$$

where $\mathcal{U}_*$ (the *input space* for Σ_*) and $\mathcal{Y}_*$ (the *output space* for Σ_*) are Hilbert spaces.

The system equations associated with an SNMLS of adjoint form Σ_* involve also a choice of source-vertex cross-section $p \mapsto s_p$ as in (2.10) and are given by

$$\Sigma_* : \begin{cases} x_{*s}(w) & = \sum_{r \in R} A_{*s,r} x_{*s_{[r]}}(e_{s_{[r]},r} w) + B_{*s} u_*(w) \\ y_*(w) & = \sum_{r \in R} C_{*r} x_{*s_{[r]}}(e_{s_{[r]},r}(w) + D_* u_*(w). \end{cases} \tag{3.2}$$

The state vector $x_*(w) = \mathrm{col}_{s \in S}\, x_{*s}(w)$ takes values in the state space $\oplus_{s \in S} \mathcal{H}_{*[s]}$ with components $x_{*s}(w)$ in the auxiliary state space $\mathcal{H}_{*[s]}$ for each $s \in S$ and is required to satisfy the *compatibility condition*

$$x_{*s}(e_{s,r} w) = x_{*s'}(e_{s',r} w) \text{ for all } s, s' \in S \text{ with } [s] = [s'] \tag{3.3}$$

$$\text{and for all } r \in R \text{ and } w \in \mathcal{F}_E. \tag{3.4}$$

The adjoint input signal $u_*(w)$ takes values in $\mathcal{U}_*$ and the adjoint output signal $y_*(w)$ takes values in $\mathcal{Y}_*$. Given a positive integer N, suppose that we are given an input signal $\{u_*(w)\}_{w:\ |w| \leq N}$ on the finite horizon $\{w \in \mathcal{F}_d : |w| \leq N\}$ along with a finalization of the state $\{x_*(w)\}_{w:\ |w|=N+1}$. We can then apply the recursions in (3.2) to compute $x_*(w)$ and $y_*(w)$ for all $w \in \mathcal{F}_d$ with $|w| \leq N$. The compatibility condition (3.4) implies that the resulting solution $x_*(w)$ and $y_*(w)$ is independent of the choice of source-vertex cross-section $p \mapsto s_p$. In general we say that a triple of functions $w \mapsto (u_*(w), x_*(w), y_*(w))$ is a *trajectory of the system of adjoint form Σ_** if x_* satisfies the compatibility condition (3.4) and (u_*, x_*, y_*) satisfy the adjoint system equations (3.2) for some (and hence for any) choice of source-vertex cross-section $p \mapsto s_p$.

Given a SNMLS $\Sigma = (G, \mathcal{H}, U)$, we define the adjoint system Σ^* of Σ to be the SNMLS of adjoint form given by

$$\Sigma^* = (G, \mathcal{H}, U^*). \tag{3.5}$$

From the definition (3.2) we see that the system equations associated with Σ^* therefore have the form

$$\Sigma^* : \begin{cases} x_{*s}(w) & = \sum_{r \in R} A^*_{r,s} x_{*s_{[r]}}(e_{s_{[r]},r} w) + C^*_s u_*(w) \\ y_*(w) & = \sum_{r \in R} B^*_r x_{*s_{[r]}}(e_{s_{[r]},r} w) + D^* u_*(w). \end{cases} \tag{3.6}$$

where the adjoint state vector $x_*(w) = \mathrm{col}_{s \in S}\, x_{*s}(w)$ taking values in $\oplus_{s \in S} \mathcal{H}_{[s]}$, adjoint input signal $u_*(w)$ taking values in $\mathcal{Y}$ and adjoint output signal $y_*(w)$ taking values in $\mathcal{U}$. The defining condition of the adjoint system is given by the following Proposition. In the following statement, by a *local trajectory of the system* Σ *at the word* w we mean a function $w' \mapsto (u(w'), x(w') = \oplus_{s \in S} x_s(w'), y(w'))$ defined at least for $w' = w$ and $w' = ew$ for each $e \in E$ which satisfies the system equations (2.4) at position w. Similarly, by a *local trajectory of* Σ^* *at* w we mean a function $w' \mapsto (u_*(w'), x_*(w') = \oplus_{s \in S} x_{*s}(w'), y_*(w'))$ defined at least for $w' = w$ and $w' = ew$ for each $e \in E$ which satisfies the compatibility condition (3.4) and the adjoint system equations (3.6) at w. With these notions we avoid the issue of whether a local trajectory (of Σ or Σ^*) necessarily extends to a global trajectory.

Proposition 3.1. *Suppose that we are given a SNMLS* $\Sigma = (G, \mathcal{H}, U)$ *with adjoint system* $\Sigma^* = (G, \mathcal{H}, U^*)$.

1. *The adjoint pairing relation*

$$\sum_{r \in R} \langle x_{s_{[r]}}(e_{s_{[r]},r} w), x_{*s_{[r]}}(e_{s_{[r]},r} w) \rangle_{\mathcal{H}_{[r]}} + \langle y(w), u_*(w) \rangle_{\mathcal{Y}}$$

$$= \sum_{s \in S} \langle x_s(w), x_{*s}(w) \rangle_{\mathcal{H}_{[s]}} + \langle u(w), y_*(w) \rangle_{\mathcal{U}} \tag{3.7}$$

holds for any trajectory (u, x, y) *of* Σ *and any trajectory* (u_*, x_*, y_*) *of* Σ^*.

2. *Conversely, if a given function*

$$w \mapsto (u(w), x(w), y(w)) \in \mathcal{U} \times (\oplus_{s \in S} \mathcal{H}_{[s]}) \times \mathcal{Y}$$

satisfies the adjoint pairing relation (3.7) *with respect to every local trajectory* $(u_*(w), x_*(w), y_*(w))$ *of* Σ^* *at each* $w \in \mathcal{F}_E$, *then* (u, x, y) *is a trajectory of* Σ.

3. *Conversely, if a given function*

$$w \mapsto (u_*(w), x_*(w), y_*(w)) \in \mathcal{Y} \times (\oplus_{s \in S} \mathcal{H}_{[s]}) \times \mathcal{U}$$

satisfies the adjoint pairing relation (3.7) *with respect to every local trajectory* $(u(w), x(w), y(w))$ *of* Σ *at* w *for each* $w \in \mathcal{F}_E$, *then* (u_*, x_*, y_*) *is a trajectory of* Σ^*.

Proof. Note that the system equations (2.9) for Σ can be written in vector form as

$$\begin{bmatrix} \mathrm{col}_{r \in R}\, x_{s_{[r]}}(e_{s_{[r]},r} w) \\ y(w) \end{bmatrix} = U \begin{bmatrix} \mathrm{col}_{s \in S}\, x_s(w) \\ u(w) \end{bmatrix}. \tag{3.8}$$

Similarly, in vector form, the adjoint system equations (3.6) are

$$\begin{bmatrix} \mathrm{col}_{s \in S}\, x_{*s}(w) \\ y_*(w) \end{bmatrix} = U^* \begin{bmatrix} \mathrm{col}_{r \in R}\, x_{*s_{[r]}}(e_{s_{[r]},r} w) \\ u_*(w) \end{bmatrix} \tag{3.9}$$

and the adjoint pairing relation is

$$\left\langle \begin{bmatrix} \mathrm{col}_{r\in R}\, x_{s_{[r]}}(e_{s_{[r]},r}w) \\ y(w) \end{bmatrix},\, \begin{bmatrix} \mathrm{col}_{r\in R}\, x_{*s_{[r]}}(e_{s_{[r]},r}w) \\ u_*(w) \end{bmatrix} \right\rangle_{(\oplus_{r\in R}\mathcal{H}_{[r]})\oplus\mathcal{Y}}$$

$$= \left\langle \begin{bmatrix} \mathrm{col}_{s\in S}\, x_s(w) \\ u(w) \end{bmatrix},\, \begin{bmatrix} \mathrm{col}_{s\in S}\, x_{*s}(w) \\ y_*(w) \end{bmatrix} \right\rangle_{(\oplus_{s\in S}\mathcal{H}_{[s]})\oplus\mathcal{U}}. \tag{3.10}$$

If (u, x, y) is a trajectory of Σ and (u_*, x_*, y_*) is a trajectory of Σ^*, then substitution of (3.8) and (3.9) into (3.10) shows that (3.10) holds for (u, x, y) and (u_*, x_*, y_*) by definition of the adjoint U^* of U. More precisely, if (u, x, y) is a trajectory such that (3.7) holds for any local trajectory (u_*, x_*, y_*) of Σ^* at w, then we see that

$$\left\langle \begin{bmatrix} \mathrm{col}_{r\in R}\, x_{s_{[r]}}(e_{s_{[r]},r}w) \\ y(w) \end{bmatrix},\, \begin{bmatrix} \mathrm{col}_{r\in R}\, x_{*s_{[r]}}(e_{s_{[r]},r}w) \\ u_*(w) \end{bmatrix} \right\rangle_{(\oplus_{r\in R}\mathcal{H}_{[r]})\oplus\mathcal{Y}}$$

$$= \left\langle \begin{bmatrix} \mathrm{col}_{s\in S}\, x_s(w) \\ u(w) \end{bmatrix},\, U^* \begin{bmatrix} \mathrm{col}_{r\in R}\, x_{*s_{[r]}}(e_{s_{[r]},r}w) \\ u_*(w) \end{bmatrix} \right\rangle_{(\oplus_{s\in S}\mathcal{H}_{[s]})\oplus\mathcal{U}}.$$

As $\begin{bmatrix} \mathrm{col}_{r\in R}\, x_{s_{[r]}}(e_{s_{[r]},r}w) \\ y(w) \end{bmatrix}$ can be taken to be an arbitrary element of $(\oplus_{r\in R}\mathcal{H}_{[r]})\oplus\mathcal{U}$ and the source-vertex cross-section $p \mapsto s_p$ is also arbitrary, it follows that (u, x, y) satisfies (3.8) at w. As the choice of $w \in \mathcal{F}_E$ is arbitrary, we conclude that (u, x, y) is a trajectory of Σ. A similar argument shows that (u_*, x_*, y_*) is a trajectory of Σ^* if (u_*, x_*, y_*) satisfies (3.7) against every local trajectory (u, x, y) of Σ at each $w \in \mathcal{F}_E$, and Proposition 3.1 now follows. $\qquad\square$

4. Dissipative and conservative structured multidimensional linear systems

In case U is contractive ($\|U\| \leq 1$), we say that Σ is a *dissipative SNMLS*. In this case the trajectories of Σ have the following energy dissipation property:

$$\sum_{r\in R} \|x_{s_{[r]}}(e_{s_{[r]},r}w)\|^2 - \|x(w)\|^2 \leq \|u(w)\|^2 - \|y(w)\|^2 \tag{4.1}$$

for every choice of source-vertex cross-section $p \mapsto s_p$. We say that the SNMLS is *isometric* in case the connection matrix U is isometric. In this case the dissipation inequality (4.1) is replaced with the *energy balance* relation:

$$\sum_{r\in R} \|x_{s_{[r]}}(e_{s_{[r]},r}w)\|^2 - \|x(w)\|^2 = \|u(w)\|^2 - \|y(w)\|^2 \tag{4.2}$$

for every choice of source-vertex cross-section $p \mapsto s_p$.

An interesting special case is the case where there is a *unique* source-vertex cross-section. This happens exactly when each path-connected component p of the admissible graph G contains exactly one source vertex s_p; this occurs, e.g., for the case of noncommutative Fornasini-Marchesini systems (see Example 2.1) and for

194   J.A. Ball, G. Groenewald and T. Malakorn

noncommutative Givone-Roesser systems (see Example 2.2). In this case, each edge e has the form $e_{s_{[r]},r}$ and hence can be indexed more simply by $r \in R$: $e_{s_{[r]},r} \mapsto e_r$. Then the property that $x_s(ew) = 0$ if $s \neq \mathbf{s}(e)$ translates to $x_s(e_r w) = 0$ if $s \neq s_{[r]}$. With the use of this fact we see that, when we sum (4.1) over all words w of length at most some N, the left side of the inequality telescopes and we arrive at

$$\sum_{w:\, |w|=N+1} \|x(w)\|^2 - \|x(\emptyset)\|^2 \leq \sum_{w:\, |w| \leq N} \left[\|u(w)\|^2 - \|y(w)\|^2 \right]. \qquad (4.3)$$

This can be rearranged as

$$\sum_{w:\, |w| \leq N} \|y(w)\|^2 \leq \sum_{w:\, |w| \leq N} \|y(w)\|^2 + \sum_{w:\, |w|=N+1} \|x(w)\|^2$$

$$\leq \sum_{w:\, |w| \leq N} \|u(w)\|^2 + \|x(\emptyset)\|^2, \qquad (4.4)$$

and hence, letting $N \to \infty$ gives

$$\sum_{w \in \mathcal{F}_E} \|y(w)\|^2 \leq \sum_{w \in \mathcal{F}_E} \|u(w)\|^2 + \|x(\emptyset)\|^2. \qquad (4.5)$$

In particular, if we impose zero initial condition $x(\emptyset) = 0$ and take formal Z-transform, from the fact that $\widehat{y}(z) = T_\Sigma(z) \cdot \widehat{u}(z)$ (see (2.14)) we arrive at

$$\|T_\Sigma(z)\widehat{u}(z)\|^2_{L^2(\mathcal{F}_E, \mathcal{Y})} \leq \|\widehat{u}(z)\|^2_{L^2(\mathcal{F}_E, \mathcal{U})} \text{ for all } \widehat{u}(z) \in L^2(\mathcal{F}_E, \mathcal{U}), \qquad (4.6)$$

i.e., *multiplication by T_Σ is a contraction operator from $L^2(\mathcal{F}_E, \mathcal{U})$ into $L^2(\mathcal{F}_E, \mathcal{Y})$ in case there is a unique source-vertex cross-section $p \mapsto s_p$ for G.* We shall have further discussion of this point in Remark 5.14 below.

Given a SNMLS $\Sigma = (G, \mathcal{H}, U)$, we say that Σ is a *conservative SNMLS* if the connection matrix

$$U = \begin{bmatrix} A & B \\ C & D \end{bmatrix} = \begin{bmatrix} [A_{r,s}] & [B_r] \\ [C_s] & D \end{bmatrix} : \begin{bmatrix} \oplus_{s \in S} \mathcal{H}_{[s]} \\ \mathcal{U} \end{bmatrix} \to \begin{bmatrix} \oplus_{r \in R} \mathcal{H}_{[r]} \\ \mathcal{Y} \end{bmatrix}$$

is *unitary*. In particular U is isometric, so system trajectories satisfy the energy balance relation (4.2).

Just as in the classical case, for a system-theoretic interpretation of the meaning of the adjoint U^* of U also being isometric, we need to introduce the adjoint system Σ^*. Recall the definition of the adjoint Σ^* of a SNMLS $\Sigma = (G, \mathcal{H}, U)$ given by (3.5).

Theorem 4.1. *Suppose that $\Sigma = (G, \mathcal{H}, U)$ is a SNMLS. Then Σ is conservative (i.e., U is unitary) if and only if either one of the following conditions holds:*

1. *The function $(u, x, y)\colon \mathcal{F}_E \to \mathcal{U} \times \oplus_{s \in S}\mathcal{H}_{[s]} \times \mathcal{Y}$ is a local trajectory of Σ at w if and only if the function $(y, x, u)\colon \mathcal{F}_E \to \mathcal{Y} \times \oplus_{s \in S}\mathcal{H}_{[s]} \times \mathcal{U}$ is a local trajectory of Σ^* at w.*

2. *A local trajectory (u, x, y) of Σ at w satisfies the energy balance relation*

$$\sum_{r \in R} \|x_{s_{[r]}}(e_{s_{[r]},r}w)\|^2 - \|x(w)\|^2 = \|u(w)\|^2 - \|y(w)\|^2 \tag{4.7}$$

and any local trajectory (u_*, x_*, y_*) of Σ^* at w satisfies the adjoint energy balance relation

$$\sum_{r \in R} \|x_{*s_{[r]}}(e_{s_{[r]},r}w)\|^2 + \|u_*(w)\|^2 = \sum_{s \in S} \|x_{*s}(w)\|^2 + \|y_*(w)\|^2. \tag{4.8}$$

for all source-vertex cross-sections $p \mapsto s_p$.

In particular, if $\Sigma = (G, \mathcal{H}, U)$ is a conservative SNMLS, then

1. *(u, x, y) is a trajectory of Σ if and only if (y, x, u) is a trajectory of Σ^*,*
2. *any trajectory (u, x, y) of Σ satisfies (4.7), and*
3. *any trajectory (u_*, x_*, y_*) of Σ^* satisfies (4.8).*

Proof. From the block forms (3.8) and (3.9) of the system equations for Σ and Σ^*, we see that the equivalence between (u, x, y) being a local trajectory for Σ and (y, x, u) being a local trajectory for Σ^* is in turn equivalent to $U^* = U^{-1}$, i.e., to U being unitary. Again from the system equations (3.8) and (3.9), we see that (4.7) holding for all local trajectories just means that U is isometric while (4.8) holding for all local trajectories of Σ^* just means that U^* is isometric. This essentially completes the proof of Theorem 4.1. $\qquad\square$

A useful property of dissipative (and hence in particular of conservative) SNMLSs is the possibility of interpreting the transfer function as a function acting on tuples of noncommuting contraction operators as we now explain. In general, suppose that G is an admissible graph and that we are given a formal power series $T(z) = \sum_{v \in \mathcal{F}_E} T_v z^v$ in noncommuting variables $z = (z_e : e \in E)$ indexed by the edge set E of the graph G, with coefficients T_v equal to bounded operators acting between Hilbert spaces $\mathcal{U}$ and $\mathcal{Y}$. Suppose that we are also given a collection $\delta = (\delta_e : e \in E)$ of bounded, linear operators (not necessarily commuting) on some separable infinite-dimensional Hilbert space $\mathcal{K}$ also indexed by the edge set E of G. We define an operator $T(\delta) \colon \mathcal{U} \otimes \mathcal{K} \to \mathcal{Y} \otimes \mathcal{K}$ by

$$T(\delta) := \lim_{N \to \infty} \sum_{v \in \mathcal{F}_E : |v| \leq N} T_v \otimes \delta^v$$

$$\text{where } \delta^\emptyset = I_{\mathcal{K}} \text{ and } \delta^v = \delta_{e_N} \cdots \delta_{e_1} \text{ if } v = e_N \cdots e_1. \tag{4.9}$$

whenever the limit (say, in the norm or the strong operator topology) exists.

In general there is no reason for the limit in (4.9) to exist. However, for the case that $T(z) = T_\Sigma(z)$ is the transfer function of a conservative SNMLS $\Sigma = (G, \mathcal{H}, U)$, $T(\delta)$ always makes sense for a natural class of operator-tuples $\delta = (\delta_e : e \in E)$. To state the result we first need to agree on some notation. Suppose that Σ is a conservative SNMLS with system structure matrix $Z_\Sigma(z) = \sum_{e \in E} I_{\Sigma,e} z_e$ as in (2.15). For $\delta = (\delta_e : e \in E)$ a finite collection of (not necessarily commuting) bounded linear operators on some Hilbert space $\mathcal{K}$ indexed by the edge set E, define an operator

$$Z_\Sigma(\delta)\colon \left(\oplus_{r\in R}\mathcal{H}_{[r]}\right)\otimes\mathcal{K} \to \left(\oplus_{s\in S}\mathcal{H}_{[s]}\right)\otimes\mathcal{K}$$

by $Z_\Sigma(\delta) = \sum_{e\in E} I_{\Sigma,e}\cdot\delta_e$ where $I_{\Sigma,e}\cdot\delta_e$ is given in terms of its matrix entries

$$[I_{\Sigma,e}\cdot\delta_e]_{s,r} = \begin{cases} I_{\mathcal{H}_{[s]}}\otimes\delta_e = I_{\mathcal{H}_{[r]}}\otimes\delta_e & \text{if } s=\mathbf{s}(e) \text{ and } r=\mathbf{r}(e), \\ 0 & \text{otherwise.} \end{cases} \tag{4.10}$$

Note that the definition of $I_{\Sigma,e}$ and of $Z_\Sigma(z)$ uses only the first two pieces G and $\mathcal{H}$ of the SNMLS $\Sigma = (G, \mathcal{H}, U)$. In case $\mathcal{H}_p$ is taken to be the complex numbers $\mathbb{C}$ for each path-connected component $p \in P$, we denote the associated coefficient matrices $I_{\Sigma,e}$ and the structure matrix $Z_\Sigma(z)$ simply as $I_{G,e}$ and $Z_G(z)$. Thus $I_{G,e}\colon \oplus_{r\in R}\mathbb{C} \to \oplus_{s\in S}\mathbb{C}$ with matrix entries

$$[I_{G,e}]_{s,r} = \begin{cases} 1 & \text{if } s=\mathbf{s}(e) \text{ and } r=\mathbf{r}(e), \\ 0 & \text{otherwise} \end{cases}$$

and $Z_G(z) = \sum_{e\in E} I_{G,e}z_e$. We then define a class $\mathcal{B}_G\mathcal{L}(\mathcal{K})$ of tuples $\delta = (\delta_e\colon e\in E)$ of bounded, linear operators on the Hilbert space $\mathcal{K}$ (the G-unit ball of $\mathcal{L}(\mathcal{K})^{n_E}$) (where n_E denotes the number of edges in the graph G) by

$$\mathcal{B}_G\mathcal{L}(\mathcal{K}) = \{\delta = (\delta_e\colon e\in E)\colon \delta_e\in\mathcal{L}(\mathcal{K}) \text{ for } e\in E \text{ and } \|Z_G(\delta)\| < 1\}. \tag{4.11}$$

It is easy to see that $\|Z_G(\delta)\| = \|Z_\Sigma(\delta)\|$ whenever $\Sigma = (G, \mathcal{H}, U)$ is a SNMLS with structure graph G; thus $\|Z_\Sigma(\delta)\| < 1$ for all $\delta \in \mathcal{B}_G\mathcal{L}(\mathcal{K})$.

Theorem 4.2. *Suppose that $T(z) = D + C(I - Z_\Sigma(z)A)^{-1}Z_\Sigma(z)B$ is the transfer function of a dissipative SNMLS $\Sigma = (G, \mathcal{H}, U)$ and that $\mathcal{K}$ is some other separable Hilbert space. Then for any collection $\delta = (\delta_e\colon e\in E)$ of operators in $\mathcal{B}_G\mathcal{L}(\mathcal{K})$, $T(\delta)$ as defined in (4.9) is a well-defined contraction operator (with the limit of the partial sums in (4.9) existing in the operator-norm topology) from $\mathcal{U}\otimes\mathcal{K}$ to $\mathcal{Y}\otimes\mathcal{K}$ ($\|T(\delta)\| \leq 1$), and can alternatively be expressed as*

$$T(\delta) = (D\otimes I_\mathcal{K}) + (C\otimes I_\mathcal{K})(I - Z_\Sigma(\delta)(A\otimes I_\mathcal{K}))^{-1}Z_\Sigma(\delta)(B\otimes I_\mathcal{K}). \tag{4.12}$$

Proof. A general fact is that, if

$$U' = \begin{bmatrix} A' & B' \\ C' & D' \end{bmatrix}\colon \begin{bmatrix} \mathcal{H}' \\ \mathcal{U}' \end{bmatrix} \to \begin{bmatrix} \mathcal{H}'' \\ \mathcal{Y}' \end{bmatrix}$$

is contractive and $\Delta\colon \mathcal{H}'' \to \mathcal{H}'$ is a strict contraction, then the upper feedback connection

$$\mathcal{F}_u[U', \Delta]\colon \mathcal{U}' \to \mathcal{Y}'$$

defined implicitly by

$$\mathcal{F}_u[U', \Delta]\colon u' \mapsto y' \text{ if there exist } h' \in \mathcal{H}' \text{ and } h'' \in \mathcal{H}''$$

$$\text{so that } \begin{bmatrix} A' & B' \\ C' & D' \end{bmatrix}\begin{bmatrix} h' \\ u' \end{bmatrix} = \begin{bmatrix} h'' \\ y' \end{bmatrix} \text{ and } \Delta h'' = h'$$

is well defined, moreover is contractive ($\|\mathcal{F}_u[U', \Delta]\| \leq 1$), and is given explicitly by the linear-fractional formula

$$\mathcal{F}_u[U', \Delta] = D' + C'(I - \Delta A')^{-1}\Delta B'. \tag{4.13}$$

This fact can be found in any of a number of places where linear-fractional transformations are discussed, e.g., in [57] where there is a comprehensive treatment for the control-theory context, or in Section 3 of [7] where there is a concise summary of what we are using here.

Now suppose that $\Sigma = (G, \mathcal{H}, U)$ is a dissipative SNMLS and suppose that $\delta = (\delta_e : e \in E)$ is an operator-tuple in $\mathcal{B}_G\mathcal{L}(\mathcal{K})$. We shall use a different font $\boldsymbol{\delta}_{s,s'}$ for the Kronecker delta function

$$\boldsymbol{\delta}_{s,s'} = \begin{cases} 1 & \text{if } s = s', \\ 0 & \text{otherwise} \end{cases} \tag{4.14}$$

for which we shall have use on occasion in the sequel. We apply the linear-fractional construction (4.13) to the case

$$U' = \begin{bmatrix} A \otimes I_\mathcal{K} & B \otimes I_\mathcal{K} \\ C \otimes I_\mathcal{K} & D \otimes I_\mathcal{K} \end{bmatrix} : \begin{bmatrix} (\oplus_{s \in S} \mathcal{H}_{[s]}) \otimes \mathcal{K} \\ \mathcal{U} \otimes \mathcal{K} \end{bmatrix} \to \begin{bmatrix} (\oplus_{r \in R} \mathcal{H}_{[r]}) \otimes \mathcal{K} \\ \mathcal{Y} \otimes \mathcal{K} \end{bmatrix},$$

$$\Delta = Z_\Sigma(\delta) : \left(\oplus_{r \in R} \mathcal{H}_{[r]}\right) \otimes \mathcal{K} \to \left(\oplus_{s \in S} \mathcal{H}_{[s]}\right) \otimes \mathcal{K}.$$

Note that U' is then contractive since by assumption U is contractive and that $\|\Delta\| < 1$ since $\delta \in \mathcal{B}_G\mathcal{L}(\mathcal{K})$. Hence it follows that

$$\mathcal{F}_u[U', Z_\Sigma(\delta)] = (D \otimes I_\mathcal{K}) + (C \otimes I_\mathcal{K})(I - Z_\Sigma(\delta)(A \otimes I_\mathcal{K}))^{-1} Z_\Sigma(\delta)(B \otimes I_\mathcal{K})$$

is a well-defined contraction operator from $\mathcal{U} \otimes \mathcal{K}$ into $\mathcal{Y} \otimes \mathcal{K}$.

It remains to show that $\mathcal{F}_u[U', Z_\Sigma(\delta)] = T_\Sigma(\delta)$. Verification of this identity draws upon repeated use of the product rule for tensor products

$$(A \otimes B)(C \otimes D) = (AC) \otimes (BD)$$

as we now show. Since $\|Z_\Sigma(\delta)(A \otimes I_\mathcal{K})\| \leq \|Z_\Sigma(\delta)\| \|A\| < 1$, it follows that the inverse of $I - Z_\Sigma(\delta)(A \otimes I_\mathcal{K})$ is given by the Neumann expansion

$$(I - Z_\Sigma(\delta)(A \otimes I_\mathcal{K}))^{-1} = \sum_{N=0}^{\infty} [Z_\Sigma(\delta)(A \otimes I_\mathcal{K})]^N.$$

From this we see that the (s, s') matrix entry of

$$(I - Z_\Sigma(\delta)(A \otimes I_\mathcal{K}))^{-1} : \oplus_{s \in S} \mathcal{H}_{[s]} \to \oplus_{s \in S} \mathcal{H}_{[s]}$$

198   J.A. Ball, G. Groenewald and T. Malakorn

is given by

$$[(I - Z_\Sigma(\delta)(A \otimes I_\mathcal{K}))^{-1}]_{s,s'} = \boldsymbol{\delta}_{s,s'} I_{\mathcal{H}_{[s]}} + \sum_{N=1}^{\infty} \sum_{e_N,\ldots,e_1 \in E:\, \mathbf{s}(e_N)=s}$$

$$(A_{\mathbf{r}(e_N),\mathbf{s}(e_{N-1})} \otimes \delta_{e_N}) \cdots (A_{\mathbf{r}(e_2),\mathbf{s}(e_1)} \otimes \delta_{e_2})(A_{\mathbf{r}(e_1),s'} \otimes \delta_{e_1})$$

$$= \boldsymbol{\delta}_{s,s'} I_{\mathcal{H}_{[s]}} + \sum_{N=1}^{\infty} \sum_{e_N,\ldots,e_1 \in E:\, \mathbf{s}(e_N)=s} (A_{\mathbf{r}(e_N),\mathbf{s}(e_{N-1})} \cdots A_{\mathbf{r}(e_2),\mathbf{s}(e_1)} A_{\mathbf{r}(e_1),s'}) \otimes$$

$$\otimes (\delta_{e_N} \cdots \delta_{e_2} \delta_{e_1}) \tag{4.15}$$

Note next that $C \otimes I_\mathcal{K} \colon \left(\oplus_{s \in S} \mathcal{H}_{[s]}\right) \otimes \mathcal{K} \to \mathcal{Y} \otimes \mathcal{K}$ has row matrix representation

$$C \otimes I_\mathcal{K} = \mathrm{row}_{s \in S}[C_s \otimes I_\mathcal{K}] \tag{4.16}$$

while $Z_\Sigma(\delta)(B \otimes I_\mathcal{K}) \colon \mathcal{U} \otimes \mathcal{K} \to \left(\oplus_{s' \in S} \mathcal{H}_{[s']}\right) \otimes \mathcal{K}$ has column matrix representation

$$Z_\Sigma(\delta)(B \otimes I_\mathcal{K}) = \mathrm{col}_{s' \in S}\left[\sum_{e:\, \mathbf{s}(e)=s'} B_{\mathbf{r}(e)} \otimes \delta_e \right]. \tag{4.17}$$

Using (4.15), (4.16) and (4.17), we then compute

$$(C \otimes I_\mathcal{K})(I - Z_\Sigma(\delta)(A \otimes I_\mathcal{K}))^{-1} Z_\Sigma(\delta)(B \otimes I_\mathcal{K})$$

$$= \sum_{s,s' \in S} \sum_{e \in E:\, \mathbf{s}(e)=s'} (C_s \otimes I_\mathcal{K})\left[(I - Z_\Sigma(\delta)(A \otimes I_\mathcal{K}))^{-1}\right]_{s,s'} (B_{\mathbf{r}(e)} \otimes \delta_e)$$

$$= X_1 + X_2.$$

where we have set

$$X_1 = \sum_{s \in S} \sum_{e:\, \mathbf{s}(e)=s} (C_s \otimes I_\mathcal{K})(B_{\mathbf{r}(e)} \otimes \delta_e) \tag{4.18}$$

$$X_2 = \sum_{s,s' \in S} \sum_{e:\, \mathbf{s}(e)=s'} \sum_{N=1}^{\infty} \sum_{e_N,\ldots,e_1 \in E:\, \mathbf{s}(e_N)=s} (C_s \otimes I_\mathcal{K}) \cdot$$

$$\cdot (A_{\mathbf{r}(e_N),\mathbf{s}(e_{N-1})} \cdots A_{\mathbf{r}(e_1),s'} \otimes \delta_{e_N} \cdots \delta_{e_2} \delta_{e_1})(B_{\mathbf{r}(e)} \otimes \delta_e). \tag{4.19}$$

The first term X_1 simplifies to

$$X_1 = \sum_{s \in S} \sum_{e \in E:\, \mathbf{s}(e)=s} (C_s B_{\mathbf{r}(e)}) \otimes \delta_e$$

$$= \sum_{e \in E} T_e \otimes \delta_e \tag{4.20}$$

while the second term X_2 can be simplified to

$$X_2 = \sum_{s,s'\in S} \sum_{e:\ \mathbf{s}(e)=s'} \sum_{N=1}^{\infty} \sum_{e_N,\ldots,e_1\in E:\ \mathbf{s}(e_N)=s} (C_s A_{\mathbf{r}(e_N),\mathbf{s}(e_{N-1})} \cdots$$

$$\cdots A_{\mathbf{r}(e_2),\mathbf{s}(e_1)} A_{\mathbf{r}(e_1),s'} B_{\mathbf{r}(e)}) \otimes (\delta_{e_N} \cdots \delta_{e_2}\delta_{e_1}\delta_e)$$

$$= \sum_{v\in E:\ |v|\geq 2} T_v \otimes \delta^v. \tag{4.21}$$

Combining (4.20) and (4.21) along with the identity $T_\emptyset = D$ immediately gives us the identity (4.12) as wanted. This completes the proof of Theorem 4.2. $\qquad\square$

5. Conservative SNMLS-realization of formal power series in the class $\mathcal{SA}_G(\mathcal{U},\mathcal{Y})$

Let G be a fixed admissible graph with source-vertex set S, range-vertex set R and edge set E. Theorem 4.2 suggests that we consider the class of all formal power series $T(z) = \sum_{v\in\mathcal{F}_E} T_v z^v$ having the property in the conclusion of Theorem 4.2. We view this class as a noncommutative analogue of the Schur-Agler class studied in a series of papers (see, e.g., [1, 17, 15, 6, 7, 12]).

Definition 5.1. We say that $T(z)$ is in the noncommutative Schur-Agler class $\mathcal{SA}_G(\mathcal{U},\mathcal{Y})$ (for a given admissible graph G) if, for each Hilbert space $\mathcal{K}$ and each $\delta = (\delta_e\colon e\in E) \in \mathcal{B}_G\mathcal{L}(\mathcal{K})$, the limit $T(\delta) = \lim_{N\to\infty}\sum_{v\in\mathcal{F}_d:\ |v|\leq N} T_v \otimes \delta^v$ exists (in the operator-norm topology) and defines an operator

$$T(\delta)\colon \mathcal{U}\otimes\mathcal{K} \to \mathcal{Y}\otimes\mathcal{K}$$

which is contractive

$$\|T(\delta)\| \leq 1. \tag{5.1}$$

Remark 5.2. Alpay and Kalyuzhnyĭ-Verbovetzkiĭ in [5] have shown that a given formal power series

$$T(z) = \sum_{v\in\mathcal{F}_E} T_v z^v \in \mathcal{L}(\mathcal{U},\mathcal{Y})\langle\langle z\rangle\rangle$$

belongs to the noncommutative Schur-Agler class $\mathcal{SA}_G(\mathcal{U},\mathcal{Y})$ if and only if (5.1) holds for each $\delta = (\delta_e\colon e\in E) \in \mathcal{B}_G\mathcal{L}(\mathbb{C}^N)$ for each finite $N = 1,2,3,\ldots$. The proof there is done explicitly only for the case where each component of G consists of a single source vertex and a single range vertex (the Givone-Roesser case); we expect that this result continues to hold for the case of a general admissible graph G.

Our next goal is a converse to Theorem 4.2 (see Theorem 5.3 below). For the statement we shall need some additional notation and terminology. We let $z' = (z'_e\colon e\in E)$ be a second system of noncommuting indeterminates; while $z_e z_{e'} \neq z_{e'} z_e$ and $z'_e z'_{e'} \neq z'_{e'} z'_e$ unless $e = e'$, we will use the convention that

$z_e z'_{e'} = z'_{e'} z_e$ for all $e, e' \in E$. We also shall need the convention (2.13) to give meaning to expressions of the form

$$z_e^{\prime -1} z^v z^{\prime v'} z_e^{-1} = (z^v z_e^{-1}) \cdot (z_e^{\prime -1} z^{\prime v'}) = z^{v e^{-1}} z^{\prime e^{-1} v'}.$$

For $H(z) = \sum_{v \in \mathcal{F}_E} H_v z^v$, we will use the convention that

$$H(z)^* = \left(\sum_{v \in \mathcal{F}_E} H_v z^v \right)^* := \sum_{v \in \mathcal{F}_E} H_v^* z^{v^\top} = \sum_{v \in \mathcal{F}_E} H_{v^\top}^* z^v.$$

In general let us say that a formal power series $K(z, z') = \sum_{v, v' \in \mathcal{F}_E} [K]_{v,v'} z^v z^{\prime v'}$ with coefficients $[K]_{v,v'}$ equal to operators on a Hilbert space $\mathcal{X}$ (so $K(z, z') \in \mathcal{L}(\mathcal{X})\langle\langle z, z' \rangle\rangle$) is *positive-definite* provided that

$$\sum_{v, v' \in \mathcal{F}_E} \langle [K]_{v,v'} y_{v'^\top}, y_v \rangle_{\mathcal{X}} \geq 0 \tag{5.2}$$

for all choices of $y_v \in \mathcal{X}$ with $y_v = 0$ for all but finitely many $v \in \mathcal{F}_E$. By the standard results concerning reproducing kernel Hilbert spaces ([9]), it is known that condition (5.2) is equivalent to the existence of an auxiliary Hilbert space $\mathcal{H}'$ and operators $H_v \in \mathcal{L}(\mathcal{H}', \mathcal{X})$ for each $v \in \mathcal{F}_E$ so that $[K]_{v,v'} = H_v H_{v'^\top}^*$. Equivalently we therefore have: $K(z, z') \in \mathcal{L}(\mathcal{X})\langle\langle z, z' \rangle\rangle$ *is positive-definite if and only if there exists an auxiliary Hilbert space $\mathcal{H}'$ and a formal power series $H(z) \in \mathcal{L}(\mathcal{H}', \mathcal{X})\langle\langle z \rangle\rangle$ so that $K(z, z') = H(z) H(z')^*$.* We shall be particularly interested in this concept for the case where $\mathcal{X} = \oplus_{s \in S} \mathcal{Y}$. We therefore consider a formal power series $K(z, z')$ of the form

$$K(z, z') = [K_{s,s'}(z, z')]_{s,s' \in S} \in \mathcal{L}(\oplus_{s \in S} \mathcal{Y})\langle\langle z, z' \rangle\rangle.$$

Such a $K(z, z')$ therefore is *positive-definite* if

$$\sum_{s, s' \in S} \sum_{v, v' \in \mathcal{F}_E} \langle [K_{s,s'}]_{v,v'} y_{s', v'^\top}, y_{s,v} \rangle_{\mathcal{Y}} \geq 0 \tag{5.3}$$

for all choices of $y_{s,v} \in \mathcal{Y}$ for $s \in S$ and $v \in \mathcal{F}_E$ with $y_{s,v} = 0$ for all but finitely many such s, v, or equivalently, if and only if there exist an auxiliary Hilbert space $\mathcal{H}'$ and formal power series $H_s(z) \in \mathcal{L}(\mathcal{H}', \mathcal{Y})$ for each $s \in S$ so that $K_{s,s'}(z, z') = H_s(z) H_{s'}(z')^*$.

Theorem 5.3. *Let $T(z) = \sum_{v \in \mathcal{F}_E} T_v z^v$ be a formal power series in noncommuting indeterminates $z = (z_e : e \in E)$ indexed by the edge set E of the admissible graph G with coefficients $T_v \in \mathcal{L}(\mathcal{U}, \mathcal{Y})$ for two Hilbert spaces $\mathcal{U}$ and $\mathcal{Y}$. Then the following conditions are equivalent:*

1. *$T(z)$ is in the noncommutative Schur-Agler class $\mathcal{SA}_G(\mathcal{U}, \mathcal{Y})$.*
2. *There exists a positive-definite formal power series*

$$K(z, z') = [K_{s,s'}(z, z')]_{s,s' \in S} \in \mathcal{L}(\oplus_{s \in S} \mathcal{Y})\langle\langle z, z' \rangle\rangle$$

so that

$$I_{\mathcal{Y}} - T(z)T(z')^*$$
$$= \sum_{s \in S} K_{s,s}(z, z') - \sum_{r \in R} \sum_{s,s' \in S:\ [s]=[s']=[r]} z'_{e_{s',r}} K_{s,s'}(z, z') z_{e_{s,r}}. \qquad (5.4)$$

3. *There exists a collection of Hilbert spaces $\mathcal{H}' = \{\mathcal{H}'_p\colon p \in P\}$ (where P is the set of path-connected components of the admissible graph G) and a formal power series $H(z) = \sum_{v \in \mathcal{F}_E} H_v z^v$ with coefficients $H_v \in \mathcal{L}(\oplus_{s \in S}\mathcal{H}'_{[s]}, \mathcal{Y})$ so that we have the* noncommutative Agler decomposition

$$I - T(z)T(z')^* = H(z)\left(I - Z_{G,\mathcal{H}'}(z)Z_{G,\mathcal{H}'}(z')^*\right)H(z')^* \qquad (5.5)$$

where we have set $Z_{G,\mathcal{H}'}(z) = \sum_{e \in E} I_{G,\mathcal{H}';e} z_e$ with coefficients $I_{G,\mathcal{H}';e}$ equal to operators acting from $\oplus_{r \in R}\mathcal{H}'_{[r]}$ to $\oplus_{s \in S}\mathcal{H}'_{[s]}$ determined from matrix entries $[I_{G,\mathcal{H}';e}]_{s,r}$ given by

$$[I_{G,\mathcal{H}';e}]_{s,r} = \begin{cases} I_{\mathcal{H}'_{[s]}} = I_{\mathcal{H}'_{[r]}} & \text{if } s = \mathbf{s}(e) \text{ and } r = \mathbf{r}(e), \\ 0 & \text{otherwise.} \end{cases} \qquad (5.6)$$

4. *There is a conservative SNMLS $\Sigma = (G, \mathcal{H}, U)$ with structure graph equal to the given admissible graph G so that $T(z) = T_\Sigma(z)$, i.e., so that*

$$T(z) = D + C(I - Z_\Sigma(z)A)^{-1}Z_\Sigma(z)B$$

where $U = \begin{bmatrix} A & B \\ C & D \end{bmatrix}\colon \begin{bmatrix} \oplus_{s \in S}\mathcal{H}_{[s]} \\ \mathcal{U} \end{bmatrix} \to \begin{bmatrix} \oplus_{r \in R}\mathcal{H}_{[r]} \\ \mathcal{Y} \end{bmatrix}$ is unitary and where $Z_\Sigma(z) = \sum_{e \in E} I_{\Sigma,e} z_e$ with $I_{\Sigma,e}$ as in (2.8).

Remark 5.4. We give the name *Agler decomposition* to an identity of the form (5.4) or (5.5) since representations of this type to our knowledge originate in the work of Agler (see [1]) in the context of the commutative polydisk.

Remark 5.5. We note that the paper [5] of Alpay and Kalyuzhnyĭ-Verbovetzkiĭ gives a uniqueness result for conservative realizations of rational inner formal power series in the Givone-Roesser case. We leave a systematic development of the uniqueness theory for realizations as in part (4) of Theorem 5.3 to another occasion.

Proof. The proof breaks up into several implications which need to be shown:

(2) $\Longleftrightarrow$ **(3):** Suppose that $K(z, z') = [K_{s,s'}(z, z')]_{s,s' \in S} \in \mathcal{L}(\oplus_{s \in S}\mathcal{Y})\langle\langle z, z'\rangle\rangle$ is positive definite. Then, by the remarks preceding the statement of the Theorem, $K_{s,s'}(z, z')$ has a factorization

$$K_{s,s'}(z, z') = H_s(z)H_{s'}(z')^*$$

for a formal power series $H_s(z) \in \mathcal{L}(\mathcal{H}'_{[s]}, \mathcal{Y})$ for $s \in S$.

Then, if we set $H(z) = \mathrm{row}_{s \in S}\, H_s(z) \in \mathcal{L}(\oplus_{s \in S}\mathcal{H}'_{[s]}, \mathcal{Y})\langle\langle z \rangle\rangle$, then (5.4) assumes the form

$$I - T(z)T(z')^*$$

$$= \sum_{s \in S} H_s(z)H_s(z')^* - \sum_{r \in R} \sum_{s,s' \in S\,:\,[s]=[s']=[r]} H_s(z) \cdot (1 - z_{e_s,r} z'_{e_{s'},r})I_{\mathcal{H}'} \cdot H_{s'}(z')^*$$

$$= H(z)\left(I_{\oplus_{s \in S}\mathcal{H}'_{[s]}} - Z_{G,\mathcal{H}'}(z)Z_{G,\mathcal{H}'}(z')^*\right)H(z')^*.$$

from which (5.5) follows. Conversely, if

$$H(z) = \mathrm{row}_{s \in S}\, H_s(z) \in \mathcal{L}(\oplus_{s \in S}\mathcal{H}'_{[s]}, \mathcal{Y})\langle\langle z \rangle\rangle$$

is as in (5.5), we may embed each $\mathcal{H}'_p$ into a common Hilbert space $\mathcal{H}'$ and without loss of generality assume that $\mathcal{H}'_p = \mathcal{H}'$ for each $p \in P$. We then set $K_{s,s'}(z,z') = H_s(z)H_{s'}(z')^* \in \mathcal{L}(\mathcal{Y})\langle\langle z, z' \rangle\rangle$ and $K(z,z') = [K_{s,s'}(z,z')]_{s,s' \in S} \in \mathcal{L}(\oplus_{s \in S}\mathcal{Y})\langle\langle z, z' \rangle\rangle$. Then by the factored form of

$$K(z,z') = \left(\mathrm{col}_{s \in S}\, H_s(z)\right)\left(\mathrm{col}_{s' \in S}\, H_{s'}(z')\right)^*$$

we see that $K(z,z')$ is positive-definite. Reversal of the steps above then shows that $K(z,z')$ satisfies (5.4). In this way we see that (2) is equivalent to (3) in Theorem 5.3.

(4) $\Longrightarrow$ (1): Since any conservative system is also dissipative, this follows from Theorem 4.2.

(1) $\Longrightarrow$ (2) or (3): As a first case we assume that $\dim \mathcal{Y} < \infty$. Let $\mathcal{X}$ denote the linear space $\mathcal{L}(\mathcal{Y})\langle\langle z, z' \rangle\rangle$ of all formal power series $\varphi(z,z') = \sum_{v,v' \in \mathcal{F}_E} \varphi_{v,v'} z^v z'^{v'}$ in the sets of noncommuting indeterminates $z = (z_e : e \in E)$ and $z' = (z'_e : e \in E)$ (but where $z_e z'_{e'} = z'_{e'} z_e$ for each $e, e' \in E$) with coefficients $\varphi_{v,v'}$ in the space of bounded linear operators $\mathcal{L}(\mathcal{Y})$ on the Hilbert space $\mathcal{Y}$. We define a sequence of increasing seminorms $\|\cdot\|_N$ on $\mathcal{L}(\mathcal{Y})\langle\langle z, z' \rangle\rangle$ according to the rule

$$\|\varphi(z,z')\|_N = \sup_{v,v' \in \mathcal{F}_E\,:\,|v|,|v'| \leq N} \|\varphi_{v,v'}\|. \tag{5.7}$$

Then $\mathcal{X}$ is a locally convex topological vector space in the topology induced by these seminorms and this topology is metrizable. Let $\mathcal{C}$ be the set of all formal power series $\varphi(z,z') = \sum_{v,v' \in \mathcal{F}_E} \varphi_{v,v'} z^v z'^{v'}$ in $\mathcal{L}(\mathcal{Y})\langle\langle z, z' \rangle\rangle$ such that

$$\varphi(z,z') = H(z)(I - Z_{G,\mathcal{H}'}(z)Z_{G,\mathcal{H}'}(z')^*)H(z')^* \tag{5.8}$$

for some collection of Hilbert spaces $\mathcal{H}' = \{\mathcal{H}'_p : p \in P\}$ indexed by the path-connected components P of G and for some formal power series

$$H(z) = \sum_{v \in \mathcal{F}_E} H_v z^v$$

with coefficients $H_v \in \mathcal{L}(\oplus_{s \in S}\mathcal{H}'_{[s]}, \mathcal{Y})$, where $Z_{G,\mathcal{H}'}(z) = \sum_{e \in E} I_{G,\mathcal{H}';e} z_e$ is defined as in (5.6). From the equivalence (2) $\Longleftrightarrow$ (3), we see that an equivalent condition for membership of φ in $\mathcal{C}$ is the existence of a positive-definite formal power series $K(z, z') = [K_{s,s'}(z, z')]_{s,s' \in S} \in \mathcal{L}(\oplus_{s \in S} \mathcal{Y})\langle\langle z, z' \rangle\rangle$ so that

$$\varphi(z, z') = \sum_{s \in S} K_{s,s}(z, z') - \sum_{r \in R} \sum_{s,s' \in S:\, [s]=[s']=[r]} z'_{e_{s',r}} K_{s,s'}(z, z') z_{e_s,r}. \tag{5.9}$$

When working with the decomposition (5.8), we may assume without loss of generality that $\mathcal{H}'_p$ is a fixed separable infinite-dimensional Hilbert space independent of the choice of $\varphi \in \mathcal{C}$ and of the particular representation (5.8) for a given $\varphi \in \mathcal{C}$. It is easily checked that $\mathcal{C}$ is closed under sums and multiplication by nonnegative scalars, i.e., that $\mathcal{C}$ is a cone in $\mathcal{X}$. We need to establish a few preliminary facts concerning $\mathcal{C}$.

Lemma 5.6. *Any positive-definite formal power series $\varphi(z, z')$ in $\mathcal{L}(\mathcal{Y})\langle\langle z, z' \rangle\rangle$ is also in $\mathcal{C}$.*

Proof. As φ is a positive kernel, we know that we can factor φ as

$$\varphi(z, z') = H(z)H(z')^*$$

for some $H(z) \in \mathcal{L}(\mathcal{K}, \mathcal{Y})\langle\langle z \rangle\rangle$ for some auxiliary Hilbert space $\mathcal{K}$. We must produce a formal power series $H'(z) \in \mathcal{L}(\oplus_{s \in S} \mathcal{H}'_{[s]}, \mathcal{Y})\langle\langle z \rangle\rangle$ so that

$$H(z)H(z')^* = H'(z)[I_{\oplus_{s \in S} \mathcal{H}'_{[s]}} - Z_{G,\mathcal{H}'}(z)Z_{G,\mathcal{H}'}(z')^*]H'(z')^*.$$

Let $s_0 \in S$ be any fixed choice of particular source vertex. Without loss of generality we may assume $\mathcal{H}'$ is presented in the form

$$\mathcal{H}' = \ell^2(\mathcal{F}_{E_0}, \mathcal{K}) \qquad \text{where} \qquad E_0 = \{e \in E \colon \mathbf{s}(e) = s_0\}.$$

We take $H'(z) \in \mathcal{L}(\oplus_{s \in S} \mathcal{H}'_{[s]}, \mathcal{Y})\langle\langle z \rangle\rangle$ to be of the form

$$H'(z) = H(z)K'(z) \text{ where } K'(z) = \text{row}_{s \in S}\, K'_s(z) \text{ with } K'_s(z) \in \mathcal{L}(\mathcal{H}'_{[s]}, \mathcal{K})\langle\langle z \rangle\rangle$$

where $K'_s(z)$ is given by

$$K'_s(z) = \begin{cases} \text{row}_{v \in \mathcal{F}_{E_0}}\, z^v I_{\mathcal{K}} & \text{if } s = s_0, \\ 0 & \text{otherwise.} \end{cases}$$

Then we check

$$H'(z)\left[I_{\oplus_{s \in S} \mathcal{H}'_{[s]}} - Z_{G,\mathcal{H}'}(z)Z_{G,\mathcal{H}'}(z')^*\right]H'(z')^*$$

$$= H(z)K'_{s_0}(z)\left[\left(1 - \sum_{e \in E_0} z_e z'_e\right) I_{\mathcal{H}'}\right] K'_{s_0}(z')^* H(z')^*$$

$$= H(z)\left[\left(\sum_{v \in \mathcal{F}_{E_0}} z^v z'^v I_{\mathcal{K}} - \sum_{v \in \mathcal{F}_{E_0} \setminus \{\emptyset\}} z^v z'^v\right) I_{\mathcal{K}}\right] H(z')^*$$

$$= H(z)H(z')^*$$

as wanted, and Lemma 5.6 follows. $\qquad \square$

We shall need to approximate the cone $\mathcal{C}$ by the cone $\mathcal{C}_\varepsilon$ (where $\varepsilon > 0$) defined as the set of all $\varphi \in \mathcal{L}(\mathcal{Y})\langle\langle z, z'\rangle\rangle$ having a representation

$$\varphi(z, z') = H(z)\left(I - (1+\varepsilon)^2 Z_{G,\mathcal{H}'}(z)Z_{G,\mathcal{H}'}(z')^*\right)H(z')^*$$
$$+ \sum_{e \in E} \gamma_e(z)(1 - \varepsilon^2 z_e z'_e)\gamma_e(z')^* \tag{5.10}$$

for some $H(z) \in \mathcal{L}(\oplus_{s \in S}\mathcal{H}'_{[s]}, \mathcal{Y})\langle\langle z\rangle\rangle$ and some $\gamma_e(z) \in \mathcal{L}(\mathcal{H}', \mathcal{Y})\langle\langle z\rangle\rangle$ for $e \in E$. Equivalently, just as in the proof of $(2) \iff (3)$ (Step 1 above), we see that, in terms of positive-definite formal power series, $\mathcal{C}_\varepsilon$ can be defined as the set of all $\varphi \in \mathcal{L}(\mathcal{Y})\langle\langle z, z'\rangle\rangle$ having a representation

$$\varphi(z, z') = \sum_{s \in S} K_{s,s}(z, z') - (1+\varepsilon)^2 \sum_{r \in R} \sum_{s,s' \in S:\ [s]=[s']=[r]} z'_{e_{s',r}} K_{s,s'}(z, z') z_{e_{s,r}}$$
$$+ \sum_{e \in E} \Gamma_e(z, z') - \varepsilon^2 \sum_{e \in E} z'_e \Gamma_e(z, z') z_e \tag{5.11}$$

for some positive-definite formal power series

$$K(z, z') = [K_{s,s'}(z, z')]_{s,s' \in S} \in \mathcal{L}(\oplus_{s \in S}\mathcal{Y})\langle\langle z, z'\rangle\rangle$$

and some positive-definite formal power series $\Gamma_e(z, z')$ in $\mathcal{L}(\mathcal{Y})\langle\langle z, z'\rangle\rangle$ for each $e \in E$.

Lemma 5.7. *Assume that $\varphi \in \mathcal{L}(\mathcal{Y})\langle\langle z, z'\rangle\rangle$ is in the cone $\mathcal{C}_\varepsilon$ for all $\varepsilon > 0$ sufficiently small. Then $\varphi \in \mathcal{C}$, i.e., φ has a representation* (5.8) *or equivalently* (5.9).

Proof. The assumption is that, for all $\varepsilon > 0$ sufficiently small, there is a positive-definite formal power series $K_\varepsilon(z, z') = [K_{\varepsilon,s,s'}(z, z')]_{s,s' \in S}$ in $\mathcal{L}(\oplus_{s \in S}\mathcal{Y})\langle\langle z, z'\rangle\rangle$ and a positive-definite formal power series $\Gamma_{\varepsilon,e}(z, z')$ in $\mathcal{L}(\mathcal{Y})\langle\langle z, z'\rangle\rangle$ so that (5.11) holds (with $K_\varepsilon(z, z')$ in place of $K(z, z')$ and $\Gamma_{\varepsilon,e}$ in place of Γ_e) for each $e \in E$. In particular for the $(\emptyset, \emptyset)$-coefficient we get

$$\varphi_{\emptyset,\emptyset} = \sum_{s \in S}[K_{\varepsilon,s,s}]_{\emptyset,\emptyset} + \sum_{e \in E}[\Gamma_{\varepsilon,e}]_{\emptyset,\emptyset}.$$

Hence $[K_{\varepsilon,s,s}]_{\emptyset,\emptyset}$ and $[\Gamma_{\varepsilon,e}]_{\emptyset,\emptyset}$ are all uniformly bounded as ϵ tends to 0. By using the positive-definiteness of $K_\varepsilon(z, z')$, we see that $\|[K_{\varepsilon,s,s'}]_{\emptyset,\emptyset}\|$ is uniformly bounded as ε tends to zero for all $s, s' \in S$ as well. More generally, computation of the (v, v') coefficient of φ from (5.11) yields

$$\varphi_{v,v'} = \sum_{s \in S}[K_{\varepsilon,s,s}]_{v,v'} - (1+\varepsilon)^2 \sum_{r \in R} \sum_{s,s' \in S:\ [s]=[s']=[r]} [K_{\varepsilon,s,s'}]_{ve_{s,r}^{-1},e_{s',r}^{-1}v'}$$
$$+ \sum_{e \in E}[\Gamma_{\varepsilon,e}]_{v,v'} - \varepsilon^2 \sum_{e \in E}[\Gamma_{\varepsilon,e}]_{ve^{-1},e^{-1}v'}. \tag{5.12}$$

Here we are using (2.3) to define words of the form ve^{-1} or $e^{-1}v$ with the convention that the coefficient is taken to be equal to zero if any of its indices is an undefined word. Inductively assume that there is a uniform bound on $\|[K_{\varepsilon,s,s'}]_{w,w'}\|$ for all words $w, w' \in \mathcal{F}_E$ having length at most N. From (5.12) we can then see that $\|[K_{\varepsilon,s,s}]_{v,v}\|$ is uniformly bounded for all $v \in \mathcal{F}_E$ with $|v| = N + 1$. Using the positive-definiteness of $K(z, z')$, we then see that this leads to a uniform bound for $\|[K_{\varepsilon,s,s'}]_{v,v'}\|$ for all $v, v' \in \mathcal{F}_E$ of length at most $N + 1$ as ε tends to zero. A similar inductive argument gives that $\|[\Gamma_{\varepsilon,e}]_{v,v'}\|$ is uniformly bounded as ε tends to zero for all words v, v' with length $|v|, |v'|$ at most some $N < \infty$.

Since we are assuming that $\mathcal{Y}$ is finite-dimensional, it follows that bounded subsets of $\mathcal{L}(\mathcal{Y})$ are precompact in the operator-norm topology. By this fact combined with a Cantor diagonalization procedure, there exists a sequence of numbers $\varepsilon_n > 0$ tending to zero such that the limits

$$\lim_{n \to \infty} [K_{\varepsilon_n,s,s'}]_{v,v'} = [K_{s,s'}]_{v,v'}, \qquad \lim_{n \to \infty} [\Gamma_{\varepsilon_n,e}]_{v,v'} = [\Gamma_e]_{v,v'}$$

all exist in the operator-norm topology of $\mathcal{L}(\mathcal{Y})$. We then take limits in (5.12) to deduce that

$$\varphi_{v,v'} = \sum_{s \in S} [K_{s,s}]_{v,v'} - \sum_{r \in R} \sum_{s,s' \in S:\, [s]=[s']=[r]} [K_{s,s'}]_{ve_{s,r}^{-1},e_{s',r}^{-1}v'} + \sum_{e \in E} [\Gamma_e]_{v,v'} \quad (5.13)$$

and hence

$$\varphi(z, z') = \sum_{s \in S} K_{s,s}(z, z') - \sum_{r \in R} \sum_{s,s' \in S:\, [s]=[s']=[r]} z'_{e_{s',r}} K_{s,s'}(z, z') z_{e_{s,r}} + \sum_{e \in E} \Gamma_e(z, z')$$
$$(5.14)$$

with $K_{s,s'}(z, z')$ and $\Gamma_e(z, z')$ given by

$$K_{s,s'}(z, z') = \sum_{v,v' \in \mathcal{F}_E} [K_{s,s'}]_{v,v'} z^v z'^{v'}, \qquad \Gamma_e(z, z') = \sum_{v,v' \in \mathcal{F}_E} [\Gamma_e]_{v,v'} z^v z'^{v'}.$$

As $K_{\varepsilon_n}(z, z') = [K_{\varepsilon_n,s,s'}(z, z')]_{s,s' \in S}$ and $\Gamma_{\varepsilon_n,e}(z, z')$ are positive-definite for each fixed n, we know that

$$\sum_{s,s' \in S} \sum_{v,v' \in \mathcal{F}_E} \langle [K_{\varepsilon_n,s,s'}]_{v,v'} y_{s',v'^\top}, y_{s,v} \rangle_\mathcal{Y} \geq 0, \qquad \sum_{v,v' \in \mathcal{F}_E} \langle [\Gamma_{\varepsilon_n,e}]_{v,v'} g_{v'^\top}, g_v \rangle_\mathcal{Y} \geq 0$$
$$(5.15)$$

for all finitely supported $\mathcal{Y}$-valued functions $(s, v) \mapsto y_{s,v}$ and $s \mapsto g_v$. We may then take the limits as $n \to \infty$ in (5.15) to get

$$\sum_{s,s' \in S} \sum_{v,v' \in \mathcal{F}_E} \langle [K_{s,s'}]_{v,v'} y_{s',v'^\top}, y_{s,v} \rangle_\mathcal{Y} \geq 0, \qquad \sum_{v,v' \in \mathcal{F}_E} \langle [\Gamma_e]_{v,v'} g_{v'^\top}, g_v \rangle_\mathcal{Y} \geq 0 \quad (5.16)$$

from which we see that $K(z, z') = [K_{s,s'}(z, z')]_{s,s' \in S}$ and $\Gamma_e(z, z')$ for $e \in E$ are positive-definite formal power series as well. By Lemma 5.6, for each $e \in E$ the formal power series $\Gamma_e(z, z')$ is therefore in the cone $\mathcal{C}$. As the difference in the first two terms on the right-hand side of (5.14) is clearly in $\mathcal{C}$ by the characterization (5.9) for $\mathcal{C}$ and $\mathcal{C}$ is closed under addition, it follows that $\varphi \in \mathcal{C}$ as asserted. Lemma 5.7 now follows. $\qquad\square$

Lemma 5.8. *If $\varphi(z, z') \in \mathcal{L}(\mathcal{Y})\langle\langle z, z' \rangle\rangle$ is a positive-definite formal power series and if $\varepsilon > 0$, then:*

1. *$\varphi \in \mathcal{C}_\varepsilon$, and*
2. *for each $e \in E$, the kernel $\widetilde{\varphi}(z, z') := \varphi(z, z') - \varepsilon^2 z'_e \varphi(z, z') z_e$ is also in $\mathcal{C}_\varepsilon$.*

Proof. As $\varphi(z, z')$ is positive-definite, we have a factorization

$$\varphi(z, z') = H(z)H(z')^*$$

for some $H(z) \in \mathcal{L}(\mathcal{K}, \mathcal{Y})\langle\langle z \rangle\rangle$. To show that $\varphi(z, z') \in \mathcal{C}_\varepsilon$ it suffices to produce a representation (5.10) for φ with $\gamma_{\varepsilon,e}(z) = 0$ for each $e \in E$. As in the proof of Lemma 5.6, to produce a representation of this latter form it suffices to produce a formal power series $H'(z) \in \mathcal{L}(\oplus_{s \in S} \mathcal{H}'_{[s]}, \mathcal{Y})\langle\langle z \rangle\rangle$ with

$$\varphi(z, z') = H(z)H(z')^* = H'(z)(1 - (1 + \varepsilon)^2 Z_{G,\mathcal{H}'}(z) Z_{G,\mathcal{H}'}(z')^*)H'(z')^*. \quad (5.17)$$

For this purpose we assume that $\mathcal{H}'$ is presented as $\mathcal{H}' = \ell^2(\mathcal{F}_{E_0}, \mathcal{K})$ where $E_0 = \{e \in E \colon \mathbf{s}(e) = s_0\}$ where s_0 is some fixed source vertex $s_0 \in S$. We then take $H'(z)$ to be of the form $H'(z) = H(z)K'(z)$ where

$$K'(z) = \mathrm{row}_{s \in S}\, K'_s(z) \in \mathcal{L}(\oplus_{s \in S} \mathcal{H}'_{[s]}, \mathcal{K})\langle\langle z \rangle\rangle$$

is given by

$$K'_s(z) = \begin{cases} \mathrm{row}_{v \in \mathcal{F}_{E_0}} (1 + \varepsilon)^{|v|} z^v I_{\mathcal{K}} & \text{if } s = s_0, \\ 0 & \text{otherwise.} \end{cases}$$

Then a direct computation as in the proof of Lemma 5.6 gives

$$H'(z)(I_{\oplus_{s \in S} \mathcal{H}'_{[s]}} - (1 + \varepsilon)^2 Z_{G,\mathcal{H}'}(z) Z_{G,\mathcal{H}'}(z')^*)H'(z')^*$$

$$= H(z)K'_{s_0}(z)\left[\left(1 - (1 + \varepsilon)^2 \sum_{e \in E_0} z_e z'_e\right) I_{\mathcal{H}'}\right] K'_{s_0}(z')^* H(z')^*$$

$$= H(z)\left[\left(\sum_{v \in \mathcal{F}_{E_0}} (1 + \varepsilon)^{2|v|} z^v z'^v I_{\mathcal{K}} - \sum_{v \in \mathcal{F}_{E_0} \setminus \{\emptyset\}} (1 + \varepsilon)^{2|v|} z^v z'^v\right) I_{\mathcal{K}}\right] H(z')^*$$

$$= H(z)H(z')^* = \varphi(z, z'),$$

as wanted, and part (1) of Lemma 5.8 follows.

For the second assertion, use the characterization (5.11) for membership in $\mathcal{C}_\varepsilon$ with $K_{s,s'}(z, z') = 0$ for all $s, s' \in S$ and with $\Gamma_{e'}(z, z') = 0$ for $e' \neq e$ and $\Gamma_e(z, z') = \varphi(z, z')$. $\qquad\square$

Lemma 5.9. *For each $\varepsilon > 0$, the cone $\mathcal{C}_\varepsilon$ is closed as a subspace of*

$$\mathcal{X} = \mathcal{L}(\mathcal{Y})\langle\langle z, z' \rangle\rangle$$

with the locally convex topology induced by the sequence of seminorms $\|\cdot\|_N$ given by (5.7) for $N = 1, 2, \ldots$.

Proof. Suppose that $\{\varphi_n\}_{n=1,2,\ldots}$ is a sequence of elements of $\mathcal{C}_\varepsilon$ converging to $\varphi \in \mathcal{X}$ in the locally convex topology of $\mathcal{X}$. By the characterization (5.11) we have the existence of positive-definite formal power series

$$K_n(z, z') = [K_{n;s,s'}(z, z')]_{s,s' \in S} \in \mathcal{L}(\oplus_{s \in S} \mathcal{Y})\langle\langle z, z' \rangle\rangle, \; \Gamma_{n,e}(z, z') \in \mathcal{L}(\mathcal{Y})\langle\langle z, z' \rangle\rangle$$

so that the representation

$$\varphi_n(z,z') = \sum_{s\in S} K_{n;s,s}(z,z') - (1+\varepsilon)^2 \sum_{r\in R} \sum_{s,s'\in S:\,[s]=[s']=[r]} z'_{e_{s',r}} K_{n;s,s'}(z,z') z_{e_{s,r}}$$

$$+ \sum_{e\in E} \Gamma_{n,e}(z,z') - \varepsilon^2 \sum_{e\in E} z'_e \Gamma_{n,e}(z,z') z_e \tag{5.18}$$

holds for each $n = 1,2,\ldots$. In terms of coefficients we then have

$$[\varphi_n]_{v,v'} = \sum_{s\in S} [K_{n;s,s}]_{v,v'} - (1+\varepsilon)^2 \sum_{r\in R} \sum_{s,s'\in S:\,[s]=[s']=[r]} [K_{n;s,s'}]_{ve_{s,r}^{-1},e_{s',r}^{-1}v'}$$

$$+ \sum_{e\in E} [\Gamma_{n,e}]_{v,v'} - \varepsilon^2 \sum_{e\in E} [\Gamma_{n,e}]_{ve^{-1},e^{-1}v'}. \tag{5.19}$$

By assumption $[\varphi_n]_{v,v'}$ converges in the operator-norm of $\mathcal{L}(\mathcal{Y})$ to $[\varphi]_{v,v'}$ as $n \to \infty$. An inductive argument on the length of words combined with the positive-definiteness of $K_n(z,z')$ and $\Gamma_{n,e}(z,z')$ as in the proof of Lemma 5.7 can now be used to show that $\|[K_{n;s,s'}]_{v,v'}\|$ and $\|[\Gamma_{n,e}]_{v,v'}\|$ remain uniformly bounded as $n \to \infty$ for each $v, v' \in \mathcal{F}_E$ and $e \in E$. Since we are assuming that $\dim \mathcal{Y} < \infty$, a compactness argument together with a Cantor diagonalization argument (as in the proof of Lemma 5.7) can be used to show that there exists a subsequence $n_1 < n_2 < n_3 < \ldots$ such that the limits

$$\lim_{k\to\infty} [K_{n_k;s,s'}]_{v,v'} = [K_{s,s'}]_{v,v'}, \qquad \lim_{k\to\infty} [\Gamma_{n_k,e}]_{v,v'} = [\Gamma_e]_{v,v'}$$

all exist in $\mathcal{L}(\mathcal{Y})$-norm for each $s, s' \in S$, $e \in E$, and $v, v' \in \mathcal{F}_E$. We may then take limits in (5.19) to conclude that

$$[\varphi]_{v,v'} = \sum_{s\in S} [K_{s,s}]_{v,v'} - (1+\varepsilon)^2 \sum_{r\in R} \sum_{s,s'\in S:\,[s]=[s']=[r]} [K_{s,s'}]_{ve_{s,r}^{-1},e_{s',r}^{-1}v'}$$

$$+ \sum_{e\in E} [\Gamma_e]_{v,v'} - \varepsilon^2 \sum_{e\in E} [\Gamma_e]_{ve^{-1},e^{-1}v'}. \tag{5.20}$$

If we then set

$$K(z,z') = [K_{s,s'}(z,z')]_{s,s'\in S} \text{ with } K_{s,s'}(z,z') = \sum_{v,v'\in\mathcal{F}_E} [K_{s,s'}(z,z')]_{v,v'} z^v z'^{v'},$$

$$\Gamma_e(z,z') = \sum_{v,v'\in\mathcal{F}_E} [\Gamma_e]_{v,v'} z^v z'^{v'},$$

we conclude that

$$\varphi(z,z') = \sum_{s\in S} K_{s,s}(z,z') - (1+\varepsilon)^2 \sum_{r\in R} \sum_{s,s'\in S:\,[s]=[s']=[r]} z'_{e_{s',r}} K_{s,s'}(z,z') z_{e_{s,r}}$$

$$+ \sum_{e\in E} \Gamma_e(z,z') - \varepsilon^2 \sum_{e\in E} z'_e \Gamma_e(z,z') z_e. \tag{5.21}$$

Furthermore, as $K_{s,s'}(z,z')$ is the coefficientwise limit of $K_{n,s,s'}(z,z')$ where each $[K_{n;s,s'}]_{s,s'\in S}$ is positive definite and $\Gamma_e(z,z')$ is the coefficientwise limit of

$\Gamma_{n,e}(z, z')$ which is positive definite, it follows as in the proof of Lemma 5.7 that $K(z, z')$ and $\Gamma_e(z, z')$ for each $e \in E$ are positive definite. The identity (5.21) then shows that $\varphi(z, z')$ satisfies the criterion (5.11) for membership in $\mathcal{C}_\varepsilon$ as wanted, and Lemma 5.9 follows. $\qquad\square$

We are now ready to commence the proof of $(1) \implies (2)$ in Theorem 5.3 for the case where $\dim \mathcal{Y} < \infty$. Suppose that we are given a formal power series $T(z)$ which is in the Schur-Agler class $\mathcal{SA}_G(\mathcal{U}, \mathcal{Y})$. The issue is to show that $\varphi_T(z, z') := I_{\mathcal{Y}} - T(z)T(z')^*$ is in $\mathcal{C}$. By Lemma 5.7, it suffices to show that φ_T is in $\mathcal{C}_\varepsilon$ for all $\varepsilon > 0$ small enough. Recall the notation $\mathcal{X}$ for the topological linear space $\mathcal{L}(\mathcal{Y})\langle\langle z, z' \rangle\rangle$ with the locally convex topology of norm-convergence of power-series coefficients. By the Hahn-Banach separation principle (apply the contrapositive version of part (b) of Theorem 3.4 in [54] with $X = \mathcal{X}$, $A = \{\varphi_T\}$, and $B = \mathcal{C}_\varepsilon$), it suffices to show: *for fixed $\varepsilon > 0$ and for any continuous linear functional* $\mathbf{L} \colon \mathcal{X} \to \mathbb{C}$ *such that* $\mathfrak{R}\mathbf{L}(\varphi) \geq 0$ *for all* $\varphi \in \mathcal{C}_\varepsilon$, *it follows that* $\mathfrak{R}\mathbf{L}(\varphi_T) \geq 0$. Here $\mathfrak{R}$ denotes "real part".

Fix $\varepsilon > 0$ and let $\mathbf{L}$ be any continuous linear functional on $\mathcal{X}$ with $\mathfrak{R}\mathbf{L}|_{\mathcal{C}_\varepsilon} \geq 0$. Define $\mathbf{L}_1 \colon \mathcal{X} \to \mathbb{C}$ by

$$\mathbf{L}_1(\varphi) = \frac{1}{2}\left(\mathbf{L}(\varphi) + \overline{\mathbf{L}(\check\varphi)}\right) \tag{5.22}$$

where we have set

$$\check\varphi(z, w) = \varphi(w, z)^*.$$

Note that $\mathbf{L}_1(\varphi) = \mathfrak{R}\mathbf{L}(\varphi)$ in case $\check\varphi = \varphi$. We define a sesquilinear form $\langle \cdot, \cdot \rangle_{\mathbf{L}}$ on the space $\mathcal{H}_0 := \mathcal{L}(\mathcal{Y}, \mathbb{C})\langle\langle z' \rangle\rangle$ according to the formula

$$\langle f, g \rangle_{\mathbf{L}} = \mathbf{L}_1(g(z)^* f(z')). \tag{5.23}$$

Note that any formal power series φ of the form $\varphi(z, z') = f(z)f(z')^*$ has the property that $\check\varphi = \varphi$; by part (1) of Lemma 5.8, any such φ is in $\mathcal{C}_\varepsilon$. We conclude that

$$\langle f, f \rangle_{\mathbf{L}} = \mathfrak{R}\mathbf{L}(f(z)^* f(z')) \geq 0 \text{ for all } f \in \mathcal{H}_0.$$

We may thus identify elements of zero norm and then take a completion in the $\mathbf{L}$-norm to get a Hilbert space $\mathcal{H}_{\mathbf{L}}$.

We next seek to define operators δ_e for each $e \in E$ on $\mathcal{H}_{\mathbf{L}}$ so that δ_e^* is given by

$$\delta_e^* \colon f(z) \mapsto z_e f(z) \text{ for } f \in \mathcal{H}_0. \tag{5.24}$$

By part (2) of Lemma 5.8 we know that the kernel $f(z)^*(1 - \varepsilon^2 z_e z_e')f(z')$ belongs to $\mathcal{C}_\varepsilon$, and hence

$$\|f\|_{\mathcal{H}_{\mathbf{L}}}^2 - \varepsilon^2 \|\delta_e^* f\|_{\mathcal{H}_{\mathbf{L}}}^2 = \mathfrak{R}\mathbf{L}\left(f(z)^*(1 - \varepsilon^2 z_e z_e')f(z')\right) \geq 0 \text{ for } f \in \mathcal{H}_0.$$

Hence δ_e extends to a bounded operator on all of $\mathcal{H}_{\mathbf{L}}$ with $\|\delta_e\| = \|\delta_e^*\| \leq 1/\varepsilon$ for each $e \in E$.

It is then easy to see that the operator $Z_{G, \mathcal{H}_{\mathbf{L}}}(\delta)^* \colon (\oplus_{s \in S} \mathcal{H}_{\mathbf{L}}) \to (\oplus_{r \in R} \mathcal{H}_{\mathbf{L}})$ is given by multiplication by $Z_{G, \mathcal{H}_{\mathbf{L}}}(z')^*$ on the left:

$$Z_{G, \mathcal{H}_{\mathbf{L}}}(\delta)^* \colon f(z') \mapsto Z_{G, \mathcal{H}_{\mathbf{L}}}(z')^* f(z') \text{ for } f \in \oplus_{s \in S} \mathcal{H}_0.$$

Note that an element $f \in \oplus_{s \in S} \mathcal{H}_0$ can be viewed as an element of the space $\mathcal{L}(\mathcal{Y}, \oplus_{s \in S} \mathbb{C})\langle\langle z' \rangle\rangle$. The $(\oplus_{s \in S} \mathcal{H}_\mathbf{L})$-norm of an element $f = \oplus_{s \in S} f_s \in \oplus_{s \in S} \mathcal{H}_0$ can be computed as follows:

$$\|f\|^2_{\oplus_{s \in S}\mathcal{H}_\mathbf{L}} = \sum_{s \in S} \|f_s\|^2_{\mathcal{H}_\mathbf{L}} = \sum_{s \in S} \Re\mathbf{L}\left(f_s(z)^* f_s(z')\right) = \Re\mathbf{L}\left(f(z)^* f(z')\right).$$

Similarly,

$$\|Z_{G,\mathcal{H}_\mathbf{L}}(\delta)^* f\|^2_{\oplus_{r \in R}\mathcal{H}_\mathbf{L}} = \Re\mathbf{L}\left(f(z)^* Z_{G,\mathcal{H}_\mathbf{L}}(z) Z_{G,\mathcal{H}_\mathbf{L}}(z')^* f(z')\right).$$

We may then compute

$$\|f\|^2_{\oplus_{s \in S}\mathcal{H}_\mathbf{L}} - (1+\epsilon)^2 \|Z_{G,\mathcal{H}_\mathbf{L}}(\delta)^* f\|^2_{\oplus_{r \in R}\mathcal{H}_\mathbf{L}}$$
$$= \Re\mathbf{L}\left(f(z)^*(I_{\oplus_{s \in S}\mathbb{C}} - (1+\varepsilon)^2 Z_{G,\mathcal{H}_\mathbf{L}}(z) Z_{G,\mathcal{H}_\mathbf{L}}(z')^*)f(z')\right). \tag{5.25}$$

Clearly, $\varphi(z, z')$ given by

$$\varphi(z, z') := f(z)^*(I_{\oplus_{s \in S}\mathbb{C}} - (1+\varepsilon)^2 Z_{G,\mathcal{H}_\mathbf{L}}(z) Z_{G,\mathcal{H}_\mathbf{L}}(z')^*)f(z')$$

is in the cone $\mathcal{C}_\varepsilon$: simply take $\gamma_e(z) = 0$ for all $e \in E$ in the defining representation (5.10) for elements of $\mathcal{C}_\varepsilon$. From (5.25) and the assumption that $\Re\mathbf{L}$ is nonnegative on $\mathcal{C}_\varepsilon$, we therefore deduce that

$$\|Z_{G,\mathcal{H}_\mathbf{L}}(\delta)\| = \|Z_{G,\mathcal{H}_\mathbf{L}}(\delta)^*\| \leq \frac{1}{1+\varepsilon} < 1.$$

From our assumption that $T(z) \in \mathcal{SA}_G(\mathcal{U}, \mathcal{Y})$, we deduce that $\|T(\delta)\| \leq 1$.

If we are in the scalar-valued case $\mathcal{U} = \mathcal{Y} = \mathbb{C}$, then we see from the form (5.24) for the action of δ_e^* and from the continuity of $\mathbf{L}$ that $T(\delta)^*$ is given by

$$T(\delta)^*: f(z') \mapsto T(z')^* f(z')$$

with

$$\|T(\delta)^* f\|^2 = \Re\mathbf{L}\left(f(z)^* T(z) T(z')^* f(z')\right).$$

As $\|T(\delta)^*\| \leq 1$ we therefore have

$$0 \leq \|1\|^2_{\mathcal{H}_\mathbf{L}} - \|T(\delta)^*(1)\|^2_{\mathcal{H}_\mathbf{L}} = \Re\mathbf{L}\left(I_\mathcal{Y} - T(z) T(z')^*\right) = \Re\mathbf{L}(\varphi_T(z, z'))$$

as wanted.

The general case is a little more intricate. For $\Phi \in \mathcal{L}(\mathcal{U}, \mathcal{Y})$ and $v \in \mathcal{F}_E$, the tensor product operator $\delta^{*v^\top} \otimes \Phi^*$ acts on an element $f(z') \otimes y$ of $\mathcal{H}_\mathbf{L} \otimes \mathcal{Y}$. We assume that the formal power series $f = \sum_{v \in \mathcal{F}_E} f_v z'^v$ consists only of its constant term (so $f_v = 0$ for $v \neq \emptyset$ and $f(z') = \ell$ where $\ell \in \mathcal{L}(\mathcal{Y}, \mathbb{C})$ is a linear functional on $\mathcal{Y}$). We compute the $(\mathcal{H}_\mathbf{L} \otimes \mathcal{U})$-inner product of $(\delta^{*v^\top} \otimes \Phi^*)(\ell \otimes y)$ against another

210   J.A. Ball, G. Groenewald and T. Malakorn

such object $(\delta^{*v'^{\top}} \otimes \Phi'^*)(\ell' \otimes y')$ as follows:

$$
\begin{aligned}
\langle (\delta^{*v'^{\top}} &\otimes \Phi'^*)(\ell' \otimes y'), \quad (\delta^{*v^{\top}} \otimes \Phi^*)(\ell \otimes y) \rangle_{\mathcal{H}_{\mathbf{L}} \otimes \mathcal{U}} \\
&= \langle z'^{v'^{\top}} \ell' \otimes \Phi'^* y', z'^{v^{\top}} \ell \otimes \Phi^* y \rangle_{\mathcal{H}_{\mathbf{L}} \otimes \mathcal{U}} \\
&= \langle z'^{v'^{\top}} \ell', z'^{v^{\top}} \ell \rangle_{\mathcal{H}_{\mathbf{L}}} \cdot \langle \Phi'^* y', \Phi^* y \rangle_{\mathcal{U}} \\
&= \mathbf{L}_1 \left(z^v z'^{v'^{\top}} \ell^* \cdot \langle \Phi \Phi'^* y', y \rangle_{\mathcal{Y}} \cdot \ell' \right) \\
&= \mathbf{L}_1 \left(\ell^* y^* (\Phi z^v)(\Phi'^* z'^{v'^{\top}}) y' \ell' \right).
\end{aligned}
\tag{5.26}
$$

Here we have viewed the vector $y \in \mathcal{Y}$ as the operator $y \colon \alpha \mapsto \alpha y$ from $\mathbb{C}$ to $\mathcal{Y}$ with adjoint operator $y^* \colon \mathcal{Y} \to \mathbb{C}$ given by $y^* \colon y'' \mapsto \langle y'', y \rangle_{\mathcal{Y}} \in \mathbb{C}$. In this way, the inner product $\langle \Phi \Phi'^* y', y \rangle_{\mathcal{Y}}$, when viewed as an operator on $\mathbb{C}$, can be written as the operator composition

$$
\langle \Phi \Phi'^* y', y \rangle_{\mathcal{Y}} = y^* \Phi \Phi'^* y' \colon \mathbb{C} \to \mathbb{C}.
$$

By linearity we can generalize (5.26) to

$$
\langle G'(\delta)^* (\ell' \otimes y'), G(\delta)^* (\ell \otimes y) \rangle_{\mathcal{H}_{\mathbf{L}} \otimes \mathcal{U}} = \mathbf{L}_1 \left(\ell^* y^* G(z) G'(z')^* y' \ell' \right)
\tag{5.27}
$$

for any polynomials $G(z'), G'(z')$ in the noncommuting indeterminates z' with coefficients in $\mathcal{L}(\mathcal{U}, \mathcal{Y})$ $(G, G' \in \mathcal{L}(\mathcal{U}, \mathcal{Y})\langle z' \rangle)$. More generally, by the assumed continuity of $\mathbf{L}$ on $\mathcal{X}$, (5.27) continues to hold if G and G' are formal power series in $\mathcal{L}(\mathcal{U}, \mathcal{Y})\langle\langle z' \rangle\rangle$ for which $G(\delta)$ and $G'(\delta)$ are defined.

We now apply (5.27) to the case where $G = G' = T \in \mathcal{SA}_G(\mathcal{U}, \mathcal{Y})$ and where $(\ell')^* = y' = y_j$ and $\ell^* = y = y_i$, where $y_1, y_2, \ldots, y_M$ is an orthonormal basis for $\mathcal{Y}$, to get

$$
\langle T(\delta)^* (y_j^* \otimes y_j), T(\delta)^* (y_i^* \otimes y_i) \rangle_{\mathcal{H}_{\mathbf{L}} \otimes \mathcal{U}} = \mathbf{L}_1 \left(y_i y_i^* T(z) T(z')^* y_j y_j^* \right).
$$

Summing over $i, j = 1, \ldots, M$ then gives

$$
\left\| T(\delta)^* \left(\sum_{j=1}^{M} y_j^* \otimes y_j \right) \right\|_{\mathcal{H}_{\mathbf{L}} \otimes \mathcal{U}}^2 = \sum_{i,j=1,\ldots,M} \mathbf{L}_1 \left(y_i y_i^* T(z) T(z')^* y_j y_j^* \right)
$$

$$
= \Re \mathbf{L} \left(T(z) T(z')^* \right).
\tag{5.28}
$$

Moreover, we compute

$$
\langle y_j^* \otimes y_j, y_i^* \otimes y_i \rangle_{\mathcal{H}_{\mathbf{L}} \otimes \mathcal{Y}} = \langle y_j^*, y_i^* \rangle_{\mathcal{H}_{\mathbf{L}}} \cdot \langle y_j, y_i \rangle_{\mathcal{Y}} = \delta_{i,j} \mathbf{L}_1 \left(y_i y_i^* \right).
$$

Summing this over $i, j = 1, \ldots, M$ then gives

$$
\left\| \sum_{j=1}^{M} y_j^* \otimes y_j \right\|_{\mathcal{H}_{\mathbf{L}} \otimes \mathcal{Y}}^2 = \sum_{j=1}^{M} \mathbf{L}_1 \left(y_j y_j^* \right) = \Re \mathbf{L} \left(I_{\mathcal{Y}} \right).
\tag{5.29}
$$

Using that $\|T(\delta)\| \le 1$ and combining (5.28) and (5.29) then gives

$$0 \le \left\| \sum_{j=1}^{M} y_j^* \otimes y_j \right\|_{\mathcal{H}_{\mathbf{L}} \otimes \mathcal{Y}}^2 - \left\| T(\delta)^* \left(\sum_{j=1}^{M} y_j^* \otimes y_j \right) \right\|_{\mathcal{H}_{\mathbf{L}} \otimes \mathcal{U}}^2$$
$$= \Re \mathbf{L}(I_{\mathcal{Y}} - T(z)T(z')^*) = \Re \mathbf{L}(\varphi_T(z, z')) \tag{5.30}$$

as wanted.

This completes the proof of (1) $\implies$ (2) or (3) for the case that $\dim \mathcal{Y} < \infty$.

We now consider the case of a general separable Hilbert output space $\mathcal{Y}$. Let $y_1, y_2, \ldots, y_M, \ldots$ be an orthonormal basis for $\mathcal{Y}$ and let $P_M \colon \mathcal{Y} \to \mathcal{Y}$ be the orthogonal projection onto the closed span $\mathcal{Y}_M$ of $\{y_1, \ldots, y_M\}$. Suppose that the formal power series $T(z) \in \mathcal{L}(\mathcal{U}, \mathcal{Y})\langle\langle z \rangle\rangle$ is in the noncommutative Schur-Agler class $\mathcal{SA}_G(\mathcal{U}, \mathcal{Y})$. Then clearly $P_M T(z) \in \mathcal{L}(\mathcal{U}, \mathcal{Y}_M)\langle\langle z \rangle\rangle$ is in the noncommutative Schur-Agler class $\mathcal{SA}_G(\mathcal{U}, \mathcal{Y}_M)$. Hence, by the special case of (1) $\implies$ (2) or (3) already proved,

$$\varphi_{T,M} = P_M(I_{\mathcal{Y}} - T(z)T(z')^*)P_M$$

has a representation of the form

$$\varphi_{T,M}(z, z') = \sum_{s \in S} K_{M;s,s}(z, z') - \sum_{r \in R} \sum_{s,s' \in S \colon [s]=[s']=[r]} z'_{e_{s',r}} K_{M;s,s'}(z, z') z_{e_{s,r}} \tag{5.31}$$

for a positive-definite formal power series

$$K_M(z, z') = [K_{M;s,s'}(z, z')]_{s,s' \in S} \in \mathcal{L}(\oplus_{s \in S} \mathcal{Y}_M)\langle\langle z, z' \rangle\rangle.$$

In terms of power-series coefficients, we therefore have

$$[\varphi_{T,M}]_{v,v'} = \sum_{s \in S} [K_{M;s,s}]_{v,v'} - \sum_{r \in R} \sum_{s,s' \in S \colon [s]=[s']=[r]} [K_{M;s,s'}]_{v e_{s,r}^{-1}, e_{s',r}^{-1} v'}. \tag{5.32}$$

By construction

$$[\varphi_{T,M}]_{v,v'} = P_M[\varphi_T]_{v,v'} P_M \tag{5.33}$$

and hence

$$\|[\varphi_{T,M}]_{v,v'}\| \le \|[\varphi_T]_{v,v'}\| \text{ for all } v, v' \in \mathcal{F}_E \text{ and } M = 1, 2, \ldots. \tag{5.34}$$

The uniform estimate (5.34) combined with an inductive argument on the length of words (as in the proof of Lemma 5.7) implies that $\|[K_{M;s,s'}]_{v,v'}\|$ is uniformly bounded in the operator norm of $\mathcal{L}(\mathcal{Y})$ as $M \to \infty$ for each $v, v' \in \mathcal{F}_E$. Furthermore, $\mathcal{L}(\mathcal{Y})$ carries a weak-$*$ topology as the dual space of the trace-class operators $\mathcal{L}_1(\mathcal{Y})$ under the duality pairing induced by the trace (see [28, Theorem 19.2 page 94]). By Alaoglu's Theorem (see [55, Theorem 10.3 page 174]), norm-bounded sets in $\mathcal{L}(\mathcal{Y})$ are precompact in the weak-$*$ topology. Moreover (see [32, Theorem 1 page 426]), since the predual $\mathcal{L}_1(\mathcal{Y})$ of $\mathcal{L}(\mathcal{Y})$ is separable, it follows that the weak-$*$ topology on bounded subsets of $\mathcal{L}(\mathcal{Y})$ is metrizable. These observations combined with another Cantor diagonalization procedure allow us to conclude that there exists a subsequence $M_{n_k} \to \infty$ so that

$$\text{weak-}* \ \lim_{k \to \infty} K_{M_k;v,v'} = K_{v,v'} \tag{5.35}$$

exists for each $v, v' \in \mathcal{F}_E$. Furthermore, a consequence of (5.33) is that

$$\text{weak-}* \ \lim_{M \to \infty} [\varphi_{T,M}]_{v,v'} = [\varphi_T]_{v,v'}. \tag{5.36}$$

Using (5.35) and (5.36) to take weak-$*$ limits in (5.32), we get

$$[\varphi_T]_{v,v'} = \sum_{s \in S} [K_{s,s}]_{v,v'} - \sum_{r \in R} \sum_{s,s' \in S \colon [s]=[s']=[r]} [K_{s,s'}]_{ve_{s,r}^{-1}, e_{s',r}^{-1} v'}$$

or, in terms of formal power series,

$$\varphi_T(z, z') = \sum_{s \in S} K_{s,s}(z, z') - \sum_{r \in R} \sum_{s,s' \in S \colon [s]=[s']=[r]} z'_{e_{s',r}} K_{s,s'}(z, z') z_{e_{s,r}}, \tag{5.37}$$

where we set $K_{s,s'}(z, z') = \sum_{v,v' \in \mathcal{F}_E} [K_{s,s'}]_{v,v'} z^v z'^{v'}$. Furthermore, using the characterization (5.3) it is easy to see from (5.35) that $K(z, z')$ is a positive-definite formal power series since each $K_{M_k}(z, z')$ is positive-definite. We have thus verified that φ_T has a representation (5.4) as wanted. This completes the proof of (1) $\implies$ (2) or (3).

(3) $\implies$ **(4):** Suppose that $T(z) \in \mathcal{L}(\mathcal{U}, \mathcal{Y})\langle\langle z \rangle\rangle$ is such that

$$I - T(z)T(z')^* = H(z)(I - Z_{G,\mathcal{H}'}(z)Z_{G,\mathcal{H}'}(z')^*)H(z')^* \tag{5.38}$$

for some $H(z) \in \mathcal{L}(\oplus_{s \in S} \mathcal{H}'_{[s]}, \mathcal{Y})\langle\langle z \rangle\rangle$. Thus $H(z)$ has a row matrix representation

$$H(z) = \text{row}_{s \in S} H_s(z)$$

where $H_s(z) \in \mathcal{L}(\mathcal{H}'_{[s]}, \mathcal{Y})\langle\langle z \rangle\rangle$. Write the coefficient of $H_s(z)$ for the word $v \in \mathcal{F}_E$ as $[H_s]_v$. Given two words $v, v' \in \mathcal{F}_E$, equating coefficients of $z^v z'^{v'^\top}$ in (5.38) gives

$$\delta_{v,\emptyset} \delta_{v',\emptyset} I_{\mathcal{Y}} - T_v T_{v'}^*$$

$$= \sum_{s \in S} [H_s]_v ([H_s]_{v'})^* - \sum_{r \in R} \left(\sum_{s \colon [s]=[r]} [H_s]_{ve_{s,r}^{-1}} \right) \cdot \left(\sum_{s' \colon [s']=[r]} ([H_{s'}]_{v'e_{s',r}^{-1}})^* \right).$$

Rewrite this identity in the form

$$\sum_{r \in R} \left(\sum_{s \colon [s]=[r]} [H_s]_{ve_{s,r}^{-1}} \right) \cdot \left(\sum_{s' \colon [s']=[r]} ([H_{s'}]_{v'e_{s',r}^{-1}})^* \right) + \delta_{v,\emptyset} \delta_{v',\emptyset} I_{\mathcal{Y}}$$

$$= \sum_{s \in S} [H_s]_v ([H_s]_{v'})^* + T_v T_{v'}^*. \tag{5.39}$$

As a consequence of (5.39) we see that the map $V' \colon \mathcal{D}_{V'} \to \mathcal{R}_{V'}$ defined by

$$V' \colon \begin{bmatrix} \text{col}_{r \in R} \sum_{s \colon [s]=[r]} ([H_s]_{ve_{s,r}^{-1}})^* \\ \delta_{v,\emptyset} I_{\mathcal{Y}} \end{bmatrix} y \mapsto \begin{bmatrix} \text{col}_{s \in S} ([H_s]_v)^* \\ T_v^* \end{bmatrix} y \tag{5.40}$$

extends by linearity and limits to define a unitary transformation from

$$\mathcal{D}_{V'} := \text{closed span} \left\{ \begin{bmatrix} \text{col}_{r \in R} \sum_{s:\,[s]=[r]} ([H_s]_{ve_{s,r}^{-1}})^* \\ \delta_{v,\emptyset} I_{\mathcal{Y}} \end{bmatrix} y : v \in \mathcal{F}_E, \quad y \in \mathcal{Y} \right\}$$

onto

$$\mathcal{R}_{V'} := \text{closed span} \left\{ \begin{bmatrix} \text{col}_{s \in S} ([H_s]_v)^* \\ T_v^* \end{bmatrix} y : v \in \mathcal{F}_E, \quad y \in \mathcal{Y} \right\}.$$

Extend V' to a unitary transformation V of the form

$$V = \begin{bmatrix} A^* & C^* \\ B^* & D^* \end{bmatrix} : \begin{bmatrix} \text{col}_{r \in R} \mathcal{H}_{[r]} \\ \mathcal{Y} \end{bmatrix} \to \begin{bmatrix} \text{col}_{s \in S} \mathcal{H}_{[s]} \\ \mathcal{U} \end{bmatrix}$$

where, for $p \in P$, $\mathcal{H}_p \supset \mathcal{H}_p'$. Set $\mathcal{H}$ equal to the collection $\{\mathcal{H}_p : p \in P\}$. Putting the pieces together, we have that

$$\Sigma = (G, \mathcal{H}, U = V^*)$$

is a conservative SNMLS.

We next verify that $T(z)$ is the transfer function $T(z) = T_\Sigma(z)$ of the conservative SNMLS Σ constructed as above. Indeed, since V extends V' we see from (5.40) that

$$\begin{bmatrix} A^* & C^* \\ B^* & D^* \end{bmatrix} : \begin{bmatrix} \text{col}_{r \in R} \sum_{s:\,[s]=[r]} ([H_s]_{ve^{-1}})^* \\ \delta_{v,\emptyset} I_{\mathcal{Y}} \end{bmatrix} y \mapsto \begin{bmatrix} \text{col}_{s \in S} ([H_s]_v)^* \\ T_v^* \end{bmatrix} y \qquad (5.41)$$

for all $v \in \mathcal{F}_E$ and $y \in \mathcal{Y}$. If we multiply both sides by $z^{v^\top}$ and sum over all $v \in \mathcal{F}_E$ and cancel off the common factor y we get the formal power series identity

$$\begin{bmatrix} A^* & C^* \\ B^* & D^* \end{bmatrix} : \begin{bmatrix} Z_\Sigma(z')^* & 0 \\ 0 & I_{\mathcal{Y}} \end{bmatrix} \begin{bmatrix} H(z')^* \\ I_{\mathcal{Y}} \end{bmatrix} \mapsto \begin{bmatrix} H(z')^* \\ T(z')^* \end{bmatrix}. \qquad (5.42)$$

In particular, the top block component of (5.42) gives

$$A^* Z_\Sigma(z')^* H(z')^* + C^* = H(z')^*$$

from which we get

$$H(z')^* = (I - A^* Z_\Sigma(z')^*)^{-1} C^*.$$

Substituting this into the equality of the bottom block components of (5.42) then gives

$$T(z')^* = B^* Z_\Sigma(z')^* (I - A^* Z_\Sigma(z')^*)^{-1} C^* + D^*$$

and hence

$$T(z) = D + C(I - Z_\Sigma(z)A)^{-1} Z_\Sigma(z) B$$

and we have verified that $T(z) = T_\Sigma(z)$ as wanted. This completes the proof of $(3) \implies (4)$ in Theorem 5.3.

As we have now verified $(2) \iff (3)$, $(4) \implies (1)$, $(1) \implies (2)$ or (3) and $(3) \implies (4)$, the proof of all of Theorem 5.3 is now complete. $\qquad\qquad \square$

Remark 5.10. It is possible to give an elementary direct proof of (4) $\implies$ (3) in Theorem 5.3. Assume that the formal power series $T(z)$ is realized as $T(z) = T_\Sigma(z)$ for a conservative SNMLS $\Sigma = (G, \mathcal{H}, U)$. In particular

$$
U = \begin{bmatrix} A & B \\ C & D \end{bmatrix} : \begin{bmatrix} \oplus_{s \in S} \mathcal{H}_{[s]} \\ \mathcal{U} \end{bmatrix} \rightarrow \begin{bmatrix} \oplus_{r \in R} \mathcal{H}_{[r]} \\ \mathcal{Y} \end{bmatrix}
$$

is unitary, and hence we have the relations

$$
BB^* = I - AA^*, \qquad BD^* = -AC^*, \qquad DB^* = -CA^*, \qquad DD^* = I - CC^*.
\tag{5.43}
$$

Then we compute

$$
\begin{aligned}
I - T(z)T(z')^* &= I - \big(D + C(I - Z_\Sigma(z)A)^{-1}Z_\Sigma(z)B\big) \cdot \\
&\quad \cdot \big(D^* + B^* Z_\Sigma(z')^*(I - A^* Z_\Sigma(z')^*)^{-1}C^*\big) \\
&= I - DD^* - C(I - Z_\Sigma(z)A)^{-1}Z_\Sigma(z)BD^* - DB^* Z_\Sigma(z')^*(I - A^* Z_\Sigma(z')^*)^{-1}C^* \\
&\quad - C(I - Z_\Sigma(z)A)^{-1}Z_\Sigma(z)BB^* Z_\Sigma(z')^*(I - A^* Z_\Sigma(z')^*)^{-1}C^* \\
&= CC^* + C(I - Z_\Sigma(z)A)^{-1}Z_\Sigma(z)AC^* + CA^* Z_\Sigma(z')^*(I - A^* Z_\Sigma(z')^*)^{-1}C^* \\
&\quad - C(I - Z_\Sigma(z)A)^{-1}Z_\Sigma(z)(I - AA^*)Z_\Sigma(z')^*(I - A^* Z_\Sigma(z')^*)^{-1}C^* \\
&\qquad \text{(where we used (5.43))} \\
&= C(I - Z_\Sigma(z)A)^{-1}\,[I - Z_\Sigma(z)Z_\Sigma(z')^*]\,(I - A^* Z_\Sigma(z')^*)^{-1}C^* \quad \text{(by algebra)}
\end{aligned}
$$

and (3) follows with

$$
H(z) = C(I - Z_\Sigma(z)A)^{-1} \in \mathcal{L}(\oplus_{s \in S}\mathcal{H}_{[s]}, \mathcal{Y})\langle\langle z \rangle\rangle.
$$

Note that this computation uses only that U is coisometric; we conclude that if $T(z)$ has a realization as the transfer function of a SNMLS $\Sigma = (G, \mathcal{H}, U)$ with coisometric connection matrix U, then T also has such a realization with unitary connection matrix U.

By a completely parallel computation, one can verify that, whenever U is isometric,

$$
I - T(z)^*T(z') = G(z)^*\,(I - Z_{G,\mathcal{H}'}(z)^* Z_{G,\mathcal{H}'}(z'))\,G(z')
\tag{5.44}
$$

for some formal power series

$$
G(z') = \mathrm{col}_{r \in R}\, G_r(z') \in \mathcal{L}(\mathcal{U}, \oplus_{r \in R}\mathcal{H}''_{[r]})\langle\langle z' \rangle\rangle
$$

and some collection of Hilbert spaces $\mathcal{H}'' = \{\mathcal{H}''_p : p \in P\}$. Under the correspondence

$$
\widetilde{K}(z, z') = (\mathrm{col}_{r \in R}\, G_r(z)^*) \cdot (\mathrm{row}_{r \in R}\, G_r(z'))
$$

it is easy to see that (5.44) can be equivalently expressed as

$$
I - T(z)^*T(z') = \sum_{r \in R} \widetilde{K}_{r,r}(z, z') - \sum_{s \in S} \sum_{r,r' \in R:\, [r]=[r']=[s]} z'_{e_{s,r'}} \widetilde{K}_{r,r'}(z, z') z_{e_{s,r}}
\tag{5.45}
$$

for some positive-definite formal power series

$$\widetilde{K}(z, z') = [\widetilde{K}_{r,r'}(z, z')]_{r,r' \in R} \in \mathcal{L}(\oplus_{r \in R} \mathcal{H}''_{[r]})\langle\langle z, z' \rangle\rangle.$$

Hence if U is unitary then we have both (5.5) and (5.44) for some $H(z) \in \mathcal{L}(\oplus_{s \in S} \mathcal{H}'_{[s]}, \mathcal{Y})\langle\langle z \rangle\rangle$ and some $G(z') \in \mathcal{L}(\mathcal{U}, \oplus_{r \in R} \mathcal{H}''_{[r]})\langle\langle z' \rangle\rangle$, or equivalently, both (5.4) and (5.45) for some positive-definite formal power series

$$K(z, z') \in \mathcal{L}(\oplus_{s \in S} \mathcal{H}'_{[s]})\langle\langle z, z' \rangle\rangle \text{ and } \widetilde{K}(z, z') \in \mathcal{L}(\oplus_{r \in R} \mathcal{H}''_{[r]})\langle\langle z, z' \rangle\rangle.$$

Moreover, since Theorem 4.2 is valid for dissipative SNMLSs, we see that $T_\Sigma(z)$ satisfies (1) in Theorem 5.3 if $\Sigma = (G, \mathcal{H}, U)$ with U merely contractive. We conclude that a given power series $T(z)$ can be realized as the transfer function of a dissipative SNMLS Σ (i.e., $\Sigma = (G, \mathcal{H}, U)$ with U contractive) if and only if it has a (possibly different) realization as the transfer function of a conservative SNMLS (with U unitary). Moreover, any of these characterizations is equivalent to the 2×2-block kernel decomposition

$$\begin{bmatrix} I_\mathcal{Y} - T(z)T(z')^* & T(z) - T(z') \\ T(z)^* - T(z')^* & I_\mathcal{U} - T(z)^*T(z') \end{bmatrix} = \begin{bmatrix} H(z) & 0 \\ 0 & G(z)^* \end{bmatrix} \tag{5.46}$$

$$\cdot \begin{bmatrix} I_{\oplus_{s \in S} \mathcal{H}'_{[s]}} - Z_{G,\mathcal{H}'}(z)Z_{G,\mathcal{H}'}(z')^* & Z_{G,\mathcal{H}'}(z) - Z_{G,\mathcal{H}'}(z') \\ Z_{G,\mathcal{H}'}(z)^* - Z_{G,\mathcal{H}'}(z')^* & I_{\oplus_{r \in R} \mathcal{H}'_{[r]}} - Z_{G,\mathcal{H}'}(z)^*Z_{G,\mathcal{H}'}(z') \end{bmatrix} \cdot \begin{bmatrix} H(z')^* & 0 \\ 0 & G(z') \end{bmatrix}$$

for some $H(z) \in \mathcal{L}(\oplus_{s \in S} \mathcal{H}'_{[s]}, \mathcal{Y})\langle\langle z \rangle\rangle$, $G(z') \in \mathcal{L}(\mathcal{U}, \oplus_{r \in R} \mathcal{H}''_{[r]})\langle\langle z' \rangle\rangle$ and a common collection of Hilbert spaces $\mathcal{H} = \{\mathcal{H}_p : p \in P\}$. Equivalently, under the correspondence

$$K(z, z') = \begin{bmatrix} K_{SS}(z, z') & K_{SR}(z, z') \\ K_{RS}(z, z') & K_{RR}(z, z') \end{bmatrix}$$

$$= \begin{bmatrix} \mathrm{col}_{s \in S} H_s(z) \\ \mathrm{col}_{r \in R} G_r(z)^* \end{bmatrix} \begin{bmatrix} \mathrm{row}_{s \in S} H_s(z')^* & \mathrm{row}_{r \in R} G_r(z') \end{bmatrix},$$

(5.46) is equivalent to

$$\begin{bmatrix} I_\mathcal{Y} - T(z)T(z')^* & T(z) - T(z') \\ T(z)^* - T(z')^* & I_\mathcal{U} - T(z)^*T(z') \end{bmatrix} = \begin{bmatrix} M_{SS}(z, z') & M_{SR}(z, z') \\ M_{RS}(z, z') & M_{SS}(z, z') \end{bmatrix} \tag{5.47}$$

where we have set

$$M_{SS}(z, z') = \sum_{s \in S} K_{SS;s,s}(z, z') - \sum_{r \in R} \sum_{s,s' \in S: \, [s]=[s']=[r]} z'_{e_{s',r}} K_{SS;s,s'}(z, z') z_{e_{s,r}}$$

$$M_{SR}(z, z') = \sum_{r \in R} \sum_{s \in S: \, [s]=[r]} \left(K_{SR;s,r}(z, z') z_{e_{s,r}} - z'_{e_{s,r}} K_{SR;s,r}(z, z') \right)$$

$$M_{RS}(z, z') = \sum_{s \in S} \sum_{r \in R: \, [r]=[s]} \left(K_{RS;r,s}(z, z') z_{e_{s,r}} - z'_{e_{s,r}} K_{RS;r,s}(z, z') \right)$$

$$M_{RR}(z, z') = \sum_{r \in R} K_{RR;r,r}(z, z') - \sum_{s \in S} \sum_{r,r' \in R: \, [r]=[r']=[s]} z'_{e_{s,r'}} K_{RR;r,r'}(z, z') z_{e_{s,r}}$$

$$\tag{5.48}$$

for some positive-definite formal power series

$$K(z, z') = \begin{bmatrix} K_{SS}(z, z') & K_{SR}(z, z') \\ K_{RS}(z, z') & K_{RR}(z, z') \end{bmatrix}$$

$$= \begin{bmatrix} [K_{SS;s,s'}(z, z')]_{s,s' \in S} & [K_{SR;s,r}(z, z')]_{s \in S; r \in R} \\ [K_{RS;r,s}(z, z')]_{r \in R; s \in S} & [K_{RR;r,r'}(z, z')]_{r,r' \in R} \end{bmatrix} \in \mathcal{L}\left(\begin{bmatrix} \oplus_{s \in S} \mathcal{H}'_{[s]} \\ \oplus_{r \in R} \mathcal{H}'_{[r]} \end{bmatrix} \right) \langle\langle z, z' \rangle\rangle.$$

Complete details for the commutative case appear in [7, 12].

Remark 5.11. **Fornasini-Marchesini conservative SNMLS.** In this extended remark we lay out how Theorem 5.3 specializes for the case where $G = G^{FM}$ is as in the setting of the noncommutative Fornasini-Marchesini systems explored in Example 2.1. In this case the Agler decompositions (5.4) or (5.5) for the formal power series $T(z) \in \mathcal{L}(\mathcal{U}, \mathcal{Y})\langle\langle z \rangle\rangle$ assume the forms

$$I - T(z)T(z')^* = K(z, z') - \sum_{k=1}^{d} z_k'^{-1} K(z, z') z_k^{-1}$$

$$= H(z) \cdot (1 - z_1 z_1' - \cdots - z_d z_d') I_{\mathcal{H}'} \cdot H(z')^*. \tag{5.49}$$

By Theorem 5.3 applied to the Fornasini-Marchesini case, we see that a given formal power series $T(z) \in \mathcal{L}(\mathcal{U}, \mathcal{Y})\langle\langle z \rangle\rangle$ satisfies (5.49) if and only if T is in the Fornasini-Marchesini Schur-Agler class $\mathcal{SA}_{G^{FM}}(\mathcal{U}, \mathcal{Y})$ given in this case by

$$\mathcal{SA}_{G^{FM}}(\mathcal{U}, \mathcal{Y}) = \{T(z) \in \mathcal{L}(\mathcal{U}, \mathcal{Y})\langle\langle z \rangle\rangle : \|T(\delta_1, \ldots, \delta_d)\| \leq 1$$

$$\text{for all } \delta_1, \ldots, \delta_d \in \mathcal{L}(\mathcal{K}) \text{ with } \delta_1 \delta_1^* + \cdots + \delta_d \delta_d^* \leq 1\}. \tag{5.50}$$

There has been much work of late from a number of different points of view on a noncommutative analogue of the algebra of Toeplitz operators on the unit disk (the "Cuntz-Toeplitz algebra" – see [49, 51, 52, 29, 31, 30, 20, 21]. The Cuntz-Toeplitz algebra (expressed in our notation) is the multiplier algebra $\mathcal{M}_{nc,d}$ consisting of the formal power series $T(z) = \sum_{v \in \mathcal{F}_d} T_v z^v$ with scalar coefficients $T_v \in \mathbb{C}$ such that the left multiplication operator

$$M_T \colon f(z) \mapsto T(z) \cdot f(z) \tag{5.51}$$

defines a bounded operator on the Fock space $L^2(\mathcal{F}_d)$ defined by

$$L^2(\mathcal{F}_d) = \left\{ f(z) = \sum_{v \in \mathcal{F}_d} f_v z^v \in \mathbb{C}\langle\langle z \rangle\rangle : \sum_{v \in \mathcal{F}_d} |f_v|^2 < \infty. \right\}$$

The tensor product of this space with $\mathcal{L}(\mathcal{U}, \mathcal{Y})$ is the space $\mathcal{M}_{nc,d}(\mathcal{U}, \mathcal{Y})$ consisting of formal power series

$$T(z) = \sum_{v \in \mathcal{F}_d} T_v z^v \in \mathcal{L}(\mathcal{U}, \mathcal{Y})\langle\langle z \rangle\rangle$$

for which the associated left multiplication operator as in (5.51) defines a bounded operator from $L^2(\mathcal{F}_d, \mathcal{U}) := L^2(\mathcal{F}_d) \otimes \mathcal{U}$ into $L^2(\mathcal{F}_d, \mathcal{Y}) := L^2(\mathcal{F}_d) \otimes \mathcal{Y}$. It is then natural to define a d-variable, noncommutative analogue of the Schur class $\mathcal{S}_{nc,d}(\mathcal{U},\mathcal{Y})$ by

$$\mathcal{S}_{nc,d}(\mathcal{U},\mathcal{Y}) = \left\{ T(z) \in \mathcal{L}(\mathcal{U},\mathcal{Y})\langle\langle z \rangle\rangle \colon \|M_T\|_{\mathcal{L}(L^2(\mathcal{F}_d,\mathcal{U}),L^2(\mathcal{F}_d,\mathcal{Y}))} \leq 1 \right\} \qquad (5.52)$$

where $M_T\colon L^2(\mathcal{F}_d,\mathcal{U}) \to L^2(\mathcal{F}_d,\mathcal{Y})$ is as in (5.51). As pointed out in [21], the condition that $T \in \mathcal{S}_{nc,d}(\mathcal{U},\mathcal{Y})$ can also be expressed as the positivity of a certain kernel

$$k_{\mathcal{F}_d}(z,z')I_{\mathcal{Y}} - T(z)(k_{\mathcal{F}_d}(z,z')I_{\mathcal{U}})T(z')^* = H(z)H(z')^* \qquad (5.53)$$

for some $H(z) \in \mathcal{L}(\mathcal{H}',\mathcal{Y})\langle\langle z \rangle\rangle$, where we have set $k_{\mathcal{F}_d}$ equal to the noncommutative Szegö kernel

$$k_{\mathcal{F}_d}(z,z') := \sum_{v \in \mathcal{F}_d} z^v z'^{v^\top}. \qquad (5.54)$$

It turns out that the Schur-Agler class $\mathcal{S}\mathcal{A}_{GFM}(\mathcal{U},\mathcal{Y})$ and the d-variable noncommutative Schur class $\mathcal{S}_{nc,d}(\mathcal{U},\mathcal{Y})$ are identical, as explained in the following Proposition.

Proposition 5.12. *Let $T(z) = \sum_{v \in \mathcal{F}_d} T_v z^v \in \mathcal{L}(\mathcal{U},\mathcal{Y})\langle\langle z \rangle\rangle$ be a formal power series with coefficients in $\mathcal{L}(\mathcal{U},\mathcal{Y})$. Then $T(z)$ satisfies (5.49) for some $H(z)$ in $\mathcal{L}(\mathcal{H},\mathcal{Y})\langle\langle z \rangle\rangle$ if and only if $T(z)$ satisfies (5.53) for the same $H(z)$ in $\mathcal{L}(\mathcal{H},\mathcal{Y})\langle\langle z \rangle\rangle$. Thus the Fornasini-Marchesini Schur-Agler class $\mathcal{S}\mathcal{A}_{GFM}(\mathcal{U},\mathcal{Y})$ defined as in (5.50) is identical to the noncommutative Schur class $\mathcal{S}_{nc,d}(\mathcal{U},\mathcal{Y})$ defined as in (5.52).*

Proof. By Theorem 5.3 we know that (5.5) characterizes the Fornasini-Marchesini Schur-Agler class $\mathcal{S}\mathcal{A}_{GFM}(\mathcal{U},\mathcal{Y})$ and by the result from [21] mentioned above we know that (5.53) characterizes the noncommutative Schur class $\mathcal{S}_{nc,d}(\mathcal{U},\mathcal{Y})$; hence the second assertion in Proposition 5.12 is an immediate consequence of the first.

Assume now that $T(z)$ satisfies (5.49) for some $H(z) \in \mathcal{L}(\mathcal{H},\mathcal{Y})\langle\langle z \rangle\rangle$. Multiplication of (5.49) on the left by $z'^{v^\top}$ and then on the right by z^v followed by the summation over all $v \in \mathcal{F}_d$ leads to (5.53). Conversely, multiplication of (5.53) on the left by z'_e and on the right by z_e followed by the sum over all $e \in E$ leads to (5.49). This completes the proof of Proposition 5.12. $\qquad\square$

For the Fornasini-Marchesini special case (where $G = G^{FM}$), it turns out that the content of Theorem 5.3 can be gleaned from various pieces already existing in the literature. Specifically, we have:

(4) $\Longrightarrow$ (1): (This amounts to Theorem 4.2 specialized to the Fornasini-Marchesini case.) As in Remark 5.10, we see that (4) $\Longrightarrow$ (3) with $H(z) = C(I - Z_\Sigma(z)A)^{-1}$, so we may assume that $T(z)$ has a Fornasini-Marchesini Agler decomposition (5.49). By Proposition 5.12, an equivalent condition is that $T(z)$ is in the noncommutative Schur class $\mathcal{S}_{nc,d}(\mathcal{U},\mathcal{Y})$ (see (5.52)), i.e., $\|M_T\| \leq 1$. Note that M_T amounts to $T(S) = \lim_{r\uparrow 1} T(rS)$ where $S = (S_1,\ldots,S_d)$ is the d-tuple of creation operators on $L^2(\mathcal{F}_d,\mathbb{C})$:

$$S_k\colon f(z) \mapsto z_k f(z) \quad \text{for } k = 1,\ldots,d. \qquad (5.55)$$

Note next that, given a d-tuple $\delta = (\delta_1, \ldots, \delta_d)$ on a Hilbert space $\mathcal{K}$, then $Z_{GFM}(\delta)$ amounts to the operator-block row matrix

$$Z_{GFM}(\delta) = \begin{bmatrix} \delta_1 & \cdots & \delta_d \end{bmatrix} : \oplus_{j=1}^d \mathcal{K} \to \mathcal{K},$$

and hence the class $\mathcal{B}_{GFM}\mathcal{L}(\mathcal{K})$ consists of *strict row contractions*, i.e., $\delta = (\delta_1, \ldots, \delta_d)$ with $\| \begin{bmatrix} \delta_1 & \cdots & \delta_d \end{bmatrix} \| < 1$. It is known (see [27, 48, 53]) that any strict row contraction $\delta = (\delta_1, \ldots, \delta_d)$ dilates to a row shift of some multiplicity, i.e., one can embed $\mathcal{K}$ as a subspace of $L^2(\mathcal{F}_d, \mathcal{E})$ for some auxiliary Hilbert space $\mathcal{E}$ in such a way that

$$\delta^v = P_{\mathcal{K}}(S \otimes I_{\mathcal{E}})^v|_{\mathcal{K}} \text{ for } v \in \mathcal{F}_d.$$

But then we have $T(\delta) = P_{\mathcal{Y} \otimes \mathcal{K}} T(S \otimes I_{\mathcal{E}})|_{\mathcal{U} \otimes \mathcal{K}}$. From the fact that $\|T(S \otimes I_{\mathcal{E}})\| = \|T(S)\| \leq 1$, we conclude that $\|T(\delta)\| \leq 1$. Alternatively, once we have established that $\|M_T\| \leq 1$, we may apply von Neumann's inequality for the noncommutative ball setting (see [50, 53]) to conclude that $\|T(\delta)\| \leq 1$; in fact, a natural way to prove von Neumann's inequality is as an application of dilation theory as sketched above. Via either way, we have verified (4) $\Longrightarrow$ (3) $\Longrightarrow$ (1) in Theorem 5.3 for the Fornasini-Marchesini case.

(1) $\Longrightarrow$ **(3):** Suppose now that $T(z) \in \mathcal{L}(\mathcal{U}, \mathcal{Y})\langle\langle z \rangle\rangle$ satisfies condition (1) in Theorem 5.3 (specialized to the Fornasini-Marchesini case), i.e., $T(z)$ is in the Fornasini-Marchesini Schur-Agler class $\mathcal{SA}_{GFM}(\mathcal{U}, \mathcal{Y})$ given by (5.50). In particular, we have that $\|T(rS)\| \leq 1$ for each $r < 1$ where $S = (S_1, \ldots, S_d)$ is the d-tuple of row shifts on $L^2(\mathcal{F}_d, \mathbb{C})$ as in (5.55). By letting $r \to 1$ we see that $\|T(S)\| \leq 1$. As observed in the previous paragraph, $T(S) = M_T$ and we see that $T(z) \in \mathcal{S}_{nc,d}(\mathcal{U}, \mathcal{Y})$. Again by Proposition 5.12, equivalently $T(z)$ satisfies (5.49) for some $H(z) \in \mathcal{L}(\mathcal{U}, \mathcal{Y})\langle\langle z \rangle\rangle$. In this way we have verified (1) $\Longrightarrow$ (3) in Theorem 5.3 for the Fornasini-Marchesini case.

(3) $\Longrightarrow$ **(4):** Assume now that $T(z)$ satisfies condition (3) in Theorem 5.3 specialized to the Fornasini-Marchesini case (5.49), i.e., that $T(z)$ satisfies (5.49) for some $H(z) \in \mathcal{L}(\oplus_{s \in S}\mathcal{H}'_{[s]}, \mathcal{Y})\langle\langle z \rangle\rangle$. By Proposition 5.12 an equivalent assumption is that $T(z)$ is in the noncommutative Schur class $\mathcal{S}_{nc,d}(\mathcal{U}, \mathcal{Y})$ given by (5.52), i.e., $\|M_T\| \leq 1$. In case $I - T(z)^*T(z')$ has 0 maximal factorable minorant, the fact that $T(z)$ has a realization $T(z) = T_\Sigma(z)$ for a Fornasini-Marchesini conservative SNMLS $\Sigma = (G^{FM}, \mathcal{H}, U)$ follows from the work of Popescu (see [49]), where a Sz.-Nagy-Foiaş model theory for row contractions is developed. By later results obtained in [20], it follows that in fact *any* noncommutative Schur class formal power series $T(z) \in \mathcal{S}_{nc,d}(\mathcal{U}, \mathcal{Y})$ can be realized as $T(z) = T_\Sigma(z)$ for a Fornasini-Marchesini conservative SNMLS Σ, i.e., the restriction that $I - T(z)^*T(z')$ have 0 maximal factorable minorant in the Popescu result can be removed. The result from [20] used functional models for representations of the Cuntz algebra (see [19]) to extend the model theory from [49] to the case of a general completely-nonunitary row contraction. In this way we have an alternate verification of (3) $\Longrightarrow$ (4) in Theorem 5.3 for the Fornasini-Marchesini case.

Finally we mention that the proof of (3) $\implies$ (4) presented here, but specialized to the Fornasini-Marchesini case, is presented in [21].

Remark 5.13. When specialized to the special case of Givone-Roesser conservative noncommutative systems (see Example 2.2), Theorem 5.3 can be viewed as a noncommutative analogue of the realization result of Agler [1] (see also [17]). Similarly, the specialization of Theorem 5.3 to the case of full-structured conservative noncommutative multidimensional systems (Example 2.3) can be viewed as a noncommutative extension of the realization result of [12] (see also [7, 6]) for the special case of Cartan domains of Type I.

Remark 5.14. Let us say that the admissible graph G^{RS} is a *row-sum* graph if each path-connected component p of G^{RS} contains exactly one source vertex s_p. As the name suggests, the associated structure matrix $Z_{G^{RS}}(z)$ is the direct sum of Fornasini-Marchesini structure graphs:

$$Z_{G^{RS}}(z) = \begin{bmatrix} z_{1,1} & \cdots & z_{1,d_1} & & & & & \\ & & & z_{2,1} & \cdots & z_{2,d_2} & & \\ & & & & & & \ddots & \\ & & & & & & & z_{K,1} & \cdots & z_{K,d_K} \end{bmatrix}.$$

Equivalently, row-sum graphs are exactly the admissible graphs for which there is a unique source-vertex cross-section $p \mapsto s_p$ (see (2.10)). The result Theorem 4.2 combined with the observation (4.6) gives the following: *if G^{RS} is a row-sum graph and $\Sigma = (G^{RS}, \mathcal{H}, U)$ is a conservative (or dissipative) system with structure graph G^{RS}, then the associated transfer function $T_\Sigma(z)$ is in the noncommutative Schur class $\mathcal{S}_{nc,d}(\mathcal{U}, \mathcal{Y})$.* On the other hand, from the discussion above we have seen that $\mathcal{S}_{nc,d}(\mathcal{U}, \mathcal{Y})$ coincides with the class $\mathcal{SA}_{G^{FM}}(\mathcal{U}, \mathcal{Y})$ of formal power series $T(z) \in \mathcal{L}(\mathcal{U}, \mathcal{Y})\langle\langle z \rangle\rangle$ having realization as the transfer function of a conservative SNMLS $\Sigma = (G^{FM}, \mathcal{H}, U)$ with Fornasini-Marchesini structure graph G^{FM}. By Theorem 5.3, the first class is characterized by $\|T(\delta)\| \le 1$ for any $\delta \in \mathcal{B}_{G^{RS}}\mathcal{L}(\mathcal{K})$ while the second class is characterized by $\|T(\delta)\| \le 1$ for any $\delta \in \mathcal{B}_{G^{FM}}\mathcal{L}(\mathcal{K})$. Thus it must be the case that: *given a formal power series $T(z) \in \mathcal{L}(\mathcal{U}, \mathcal{Y})\langle\langle z \rangle\rangle$ in d noncommuting indeterminates $z = (z_1, \ldots, z_d)$ and a row-sum graph G^{RS} with edge set*

$$E = \{(1,1)\ldots,(1,d_1),(2,1)\ldots,(2,d_2),\ldots,(K,1),\ldots,(K,d_K)\},$$

if $\|T(\delta)\| \le 1$ for all $\delta \in \mathcal{B}_{G^{RS}}\mathcal{L}(\mathcal{K})$, then also $\|T(\delta)\| \le 1$ for all $\delta \in \mathcal{B}_{G^{FM}}\mathcal{L}(\mathcal{K})$ (where G^{FM} is the Fornasini-Marchesini graph with edge set E). In fact one can see this result directly from the fact that

$$\mathcal{B}_{G^{FM}}\mathcal{L}(\mathcal{K}) \subset \mathcal{B}_{G^{RS}}\mathcal{L}(\mathcal{K}) \text{ if } G^{RS} \text{ is a row-sum graph.} \tag{5.56}$$

Indeed, if $\delta = (\delta_{1,1}, \ldots, \delta_{1,d_1}, \delta_{2,1}, \ldots, \delta_{2,d_2}, \ldots, \delta_{K,1}, \ldots, \delta_{K,d_K}) \in \mathcal{B}_{G^{FM}}\mathcal{L}(\mathcal{K})$, then the row matrix

$$Z_{G^{FM}}(\delta) = \begin{bmatrix} \delta_{1,1} & \cdots & \delta_{1,d_1} & \delta_{2,1} & \cdots & \delta_{2,d_2} & \cdots, \delta_{K,1} & \cdots & \delta_{K,d_K} \end{bmatrix}$$

is contractive. In particular for each $k = 1, \ldots, K$ the shorter row $\begin{bmatrix} \delta_{k,1} & \cdots & \delta_{k,d_k} \end{bmatrix}$ is contractive from which we see that the row-sum matrix

$$
Z_{G^{RS}}(\delta) = \begin{bmatrix} \delta_{1,1} & \cdots & \delta_{1,d_1} \\ & & & \delta_{2,1} & \cdots & \delta_{2,d_2} \\ & & & & & & \ddots \\ & & & & & & & \delta_{K,1} & \cdots & \delta_{K,d_K} \end{bmatrix}
$$

is contractive, i.e., $\delta \in \mathcal{B}_{G^{RS}}\mathcal{L}(\mathcal{K})$. In this way the containment (5.56) follows in a simple direct way.

Remark 5.15. One can view Theorem 5.3 as really concerning the formal power series

$$
K(z, z') = I - T(z)T(z')^* \tag{5.57}
$$

in two sets of noncommuting indeterminates $z = (z_1, \ldots, z_d)$ and $z' = (z'_1, \ldots, z'_d)$ rather than $T(z)$ itself. Expressed in this way, Theorem 5.3 says that a formal power series $K(z, z') = \sum_{v,v' \in \mathcal{F}_d} K_{v,v'} z^v z'^{v'} \in \mathcal{L}(\mathcal{Y})\langle\langle z, z'\rangle\rangle$ of the special form (5.57) has the representation

$$
K(z, z') = H(z)\left(I - Z_{G,\mathcal{H}'}(z)Z_{G,\mathcal{H}'}(z')^*\right)H(z')^* \tag{5.58}
$$

for some $H(z) \in \mathcal{L}(\mathcal{H}, \mathcal{Y})\langle\langle z\rangle\rangle$ if and only if

$$
K(\delta, \delta) = \sum_{v,v' \in \mathcal{F}_d} K_{v,v'} \otimes \delta^v (\delta^*)^{v'} \geq 0 \tag{5.59}
$$

for all operator d-tuples $\delta = (\delta_1, \ldots, \delta_d) \in \mathcal{B}_G\mathcal{L}(\mathcal{K})$. One can pose the question of obtaining results along this line without the restriction that $K(z, z')$ a priori has the special form (5.57).

In case one takes $K(z, z')$ to be a general hereditary kernel, sets $Z_{G,\mathcal{H}'}(z)$ formally equal to zero, and replaces $\mathcal{B}_G\mathcal{L}(\mathcal{K})$ by the set $\mathcal{N}$ of nilpotent d-tuples δ of matrices of arbitrary finite size (δ with $\delta^v = 0$ for $|v|$ sufficiently large), such a result appears in the recent paper of Kalyuzhnyĭ-Verbovetzkiĭ and Vinnikov (see [43]). For the special case where $K(z, z')$ is a polynomial and sets $\mathcal{B}_G\mathcal{L}(\mathcal{K})$ equal to all of $\mathcal{L}(\mathcal{K})^d$ (where $\mathcal{K}$ is taken to be any finite-dimensional Hilbert space), the Positivstellensatz of [39] gives a similar type result. For $\mathcal{B}_G\mathcal{L}(\mathcal{K})$ set equal to other types of algebraic varieties or semivarieties, see [40] and [41].

References

[1] J. Agler, On the representation of certain holomorphic functions defined on a polydisk, in *Topics in Operator Theory: Ernst D. Hellinger memorial Volume* (Ed. L. de Branges, I. Gohberg and J. Rovnyak), pp. 47-66, **OT48** Birkhäuser Verlag, Basel, 1990.

[2] J. Agler and J.E. McCarthy, Nevanlinna-Pick interpolation on the bidisk, *J. Reine Angew. Math.* **506** (1999), 191–204.

[3] J. Agler and J.E. McCarthy, Complete Nevanlinna-Pick kernels, *J. Functional Analysis*, **175** (2000), 111–124.

[4] D. Alpay, V. Bolotnikov and T. Kaptanoğlu, The Schur algorithm and reproducing kernel Hilbert spaces in the ball, *Linear Algebra Appl.* **342** (2002), 163–186.

[5] D. Alpay and D.S. Kalyuzhnyĭ-Verbovetzkiĭ, Matrix-J-unitary non-commutative rational formal power series, in this volume.

[6] C.-G. Ambrozie and J. Eschmeier, A commutant lifting theorem on analytic polyhedra, *Proceedings of Operator Theory Conference Dedicated to Prof. Wieslaw Zelazko*, Banach Center publ., Warszawa, to appear.

[7] C.-G. Ambrozie and D. Timotin. A von Neumann type inequality for certain domains in $\mathbb{C}^n$, *Proc. Amer. Math. Soc.*, **131** (2003), 859–869.

[8] A. Arias and G. Popescu, Noncommutative interpolation and Poisson transforms, *Israel J. Math.* **115** (2000), 205–234.

[9] N. Aronszajn, Theory of reproducing kernels, *Trans. Amer. Math. Soc.*, **68** (1950), 337–404.

[10] W. Arveson, Subalgebras of C^*-algebras III: multivariable operator theory, *Acta Math.*, **181** (1998), 159–228.

[11] J.A. Ball, Linear systems, operator model theory and scattering: Multivariable generalizations, in *Operator Theory and its Applications* (Ed. A.G. Ramm, P.N. Shivakumar and A.V. Strauss), **FIC25**, Amer. Math. Soc., Providence, 2000.

[12] J.A. Ball and V. Bolotnikov, Realization and interpolation for Schur-Agler class functions on domains with matrix polynomial defining function in $\mathbb{C}^n$, *J. Functional Analysis* **213** (2004), 45–87.

[13] J.A. Ball, G. Groenewald and T. Malakorn, Structured noncommutative multidimensional linear systems, *SIAM J. Control and Optimization*, to appear.

[14] J.A. Ball, G. Groenewald and T. Malakorn, Bounded real lemma for structured noncommutative multidimensional linear systems and robust control, preprint (2005).

[15] J.A. Ball, W.S. Li, D. Timotin and T.T. Trent, A commutant lifting theorem on the polydisc, *Indiana Univ. Math. J.* **48** (1999), 653–675.

[16] J.A. Ball and T. Malakorn, Multidimensional linear feedback control systems and interpolation problems for multivariable holomorphic functions, *Multidimensional Systems and Signal Processing* **15** (2004), 7–36.

[17] J.A. Ball and T. Trent, Unitary colligations, reproducing kernel Hilbert spaces and Nevanlinna–Pick interpolation in several variables, *J. Functional Analysis*, **157** (1998), no.1, 1–61.

[18] J.A. Ball, T.T. Trent and V. Vinnikov, Interpolation and commutant lifting for multipliers on reproducing kernel Hilbert spaces, in: *Operator Theory and Analysis: The M.A. Kaashoek Anniversary Volume (Workshop in Amsterdam, Nov. 1997)*, pages 89–138, **OT 122**, Birkhäuser Verlag, Basel, 2001.

[19] J.A. Ball and V. Vinnikov, Functional models for representations of the Cuntz algebra, in *Operator Theory, System Theory and Scattering Theory: Multidimensional Generalizations* (Ed. D. Alpay and V. Vinnikov), Birkhäuser Verlag OT volume, to appear.

[20] J.A. Ball and V. Vinnikov, Lax-Phillips scattering and conservative linear systems: a Cuntz-algebra multidimensional setting, *Memoir of the AMS*, to appear.

[21] J.A. Ball and V. Vinnikov, Formal reproducing kernel Hilbert spaces: the commutative and noncommutative settings, in *Reproducing Kernel Hilbert Spaces* (Ed. D. Alpay), pages 77–134, **OT 143**, Birkhäuser Verlag, Basel, 2003.

[22] C.L. Beck, On formal power series representations for uncertain systems, *IEEE Trans. Auto. Contr.* **46** No. 2 (2001), 314–319.

[23] C.L. Beck and J.C. Doyle, A necessary and sufficient minimality condition for uncertain systems, *IEEE Trans. Auto. Contr.* **44** No. 10 (1999), 1802–1813.

[24] C.L. Beck, J.C. Doyle and K. Glover, Model reduction of multidimensional and uncertain systems, *IEEE Trans. Auto. Contr.* **41** No. 10 (1996), 1466–1477.

[25] L. de Branges and J. Rovnyak, Canonical models in quantum scattering theory, in *Perturbation Theory and its Applications in Quantum Mechanics* (Ed. C.H. Wilcox), Wiley, New York, 1966, pp. 295–392.

[26] M.S. Brodskiĭ, *Triangular and Jordan Representations of Linear Operators*, Volume Thirty-Two, Translations of Mathematical Monographs, American Mathematical Society, Providence, 1971.

[27] J.W. Bunce, Models for n-tuples of noncommuting operators, *J. Functional Analysis* **57** (1984), 21–30.

[28] J.B. Conway, *A Course in Operator Theory*, Graduate Studies in Mathematics Vol. **21**, American Mathematical Society (Providence), 2000.

[29] K.R. Davidson and D.R. Pitts, The algebraic structure of non-commutative analytic Toeplitz algebras, *Math. Ann.* **311** (1998), 275–303.

[30] K.R. Davidson and D.R. Pitts, Nevanlinna–Pick interpolation for non-commutative analytic Toeplitz algebras, *Integral Equations Operator Theory* **31** (1998), no. 3, 321–337.

[31] K.R. Davidson and D.R. Pitts, Invariant subspaces and hyper-reflexivity for free semigroup algebras, *Proc. London Math. Soc.* **78** (1999), 401–430.

[32] N. Dunford and L.T. Schwartz, *Linear Operators Part I: General Theory*, Interscience Publishers, New York, 1958.

[33] S.W. Drury, A generalization of von Neumann's inequality to the complex ball, *Proc. Amer. Math. Soc.*, **68** (1978), 300–304.

[34] E. Fornasini and G. Marchesini, Doubly-indexed dynamical systems: state space models and structural properties, *Math. System Theory* **12** (1978), 59–72.

[35] D.D. Givone and R.P. Roesser, Multidimensional linear iterative circuits – general properties, *IEEE Trans. Comp.* **C-21** no. 10 (1972),1067–1073.

[36] D.D. Givone and R.P. Roesser, Minimization of multidimensional linear iterative circuits, *IEEE Trans. Comp.* **C-22** no. 7 (1973), 673–678.

[37] D. Greene, S. Richter and C. Sundberg, The structure of inner multipliers on spaces with complete Nevanlinna Pick kernels, *J. Functional Analysis* **194** no. 2 (2002), 311–331.

[38] J.W. Helton, The characteristic functions of operator theory and electrical network realization, *Indiana Univ. Math. J.* **22** (1972/73), 403–414.

[39] J.W. Helton, "Positive" noncommutative polynomials are sums of squares, *Ann. Math.* **56** (2002), 675–694.

[40] J.W. Helton and S.A. McCullough, A Positivstellensatz for noncommutative polynomials, *Trans. Amer. Math. Soc.* **356** No. 9 (2004), 3721–3737.

[41] J.W. Helton, S.A. McCullough and M. Putinar, A non-commutative Positivstellensatz on isometries, *J. Reine Angew. Math.* **568** (2004), 71–80.

[42] I. Gohberg (ed.), *I. Schur Methods in Operator Theory and Signal Processing*, **OT18** Birkhäuser Verlag, Basel-Boston, 1986.

[43] D.S. Kalyuzhnyĭ-Verbovetzkiĭ and V. Vinnikov, Non-commutative positive kernels and their matrix evaluations, *Proc. Amer. Math. Soc.*, to appear.

[44] W.-M. Lu, K. Zhou and J.C. Doyle, Stabilization of uncertain linear systems: an LFT approach, *IEEE Trans. Auto. Contr.* **41** No. 1 (1996), 50–65.

[45] S. McCullough and T.T. Trent, Invariant subspaces and Nevanlinna-Pick kernels, *J. Functional Analysis* **178** (2000), 226–249.

[46] B. Sz.-Nagy and C. Foiaş, *Harmonic Analysis of Operators on Hilbert Space*, North Holland/American Elsevier, 1970.

[47] N.K. Nikol'skiĭ, *Treatise on the Shift Operator: Spectral Function Theory*, Springer-Verlag, Berlin, 1986.

[48] G. Popescu, Models for infinite sequences of noncommuting operators, *Acta Sci. Math.* **53** (1989), 355–368.

[49] G. Popescu, Characteristic functions for infinite sequences of noncommuting operators, *J. Operator Theory* **22** (1989), 51–71.

[50] G. Popescu, Von Neumann inequality for $(B(\mathcal{H})^n)_1$, *Math. Scand.* **68** (1991), 292–304.

[51] G. Popescu, Multi-analytic operators on Fock spaces, *Math. Ann.* **303** (1995), 31–46.

[52] G. Popescu, Interpolation problems in several variables, *J. Math. Anal. Appl.* **227** (1998), 227–250.

[53] G. Popescu, Poisson transforms on some C^*-algebras generated by isometries, *J. Functional Analysis* **161** (1999), 27–61.

[54] W. Rudin, *Functional Analysis*, McGraw-Hill, New York, 1973.

[55] A.E. Taylor and D.C. Lay, *Introduction to Functional Analysis, Second Edition*, Wiley, 1980.

[56] A.T. Tomerlin, Products of Nevanlinna-Pick kernels and operator colligations, *Integral Equations Operator Theory* **38** (2000), no. 3, 350–356.

[57] K. Zhou, J.C. Doyle and K. Glover, *Robust and Optimal Control*, Prentice Hall, Upper Saddle River, New Jersey, 1996.

Joseph A. Ball
Department of Mathematics
Virginia Tech
Blacksburg, Virginia 24061-0123
e-mail: `ball@math.vt.edu`

Gilbert Groenewald
Department of Mathematics
North West University
Potchefstroom 2520, South Africa
e-mail: `wskgjg@puknet.puk.ac.za`

Tanit Malakorn
Department of Electrical and Computer Engineering
Naresuan University
Phitsanulok, 65000, Thailand
e-mail: `tanitm@nu.ac.th`

Operator Theory:
Advances and Applications, Vol. 161, 225–270
© 2005 Birkhäuser Verlag Basel/Switzerland

The Bezout Integral Operator: Main Property and Underlying Abstract Scheme

I. Gohberg, I. Haimovici, M.A. Kaashoek and L. Lerer

Abstract. For a class of entire matrix-functions a continuous analogue of the classical Bezout matrix for scalar polynomials is introduced and studied. This analogue is an integral operator with a matrix-valued kernel. The null space of this operator is explicitly expressed in terms of the common eigenvectors and common Jordan chains of the two underlying entire matrix functions. Also a refinement of the abstract scheme from [17] for defining Bezout operators is presented and analyzed. The approach of the paper is based, to a large extent, on the state space method from mathematical system theory. In particular, an important role is played by the fact that the functions involved can be represented as transfer functions of certain infinite-dimensional input output systems.

Mathematics Subject Classification (2000). Primary 47B35, 47B99, 45E10, 30D20; Secondary 33C47, 42C05, 93B15.

Keywords. Bezout operator, continuous analogue of the Bezout matrix, convolution integral operators on a finite interval, state space method.

Contents

The research of the fourth author was partially supported by a visitor fellowship of the Netherlands Organization for Scientific Research (NWO) and by the Fund for Promotion of Research at the Technion, Haifa.

1. Introduction

Let a, b, c, and d be $n \times n$ matrix functions, a and d belong to $L_1^{n \times n}[0, \omega]$, while b and c belong to $L_1^{n \times n}[-\omega, 0]$. We shall assume that the four functions a, b, c, d satisfy the following additional condition

$$\mathcal{A}(\lambda)\mathcal{B}(\lambda) = \mathcal{C}(\lambda)\mathcal{D}(\lambda), \quad \lambda \in \mathbb{C}, \tag{1.1}$$

where $\mathcal{A}$, $\mathcal{B}$, $\mathcal{C}$, $\mathcal{D}$ are the entire $n \times n$ matrix functions given by

$$\mathcal{A}(\lambda) = I_n + \int_0^\omega e^{i\lambda s} a(s)\, ds, \quad \mathcal{B}(\lambda) = I_n + \int_{-\omega}^0 e^{i\lambda s} b(s)\, ds, \tag{1.2}$$

$$\mathcal{C}(\lambda) = I_n + \int_{-\omega}^0 e^{i\lambda s} c(s)\, ds, \quad \mathcal{D}(\lambda) = I_n + \int_0^\omega e^{i\lambda s} d(s)\, ds. \tag{1.3}$$

Here I_n denotes the $n \times n$ identity matrix. Notice that in the scalar case ($n = 1$) the additional condition (1.1) is automatically fulfilled with $a = d$ and $b = c$.

Given four functions as above, we let T be the integral operator on $L_1^n[0, \omega]$ defined by

$$(T\varphi)(t) = \varphi(t) + \int_0^\omega \gamma(t, s)\varphi(s)\, ds, \quad 0 \le t \le \omega, \tag{1.4}$$

with the kernel function γ being given by

$$\gamma(t, s) \quad = \quad a(t - s) + b(t - s) +$$
$$+ \quad \int_0^{\min\{t,s\}} \big(a(t - r)b(r - s) - c(t - \omega - r)d(r + \omega - s)\big)\, dr. \tag{1.5}$$

We can now state the first main result of this paper, which shows that the operator T preserves the main property of the classical Bezout matrix.

Theorem 1.1. *Assume that condition (1.1) is satisfied. Then the dimension of the null space of the operator T defined by (1.4), (1.5) is equal to the total multiplicity of the common eigenvalues of the entire matrix functions $\mathcal{B}$ and $\mathcal{D}$.*

The definition of the total multiplicity of the common eigenvalues of the entire matrix functions $\mathcal{B}$ and $\mathcal{D}$, which involves the notion of common Jordan chains of $\mathcal{B}$ and $\mathcal{D}$, will be given at the end of Section 2.1 below. We shall also present a basis of the null space of T in terms of these common Jordan chains (Theorem 4.6).

For the scalar case and with $a = d$ and $b = c$, the above theorem, together with the description of its null space, has been proved in Section 6 of [9]. In [9] it has also been shown that for the scalar case and with $a = d$ and $b = c$ the operator T is the natural continuous analogue of the classical Bezout matrix for polynomials. For the matrix case $(n > 1)$ it is proved in [17] that this analogy with the classical Bezout matrix remains true provided condition (1.1) is satisfied. For this reason we shall refer to the operator T as the *Bezout integral operator* associated with $\{\mathcal{A}, \mathcal{C}; \mathcal{B}, \mathcal{D}\}$, and we simply write

$$T = T\{\mathcal{A}, \mathcal{C}; \mathcal{B}, \mathcal{D}\}.$$

We call (1.1) the *quasi commutativity property* of the quadruple $\{\mathcal{A}, \mathcal{C}; \mathcal{B}, \mathcal{D}\}$. Theorem 1.1 does not remain true, not even in the scalar case, when the quasi commutativity property is not satisfied (see the example at the end of Chapter 3).

Theorem 1.1 has been proved in the dissertation [16] using the general scheme for Bezout operators developed in [17]. In this paper we give a self-contained and independent proof of Theorem 1.1. Theorem 4.6, which gives explicit formulas for a basis of the null space of T, is our second main result and seems to be new. We also present and analyse a more refined version of the scheme from [17], and use this to prove Theorem 4.6.

This paper consists of four chapters including the present introduction. In the second chapter we recall the notion of total common multiplicity (that is, the total multiplicity of the common eigenvalues) of two entire matrix functions and study common spectral data, including common Jordan chains, of such functions. The latter is done by representing the functions involved as transfer functions of certain infinite-dimensional input output systems. The main result is Theorem 2.6 which identifies the total common multiplicity of two entire matrix functions in terms of certain invariant subspaces. Theorem 1.1 is proved in Chapter 3 using Theorem 2.6. In the final chapter we return to the definition of a Bezout integral operator T. We present a refinement of the scheme from [17] for defining Bezout operators, and we show how our operator T fits into this scheme. In the final section we prove our second main theorem (Theorem 4.6).

We conclude this introduction with a few remarks about the literature on the Bezout matrix and its generalizations. For the definition of the classical Bezout matrix for two scalar polynomials and a comprehensive survey of its properties and its use in various applications, we refer the reader to [20]. Getting the main property of the Bezout matrix for matrix (non-commutative) functions presented serious difficulties. Attempts to generalize the notion of the Bezout matrix to matrix polynomials by replacing scalar multiplication by the usual matrix multiplication or by the tensor (Kronecker) product did not yield a natural analogue with the same main property (see [24] for details).

The idea of involving supplementary functions $\mathcal{A}$ and $\mathcal{C}$ along with the given functions $\mathcal{B}$ and $\mathcal{D}$, such that (1.1) holds true, originates from some problems in system theory (see [1], also [6], [18]), and turned out to be successful; see, for the case of matrix and operator polynomials, [23], [24], and also Chapter 9 in the book [25]. Notice that in many problems the supplementary functions appear in a natural way or can be constructed from the given functions $\mathcal{B}$ and $\mathcal{D}$ (see, e.g., Section 5 in [10]) and Section 3 in [14]). Of course, in the commutative case one can just set $\mathcal{A} = \mathcal{D}$ and $\mathcal{C} = \mathcal{B}$.

When passing to non-polynomial matrix functions, a significant role is played by the idea of representing the matrix functions involved as transfer functions of certain input output systems. For entire scalar functions it originates from the paper [26]; see also Chapter 16 in [27]. This idea and the one of the previous paragraph were important in constructing proper analogues of the classical Bezout matrix for rational and analytic matrix functions (see, e.g., [10], [16], [17], [21], [22] and the references therein). The two ideas play also an important role in the present paper.

2. Spectral theory of entire matrix functions

In this chapter we deal with entire $n \times n$ matrix functions that are equal to the $n \times n$ identity matrix I_n at the point zero. Such a function F admits a representation of the form

$$F(\lambda) = I_n + \lambda C(I - \lambda A)^{-1} B, \quad \lambda \in \mathbb{C}. \tag{2.1}$$

Here A is a quasi-nilpotent operator on a Banach space $\mathcal{X}$, that is, A is a bounded linear operator of which the spectrum $\sigma(A)$ consists of the point zero only. Furthermore, $B : \mathbb{C}^n \to \mathcal{X}$ and $C : \mathcal{X} \to \mathbb{C}^n$ are bounded linear operators, and I is the identity operator on $\mathcal{X}$. Since $\sigma(A) = \{0\}$, the operator $I - \lambda A$ is invertible for each $\lambda \in \mathbb{C}$. Hence, both sides of (2.1) are well defined for each $\lambda \in \mathbb{C}$.

To get a representation of F as in (2.1), it is convenient to first consider the $n \times n$ matrix function $W(\lambda) = F(\lambda^{-1})$ which is defined and analytic on the set

$$\Omega = (\mathbb{C} \cup \{\infty\}) \backslash \{0\}.$$

In particular, $0 \notin \Omega$. Hence we can apply Theorem 2.5 in [5] to show that there exists a Banach space $\mathcal{X}$, a bounded linear operator A on $\mathcal{X}$ such that $\sigma(A) = \{0\}$, and bounded linear operators $B : \mathbb{C}^n \to \mathcal{X}$ and $C : \mathcal{X} \to \mathbb{C}^n$ such that

$$W(\lambda) = I_n + C(\lambda I - A)^{-1} B, \quad 0 \neq \lambda \in \mathbb{C}. \tag{2.2}$$

Here, as before, I is the identity operator on $\mathcal{X}$. Since $F(\lambda) = W(\lambda^{-1})$, from (2.2) we get (2.1) for each $0 \neq \lambda \in \mathbb{C}$. But both the left and right side of (2.1) are analytic at zero. Thus (2.1) holds for each $\lambda \in \mathbb{C}$.

One refers to the right-hand side of (2.2) as a *realization* of W. This terminology is taken from mathematical system theory, where functions of the form (2.2) appear as transfer functions of time-invariant input output systems (cf., [5], [7]). Following the system theory terminology we call the space $\mathcal{X}$ the *state space* of the realization, and the operator A in (2.2) is called the *main operator* or *state operator*. The operators B and C are called the *input operator* and *output operator*, respectively. We shall use these terms also for the operators in (2.1). In the sequel we refer to the right-hand side of (2.1) as a *co-realization*. The terms realization and co-realization will also be used when in (2.1) and (2.2) the identity matrix I_n is replaced by an arbitrary square matrix D.

This chapter, which consists of four sections, deals with the spectral properties of the functions F and W in terms of the representations (2.1) and (2.2). The first section, which has a preliminary character, reviews for analytic matrix functions the concepts of eigenvalues, corresponding eigenvectors and Jordan chains, and canonical systems of Jordan chains. In Section 2.2 we show that the representation (2.2) of W allows one to describe the eigenvalues, the corresponding eigenvectors, and canonical systems of Jordan chains corresponding to an eigenvalue of W in terms of the spectral properties of the operator $A^\times = A - BC$. Notice that the latter operator appears in a natural when one invert $W(\lambda)$. Indeed,

$$W(\lambda)^{-1} = I_n - C(\lambda I - A^\times)^{-1} B, \quad \lambda \notin \sigma(A^\times),$$

where $\sigma(A^\times)$ denotes the spectrum of $A^\times$. In Section 2.3 we use realizations to describe the common eigenvalues and common Jordan chains of two functions of the form (2.2). In the final section the results of the third section are applied to two entire matrix functions, and we use co-realizations to describe the common zero data of two such functions in operator terms.

2.1. A review of the spectral data of an analytic matrix function

Let H be an $n \times n$ matrix function, which is analytic on an open set Ω of $\mathbb{C}$. We assume H to be *regular* on Ω. The latter means that $\det H(\lambda) \not\equiv 0$ on each connected component of Ω. As usual the values of H are identified with their canonical action on $\mathbb{C}^n$. In what follows λ_0 is a point in Ω.

The point λ_0 is called an *eigenvalue* of H whenever there exists a vector $x_0 \neq 0$ in $\mathbb{C}^n$ such that $H(\lambda_0)x_0 = 0$. In that case the non-zero vector x_0 is called an *eigenvector* of H at λ_0. Note that λ_0 is an eigenvalue of H if and only if $\det H(\lambda_0) = 0$. In particular, in the scalar case, i.e., when n=1, the point λ_0 is an eigenvalue of H if and only if λ_0 is a zero of H. The *multiplicity* $\nu(\lambda_0)$ of the eigenvalue λ_0 of H is defined as the multiplicity of λ_0 as a zero of $\det H(\lambda)$. The set of eigenvectors of H at λ_0 together with the zero vector is equal to $\mathrm{Ker}\, H(\lambda_0)$.

An ordered sequence of vectors $x_0, x_1, \ldots, x_{r-1}$ in $\mathbb{C}^n$ is called a *Jordan chain of length* r of H at λ_0 if $x_0 \neq 0$ and

$$\sum_{j=0}^{k} \frac{1}{j!} H^{(j)}(\lambda_0) x_{k-j} = 0, \quad k = 0, \ldots, r-1. \tag{2.3}$$

Here $H^{(j)}(\lambda_0)$ is the jth derivative of H at λ_0. From $x_0 \neq 0$ and (2.3) it follows that λ_0 is an eigenvalue of H and x_0 a corresponding eigenvector. The converse is also true, that is, x_0 is an eigenvector of H at λ_0 if and only if x_0 is the first vector in a Jordan chain for H at λ_0.

Given an eigenvector x_0 of H at λ_0 there are in general many Jordan chains for H at λ_0 which have x_0 as their first vector. However, the fact that H is regular implies that the lengths of these Jordan chains have a finite supremum which we shall call the *rank* of the eigenvector x_0.

To organize the Jordan chains corresponding to the eigenvalue λ_0 we proceed as follows. Choose an eigenvector $x_{1,0}$ in $\operatorname{Ker} H(\lambda_0)$ such that the rank r_1 of $x_{1,0}$ is maximal, and let $x_{1,0}, \ldots, x_{1,r_1-1}$ be a corresponding Jordan chain. Next we choose among all vectors x in $\operatorname{Ker} H(\lambda_0)$, with x not a multiple of $x_{1,0}$, a vector $x_{2,0}$ of maximal rank, r_2 say, and we choose a corresponding Jordan chain $x_{2,0}, \ldots, x_{2,r_2-1}$. We proceed by induction. Assume

$$x_{1,0}, \ldots, x_{1,r_1-1}, \ldots, x_{k,0}, \ldots, x_{k,r_k-1}$$

have been chosen. Then we choose $x_{k+1,0}$ to be a vector in $\operatorname{Ker} H(\lambda_0)$ that does not belong to $\operatorname{span}\{x_{1,0}, \ldots, x_{k,0}\}$ such that $x_{k+1,0}$ is of maximal rank among all vectors in $\operatorname{Ker} H(\lambda_0)$ not belonging to $\operatorname{span}\{x_{1,0}, \ldots, x_{k,0}\}$. In this way, in a finite number of steps, we obtain a basis $x_{1,0}, x_{2,0}, \ldots, x_{p,0}$ of $\operatorname{Ker} H(\lambda_0)$ and corresponding Jordan chains

$$x_{1,0}, \ldots, x_{1,r_1-1}, x_{2,0}, \ldots, x_{2,r_2-1}, \ldots, x_{p,0}, \ldots, x_{p,r_p-1}. \tag{2.4}$$

The system (2.4) is called a *canonical system of Jordan chains* for H at λ_0. From the construction it follows that $p = \dim \operatorname{Ker} H(\lambda_0)$. Furthermore, the numbers $r_1 \geq r_2 \geq \cdots \geq r_p$ are uniquely determined by H and do not depend on the particular choices made above. They are called the *partial multiplicities* of H at λ_0. Their sum $r_1 + \cdots + r_p$ is equal to the multiplicity $\nu(\lambda_0)$.

The above definitions of eigenvalue, eigenvector and Jordan chain for H at λ_0 also make sense when H is non-regular or when H is a non-square analytic matrix function on Ω. However, in that case it may happen that the supremum of the lengths of the Jordan chains with a given first vector is not finite. On the other hand, if for each non-zero vector x_0 in $\operatorname{Ker} H(\lambda_0)$ the supremum of the lengths of the Jordan chains with x_0 as first vector is finite, then we can define a canonical set of Jordan chains for H at λ_0 in the same way as it was done above for regular analytic matrix functions.

More details on the above notions, including proofs, can be found in [15]; see also the book [13] or the appendix of [11].

Common spectral data. Next we consider two $n \times n$ matrix functions H_1 and H_2 which are analytic on an open subset Ω of $\mathbb{C}$. We also assume that either H_1 or H_2 is regular on Ω.

Let λ_0 be a point in Ω. We say that λ_0 is a *common eigenvalue* of H_1 and H_2 if there exists a vector $x_0 \neq 0$ such that $H_1(\lambda_0)x_0 = H_2(\lambda_0)x_0 = 0$. In this case we refer to x_0 as a *common eigenvector* of H_1 and H_2 at λ_0. Note that x_0 is a common eigenvector of H_1 and H_2 at λ_0 if and only if x_0 is a non-zero vector in

$$\mathrm{Ker}\, H_1(\lambda_0) \cap \mathrm{Ker}\, H_2(\lambda_0) = \mathrm{Ker} \begin{bmatrix} H_1(\lambda_0) \\ H_2(\lambda_0) \end{bmatrix}.$$

If an ordered sequence of vectors $x_0, x_1, \ldots, x_{r-1}$ is a Jordan chain for both H_1 and H_2 at λ_0, then we say that $x_0, x_1, \ldots, x_{r-1}$ is a *common Jordan chain* for H_1 and H_2 at λ_0. In other words, $x_0, x_1, \ldots, x_{r-1}$ is a common Jordan chain for H_1 and H_2 at λ_0 if and only if $x_0, x_1, \ldots, x_{r-1}$ is a Jordan chain for H at λ_0, where H is the non-square matrix function given by

$$H(\lambda) = \begin{bmatrix} H_1(\lambda) \\ H_2(\lambda) \end{bmatrix}, \qquad \lambda \in \Omega. \tag{2.5}$$

Let x_0 be a common eigenvector of H_1 and H_2 at λ_0. Since H_1 or H_2 is regular, the lengths of the common Jordan chains of H_1 and H_2 at λ_0 with initial vector x_0 have a finite supremum. In other words, if x_0 is a non-zero vector in $\mathrm{Ker}\, H(\lambda_0)$, where H is the non-square analytic matrix function defined by (2.5), then the lengths of the Jordan chains of H at λ_0 with initial vector x_0 have a finite supremum. Hence for H in (2.5) a canonical set of Jordan chains of H at λ_0 is well defined. We say that

$$x_{1,0}, \ldots, x_{1,r_1-1}, x_{2,0}, \ldots, x_{2,r_2-1}, \ldots, x_{p,0}, \ldots, x_{p,r_p-1} \tag{2.6}$$

is a *canonical set of common Jordan chains* of H_1 and H_2 at λ_0 if the chains in (2.6) form a canonical set of Jordan chains for H at λ_0, where H is defined by (2.5). Furthermore, in that case the number

$$\nu(H_1, H_2; \lambda_0) := \sum_{j=1}^{p} r_j$$

is called the *common multiplicity* of λ_0 as a common eigenvalue of the analytic matrix functions H_1 and H_2.

If the analytic matrix functions H_1 and H_2 have a finite number of common eigenvalues in Ω, then we define the *total common multiplicity* of H_1 and H_2 in Ω to be the number $\nu(H_1, H_2; \Omega)$ given by

$$\nu(H_1, H_2; \Omega) = \sum_{\lambda \in \Omega} \nu(H_1, H_2; \lambda).$$

When $\Omega = \mathbb{C}$, we simply write $\nu(H_1, H_2) = \nu(H_1, H_2; \mathbb{C})$.

The total common multiplicity $\nu(\mathcal{B},\mathcal{D})$. Let $\mathcal{B}$ and $\mathcal{D}$ be the $n \times n$ entire matrix functions defined by (1.2) and (1.3), respectively. From the definitions of these functions it follows that

$$\lim_{\Im\lambda\leq 0,\,|\lambda|\to\infty}\mathcal{B}(\lambda)=I_n,\qquad \lim_{\Im\lambda\geq 0,\,|\lambda|\to\infty}\mathcal{D}(\lambda)=I_n.$$

Thus $\mathcal{B}$ has only a finite number of eigenvalues in the closed lower half-plane, and the same is true for $\mathcal{D}$ with respect to the closed upper half-plane. We conclude that the number of common eigenvalues of $\mathcal{B}$ and $\mathcal{D}$ in $\mathbb{C}$ is finite. This allows us to define the *total common multiplicity* $\nu(\mathcal{B},\mathcal{D})$ of $\mathcal{B}$ and $\mathcal{D}$, namely:

$$\nu(\mathcal{B},\mathcal{D})=\sum_{\lambda}\nu(\mathcal{B},\mathcal{D};\lambda),$$

where the sum is taken over the common eigenvalues, and $\nu(\mathcal{B},\mathcal{D};\lambda)$ is the common multiplicity of λ as a common eigenvalue of $\mathcal{B}$ and $\mathcal{D}$.

2.2. Eigenvalues and Jordan chains in terms of realizations

Throughout this section W is an $n \times n$ matrix function which is analytic on $\mathbb{C}\backslash\{0\}$, and we assume that W is given in realized form:

$$W(\lambda)=I_n+C(\lambda I-A)^{-1}B,\quad 0\neq\lambda\in\mathbb{C}. \tag{2.7}$$

Here A, B, C, and I are as in the previous section. With the realization (2.7) we associate the operator $A^\times=A-BC$.

Since $A^\times=A-BC$ and BC is of finite rank, $A^\times$ is a finite rank perturbation of a quasi-nilpotent operator. It follows that a non-zero point λ_0 in the spectrum of $A^\times$ is an eigenvalue of finite type. Thus, if $0\neq\lambda_0\in\sigma(A^\times)$, then λ_0 is an isolated point in $\sigma(A^\times)$, and the corresponding Riesz projection $P(\lambda_0;A^\times)$ is of finite dimension (see Section II.1 in [8] for further details). In particular, the non-zero part of $\sigma(A^\times)$ consists of eigenvalues only. Recall (see Section II.2 of [8]) that $x_0,x_1,\ldots,x_{r-1}$ in $\mathcal{X}$ is called a *Jordan chain* of $A^\times$ at λ_0 if $x_0\neq 0$ and

$$A^\times x_0=\lambda_0 x_0,\quad A^\times x_j=\lambda_0 x_j+x_{j-1}\quad (j=1,\ldots,r-1). \tag{2.8}$$

In other words, in the terminology of Section 2.1, the vectors $x_0,x_1,\ldots,x_{r-1}$ form a Jordan chain of the operator $A^\times$ at λ_0 if and only if $x_0,x_1,\ldots,x_{r-1}$ is a Jordan chain of the analytic operator-valued function $\lambda I-A$ at λ_0.

The following proposition is the main result of this section.

Proposition 2.1. *Let W be given by (2.7), and put $A^\times=A-BC$. Fix $0\neq\lambda\in\mathbb{C}$. Then C maps $\mathrm{Ker}\,(\lambda_0 I-A^\times)$ in a one to one way onto $\mathrm{Ker}\,W(\lambda_0)$, and the action of the corresponding inverse map is given by $(A-\lambda_0 I)^{-1}B$. Furthermore, if $x_0,\ldots,x_{r-1}$ is a Jordan chain of $A^\times$ at λ_0, then $Cx_0,\ldots,Cx_{r-1}$ is a Jordan chain of W at λ_0, and each Jordan chain of W at λ_0 is obtained in this way.*

Proof. We shall use the fact (see [5], page 58) that the operator functions

$$\begin{bmatrix} W(\lambda) & 0 \\ 0 & I \end{bmatrix},\qquad \begin{bmatrix} \lambda I-A^\times & 0 \\ 0 & I_n \end{bmatrix} \tag{2.9}$$

are analytically equivalent on $\mathbb{C}\backslash\{0\}$. More precisely, for each $0 \neq \lambda \in \mathbb{C}$ we have the following identity

$$\begin{bmatrix} C(\lambda I - A)^{-1} & I_n \\ -(\lambda I - A)^{-1} & 0 \end{bmatrix} \begin{bmatrix} -I & B \\ 0 & I_n \end{bmatrix} \begin{bmatrix} \lambda I - A^\times & 0 \\ 0 & I_n \end{bmatrix} \tag{2.10}$$
$$= \begin{bmatrix} W(\lambda) & 0 \\ 0 & I \end{bmatrix} \begin{bmatrix} 0 & I_n \\ I & -(\lambda I - A)^{-1}B \end{bmatrix} \begin{bmatrix} I & 0 \\ -C & I_n \end{bmatrix}.$$

Notice that the first two factors in the left-hand side of (2.10) are invertible, and these factors and their inverses depend analytically on $\lambda \in \mathbb{C}\backslash\{0\}$. A similar statement holds true for the second and third factor in the right-hand side of (2.10). Thus (2.10) shows that the operator functions in (2.9) are analytically equivalent on $\mathbb{C}\backslash\{0\}$.

We first prove the statement about the Jordan chains. So, let $x_0, \ldots, x_{r-1}$ be a Jordan chain for $A^\times$ at λ_0. Put

$$x(\lambda) = x_0 + (\lambda - \lambda_0)x_1 + \cdots + (\lambda - \lambda_0)^{r-1}x_{r-1}.$$

Then $x(\lambda_0) = x_0 \neq 0$, and $(\lambda - A^\times)x(\lambda) = (\lambda - \lambda_0)^r \varphi(\lambda)$, where φ is analytic at λ_0. By applying the left-hand side (2.10) to the vector function

$$\begin{bmatrix} x(\lambda) \\ 0 \end{bmatrix},$$

we see that the function

$$\begin{bmatrix} W(\lambda) & 0 \\ 0 & I \end{bmatrix} \begin{bmatrix} -Cx(\lambda) \\ x(\lambda) + (\lambda I - A)^{-1}BCx(\lambda) \end{bmatrix}$$

must have a zero at λ_0 of order at least r. For the second component this result means that the vectors $x_0, \ldots, x_{r-1}$ are precisely equal to the first r Taylor coefficients of $-(\lambda I - A)^{-1}BCx(\lambda)$ at λ_0. In particular

$$x_0 = -(\lambda_0 I - A)^{-1}BCx_0. \tag{2.11}$$

Since $x_0 \neq 0$, formula (2.11) yields $Cx_0 \neq 0$. For the first component we have that $W(\lambda)Cx(\lambda)$ has a zero of order at least r at λ_0. Since $Cx_0 \neq 0$, this is equivalent to the statement that $Cx_0, \ldots, Cx_{r-1}$ is a Jordan chain of W at λ_0.

To prove that all Jordan chains of W at λ_0 are obtained in this way, let $y_0, \ldots, y_{r-1}$ be a Jordan chain of W at λ_0. Put

$$y(\lambda) = y_0 + (\lambda - \lambda_0)y_1 + \cdots + (\lambda - \lambda_0)^{r-1}y_{r-1}.$$

Then $y(\lambda_0) = y_0 \neq 0$ and $W(\lambda)y(\lambda) = (\lambda - \lambda_0)^r \psi(\lambda)$, where ψ is analytic at λ_0. Using the experience of the previous part of the proof, put

$$x(\lambda) = -(\lambda I - A)^{-1}By(\lambda), \tag{2.12}$$

and let $x_0, \ldots, x_{r-1}$ be the first r Taylor coefficients of $x(\lambda)$ at λ_0, that is,

$$x_k = \sum_{\alpha=0}^{k} (A - \lambda_0 I)^{-(\alpha+1)} By_{k-\alpha}, \quad k = 0, \ldots, r - 1. \tag{2.13}$$

From (2.10) it follows that the vector function

$$\begin{bmatrix} \lambda I - A^\times & 0 \\ 0 & I_n \end{bmatrix} \begin{bmatrix} I & 0 \\ C & I_n \end{bmatrix} \begin{bmatrix} (\lambda I - A)^{-1}B & I \\ I_n & 0 \end{bmatrix} \begin{bmatrix} y(\lambda) \\ 0 \end{bmatrix}$$

has a zero at λ_0 of order at least r. Using (2.12) we conclude that the same holds true for $(\lambda I - A^\times)x(\lambda)$, that is, at λ_0 the function $(\lambda I - A^\times)x(\lambda)$ has a zero of order at least r too. Since $W(\lambda)y(\lambda) = y(\lambda) - Cx(\lambda)$, the first r Taylor coefficients at λ_0 of $y(\lambda)$ and $Cx(\lambda)$ coincide. Thus

$$y_0 = Cx_0, \ldots, y_{r-1} = Cx_{r-1}. \tag{2.14}$$

From $y_0 \neq 0$ we obtain $x_0 \neq 0$. But then the fact that $(\lambda I - A^\times)x(\lambda)$ has a zero at λ_0 of order at least r is equivalent to the fact that $x_0, \ldots, x_{r-1}$ is a Jordan chain of $\lambda I - A^\times$ at λ_0. Formula (2.14) shows that C maps this chain onto the chain we started with.

Finally notice that the result about the Jordan chains specified for $r = 1$ implies the fact that C maps $\mathrm{Ker}\,(\lambda_0 I - A^\times)$ in a one to one way onto $\mathrm{Ker}\,W(\lambda_0)$. Furthermore, according to (2.11), the action of the corresponding inverse map is given by the operator $(A - \lambda_0 I)^{-1}B$. $\qquad\square$

Corollary 2.2. *Let W be given by (2.7), and put $A^\times = A - BC$. Fix $0 \neq \lambda_0 \in \sigma(A^\times)$. If $x_{1,0}, \ldots, x_{1,r_1-1}, \ldots, x_{p,0}, \ldots, x_{p,r_p-1}$ is a canonical system of Jordan chains of $A^\times$ at λ_0, then the chains*

$$Cx_{1,0}, \ldots, Cx_{1,r_1-1}, \ldots, Cx_{p,0}, \ldots, Cx_{p,r_p-1}$$

form a canonical system of Jordan chains for W at λ_0. Moreover, any canonical system of Jordan chains for W at λ_0 is obtained in this way. In particular, the multiplicity of λ_0 as an eigenvalue of W is equal to $\mathrm{rank}\,P(\lambda_0; A^\times)$, where $P(\lambda_0; A^\times)$ is the Riesz projection of $A^\times$ corresponding to λ_0.

Proof. The result follows immediately from Proposition 2.1. Indeed, notice that C maps $\mathrm{Ker}\,(\lambda_0 I - A^\times)$ in a one to one way onto $\mathrm{Ker}\,W(\lambda_0)$. Since $x_{j+1,0}$ is a vector in $\mathrm{Ker}\,(\lambda_0 I - A^\times)$ which does not belong to $\mathrm{span}\{x_{1,0}, \ldots, x_{j,0}\}$, it follows that $Cx_{j+1,0}$ is a vector in $\mathrm{Ker}\,W(\lambda_0)$ which does not belong to $\mathrm{span}\{Cx_{1,0}, \ldots, Cx_{j,0}\}$. This, together with the definition of a canonical system of Jordan chains, yields the desired result. $\qquad\square$

2.3. Common eigenvalues and common Jordan chains in terms of realizations

Throughout this section W_1 and W_2 are $n \times n$ matrix functions which are analytic on $\mathbb{C}\backslash\{0\}$ and at infinity. We assume that $W_1(\infty)$ and $W_2(\infty)$ are equal to the $n \times n$ identity matrix.

The functions W_1 and W_2 can be realized simultaneously in the following way:

$$W_1(\lambda) = I_n + C_1(\lambda I - A)^{-1}B, \quad W_2(\lambda) = I_n + C_2(\lambda I - A)^{-1}B. \tag{2.15}$$

Here A is a quasi-nilpotent operator acting on a Banach space $\mathcal{X}$, the operators C_1, C_2 act from $\mathcal{X}$ into $\mathbb{C}^n$, and B is an operator from $\mathbb{C}^n$ into $\mathcal{X}$. To get the realizations in (2.15) we apply Theorem 2.5 in [5] to the $2n \times 2n$ matrix function

$$W(\lambda) = \left[\begin{array}{cc} W_1(\lambda) & 0 \\ W_2(\lambda) & 0 \end{array} \right].$$

The zeros in the second column of $W(\lambda)$ stand for the zero $n \times n$ matrix. Since W is analytic on $\mathbb{C}\backslash\{0\}$ and at infinity, Theorem 2.5 in [5] tells us that W admits a representation

$$W(\lambda) = \hat{D} + \hat{C}(\lambda I - A)^{-1}\hat{B},$$

where A is a quasi-nilpotent operator on a Banach space $\mathcal{X}$, the operator $\hat{B}$ maps $\mathbb{C}^{2n}$ into $\mathcal{X}$, and $\hat{C}$ maps $\mathcal{X}$ into $\mathbb{C}^{2n}$. Furthermore,

$$\hat{D} = W(\infty) = \left[\begin{array}{cc} I_n & 0 \\ I_n & 0 \end{array} \right].$$

Identifying $\mathbb{C}^{2n}$ with $\mathbb{C}^n \oplus \mathbb{C}^n$, the operators $\hat{B}$ and $\hat{C}$ can be partitioned as follows:

$$\hat{B} = \left[\begin{array}{cc} B_1 & B_2 \end{array} \right] : \left[\begin{array}{c} \mathbb{C}^n \\ \mathbb{C}^n \end{array} \right] \rightarrow \mathcal{X}, \quad \hat{C} = \left[\begin{array}{c} C_1 \\ C_2 \end{array} \right] : \mathcal{X} \rightarrow \left[\begin{array}{c} \mathbb{C}^n \\ \mathbb{C}^n \end{array} \right].$$

It follows that with this choice of A, C_1, C_2 and with $B = B_1$ the identities in (2.15) are satisfied.

Our aim is to describe the common eigenvalues and common Jordan chains of W_1 and W_2 given by the realizations in (2.15). For this purpose, put

$$A_1^\times = A - BC_1, \quad A_2^\times = A - BC_2, \tag{2.16}$$

and let $\mathcal{M}$ be the largest subspace of $\mathrm{Ker}\,(C_1 - C_2)$ that is invariant under $A_1^\times$. Since $\mathcal{M} \subset \mathrm{Ker}\,(C_1 - C_2)$, the operators C_1 and C_2 coincide on $\mathcal{M}$, and hence $A_1^\times$ and $A_2^\times$ also coincide on $\mathcal{M}$. In particular, $A_2^\times$ leaves $\mathcal{M}$ invariant too. It follows that $\mathcal{M}$ is also the largest $A_2^\times$-invariant subspace contained in $\mathrm{Ker}\,(C_1 - C_2)$. In the sequel we let $A_\mathcal{M}^\times$ and $C_\mathcal{M}$ be the operators defined by

$$A_\mathcal{M}^\times = A_1^\times|_\mathcal{M} = A_2^\times|_\mathcal{M} : \mathcal{M} \to \mathcal{M}, \tag{2.17}$$

$$C_\mathcal{M} = C_1|\mathcal{M} = C_2|\mathcal{M} : \mathcal{M} \to \mathbb{C}^n. \tag{2.18}$$

By $I_\mathcal{M}$ we denote the identity operator on $\mathcal{M}$. We shall need the following lemma.

Lemma 2.3. *The non-zero part of $\sigma(A_\mathcal{M}^\times)$ consists of eigenvalues of finite type only.*

Proof. Let $\lambda_0 \neq 0$ be a point in the boundary $\partial\sigma(A_\mathcal{M}^\times)$ of $\sigma(A_\mathcal{M}^\times)$. Then λ_0 is an approximate eigenvalue of $A_\mathcal{M}^\times$, that is, there exists a sequence $m_1, m_2, \ldots$ in $\mathcal{M}$ such that $\|m_j\| = 1$ for each j and $(\lambda_0 I_\mathcal{M} - A_\mathcal{M}^\times)m_j \to 0$ for $j \to \infty$. Since $A^\times|_\mathcal{M} = A_\mathcal{M}^\times$, it follows that

$$(\lambda_0 I - A^\times)m_j = (\lambda_0 I_\mathcal{M} - A_\mathcal{M}^\times)m_j \to 0, \quad j \to \infty.$$

Hence λ_0 is also an approximate eigenvalue of $A^\times$. We conclude that

$$\partial\sigma(A_\mathcal{M}^\times)\backslash\{0\} \subset \sigma(A^\times). \tag{2.19}$$

But the non-zero part of $\sigma(A^\times)$ consists of isolated eigenvalues only. This together with (2.19) implies that the non-zero part of $\sigma(A^\times_{\mathcal{M}})$ is contained in $\sigma(A^\times)$.

Take $0 \neq \lambda_0 \in \sigma(A^\times_{\mathcal{M}})$. The result of the previous paragraph shows that λ_0 is an isolated point in $\sigma(A^\times_{\mathcal{M}})$, and hence its Riesz projection $P(\lambda_0; A^\times_{\mathcal{M}})$ is well defined. Since the resolvent sets of $A^\times$ and $A^\times_{\mathcal{M}}$ are connected, it follows that $P(\lambda_0; A^\times_{\mathcal{M}}) = P(\lambda_0; A^\times)|_{\mathcal{M}}$. But $P(\lambda_0; A^\times)$ has finite rank, and thus the same is true for $P(\lambda_0; A^\times_{\mathcal{M}})$. This proves that λ_0 is an eigenvalue of finite type for $A^\times_{\mathcal{M}}$. $\square$

The following proposition is the main result of this section.

Proposition 2.4. *Let W_1 and W_2 be given by (2.15), and let $A^\times_{\mathcal{M}}$, $C_{\mathcal{M}}$ be the operators defined by (2.17) and (2.18), respectively. Fix $0 \neq \lambda_0 \in \mathbb{C}$. Then λ_0 is a common eigenvalue of W_1 and W_2 if and only if λ_0 is an eigenvalue of $A^\times_{\mathcal{M}}$. More precisely, $C_{\mathcal{M}}$ maps $\mathrm{Ker}\,(\lambda_0 I_{\mathcal{M}} - A^\times_{\mathcal{M}})$ in a one to one way onto $\mathrm{Ker}\,W_1(\lambda_0) \cap \mathrm{Ker}\,W_2(\lambda_0)$, and the action of the corresponding inverse map is given by $(A - \lambda_0 I)^{-1} B$. Furthermore, if $x_0, \ldots, x_{r-1}$ is a Jordan chain of $A^\times_{\mathcal{M}}$ at λ_0, then $C_{\mathcal{M}} x_0, \ldots, C_{\mathcal{M}} x_{r-1}$ is a common Jordan chain of W_1 and W_2 at λ_0, and each common Jordan chain of W_1 and W_2 at λ_0 is obtained in this way.*

Proof. We first prove the statements about the Jordan chains. Let the vectors $x_0, \ldots, x_{r-1}$ form a Jordan chain of $A^\times_{\mathcal{M}}$ at λ_0. Fix $i = 1, 2$. Since $A^\times_i|_{\mathcal{M}} = A^\times_{\mathcal{M}}$, the vectors $x_0, \ldots, x_{r-1}$ also form a Jordan chain for $A^\times_i$ at λ_0. But then Proposition 2.1 implies that $C_i x_0, \ldots, C_i x_{r-1}$ is a Jordan chain for W_i at λ_0. Recall that $C_{\mathcal{M}} = C_i|_{\mathcal{M}}$ and the vectors $x_0, \ldots, x_{r-1}$ are in $\mathcal{M} \subset \mathrm{Ker}\,(C_1 - C_2)$. Thus $C_{\mathcal{M}} x_j = C_i x_j$ for $i = 1, 2$ and $j = 0, \ldots, r-1$. We conclude that $C_{\mathcal{M}} x_0, \ldots, C_{\mathcal{M}} x_{r-1}$ is a common Jordan chain of W_1 and W_2 at λ_0.

Next, let $y_0, \ldots, y_{r-1}$ be a common Jordan chain of W_1 and W_2 at λ_0. Put

$$x_k = \sum_{\alpha=0}^{k} (A - \lambda_0 I)^{-(\alpha+1)} B y_{k-\alpha}, \quad k = 0, \ldots, r-1. \tag{2.20}$$

From the proof of Proposition 2.1 (cf., formula (2.13)) we know that the vectors $x_0, \ldots, x_{r-1}$ form a Jordan chain at λ_0 for both $A^\times_1$ and $A^\times_2$, and according to formula (2.14) we have

$$y_j = C_1 x_j \quad \text{and} \quad y_j = C_2 x_j \quad (j = 0, \ldots, r-1). \tag{2.21}$$

Since $x_0, \ldots, x_{r-1}$ is a Jordan chain of $A^\times_1$, the space

$$\mathcal{N} = \mathrm{span}\{x_j \mid j = 0, \ldots, r-1\}$$

is invariant under $A^\times_1$. From (2.21) we see that the vectors $x_0, \ldots, x_{r-1}$ belong to $\mathrm{Ker}\,(C_1 - C_2)$. Thus $\mathcal{N}$ is an $A^\times_1$-invariant subspace contained in $\mathrm{Ker}\,(C_1 - C_2)$. It follows that $\mathcal{N} \subset \mathcal{M}$. We conclude that $x_0, \ldots, x_{r-1}$ is a Jordan chain of $A^\times_{\mathcal{M}}$ at λ_0 and $y_j = C_{\mathcal{M}} x_j$ for $j = 0, \ldots, r-1$, as desired.

When specified for $r = 1$, the results proved in the preceding two paragraphs imply that $C_{\mathcal{M}}$ maps $\mathrm{Ker}\,(\lambda_0 I_{\mathcal{M}} - A^\times_{\mathcal{M}})$ onto $\mathrm{Ker}\,W_1(\lambda_0) \cap \mathrm{Ker}\,W_2(\lambda_0)$. This map is also one to one because $C_{\mathcal{M}} x_0 \neq 0$ whenever x_0 is a non-zero vector in the null space $\operatorname{Ker}(\lambda_0 I_{\mathcal{M}} - A_{\mathcal{M}}^{\times})$. By taking $k = 0$ in (2.20) we see that the action of the corresponding inverse map is given by the operator $(A - \lambda_0 I)^{-1}B$. $\qquad\square$

Corollary 2.5. *Let W_1 and W_2 be given by (2.15), and let $A_{\mathcal{M}}^{\times}$, $C_{\mathcal{M}}$ be the operators defined by (2.17) and (2.18), respectively. Fix $0 \neq \lambda_0 \in \sigma(A_{\mathcal{M}}^{\times})$. If*

$$x_{1,0}, \ldots, x_{1, r_1 - 1}, \ldots, x_{p, 0}, \ldots, x_{p, r_p - 1}$$

is a canonical system of Jordan chains of $A_{\mathcal{M}}^{\times}$ at λ_0, then the chains

$$C_{\mathcal{M}} x_{1,0}, \ldots, C_{\mathcal{M}} x_{1, r_1 - 1}, \ldots, C_{\mathcal{M}} x_{p, 0}, \ldots, C_{\mathcal{M}} x_{p, r_p - 1}$$

form a canonical system of common Jordan chains of W_1 and W_2 at λ_0. Moreover, any canonical system of common Jordan chains of W_1 and W_2 at λ_0 is obtained in this way. In particular, the total common multiplicity of W_1 and W_2 at λ_0 is given by

$$\nu(W_1, W_2; \lambda_0) = \operatorname{rank} P(\lambda_0; A_{\mathcal{M}}^{\times}), \tag{2.22}$$

where $P(\lambda_0; A_{\mathcal{M}}^{\times})$ is the Riesz projection of $A_{\mathcal{M}}^{\times}$ corresponding to λ_0.

Proof. The proof follows the same line of reasoning as that of Corollary 2.2. One only has to replace the reference to Proposition 2.1 by a reference to Proposition 2.4. $\qquad\square$

Let W_1 and W_2 be given by (2.15), and let $A_{\mathcal{M}}^{\times}$ be the operator defined by (2.17). Proposition 2.4 shows that W_1 and W_2 have a finite number of common eigenvalues in $\mathbb{C}\backslash\{0\}$ if and only if the non-zero part of the spectrum of $A_{\mathcal{M}}^{\times}$ is finite. Moreover, using (2.22), we see that in that case the total common multiplicity of W_1 and W_2 in $\mathbb{C}\backslash\{0\}$ is equal to the rank of the Riesz projection (see [8]) corresponding to the non-zero part of the spectrum of $A_{\mathcal{M}}^{\times}$.

2.4. Common spectral data of entire matrix functions

In this section F_1 and F_2 are two entire $n \times n$ matrix functions which are assumed to have the value I_n at zero. The functions F_1 and F_2 can be represented simultaneously in the form

$$F_1(\lambda) = I_n + \lambda C_1(I - \lambda A)^{-1}B, \quad F_2(\lambda) = I_n + \lambda C_2(I - \lambda A)^{-1}B \tag{2.23}$$

Here A is a quasi-nilpotent operator on a Banach space $\mathcal{X}$, the operators C_1, C_2 act from $\mathcal{X}$ into $\mathbb{C}^n$, and B is an operator from $\mathbb{C}^n$ into $\mathcal{X}$. To get the co-realizations of F_1 and F_2 in (2.23) one applies the result of the second paragraph of the previous section to the matrix functions $W_1(\lambda) = F_1(\lambda^{-1})$ and $W_2(\lambda) = F_2(\lambda^{-1})$.

Theorem 2.6. *Let F_1 and F_2 be given by (2.23), and let $\mathcal{M}$ be the largest subspace contained in $\operatorname{Ker}(C_1 - C_2)$ that is invariant under $A_2^{\times} = A - BC_2$. Assume $A_2^{\times}$ is injective and $\dim \mathcal{M} < \infty$. Then F_1 and F_2 have a finite number of common eigenvalues and their total common multiplicity $\nu(F_1, F_2)$ is given by*

$$\nu(F_1, F_2) = \dim \mathcal{M}. \tag{2.24}$$

Furthermore, in terms of the common Jordan chains of F_1 and F_2 a basis of $\mathcal{M}$ can be obtained as follows. Let $z_1, \ldots, z_\ell$ be the set of distinct common eigenvalues of F_1 and F_2 in $\mathbb{C}$, and for each common eigenvalue z_ν let

$$y_{1,0}^\nu, \ldots, y_{1,r_1^{(\nu)}-1}^\nu, y_{2,0}^\nu, \ldots, y_{2,r_2^{(\nu)}-1}^\nu, \ldots, y_{p_\nu,0}^\nu, \ldots, y_{p_\nu,r_{p_\nu}^{(\nu)}-1}^\nu \tag{2.25}$$

stand for a canonical set of common Jordan chains of F_1 and F_2 at z_ν. Then the vectors

$$u_{j,k}^\nu = \sum_{\alpha=0}^{k}(I - z_\nu A)^{-(\alpha+1)} A^\alpha B y_{j,k-\alpha}^\nu, \quad k = 0, \ldots, r_j^{(\nu)} - 1, \tag{2.26}$$

$$j = 1, \ldots, p_\nu, \quad \nu = 1, \ldots, \ell,$$

form a basis of $\mathcal{M}$.

The above theorem also holds for $A_1^\times = A - BC_1$ in place of $A_2^\times = A - BC_2$. Also, notice that the operators $A_1^\times$ and $A_2^\times$ coincide on the space $\mathcal{M}$ defined in Theorem 2.6. As before (see (2.17)), we put

$$A_{\mathcal{M}}^\times = A_1^\times|_{\mathcal{M}} = A_2^\times|_{\mathcal{M}} : \mathcal{M} \to \mathcal{M}.$$

In order to prove Theorem 2.6 it is convenient first to prove the following lemma.

Lemma 2.7. *Put $W_1(\lambda) = F_1(\lambda^{-1})$ and $W_2(\lambda) = F_2(\lambda^{-1})$, and let z_0 be a common eigenvalue of F_1 and F_2. Then $z_0 \neq 0$, and $\lambda_0 = z_0^{-1}$ is a common eigenvalue of W_1 and W_2. Moreover, any non-zero common eigenvalue of W_1 and W_2 is obtained in this way, and*

$$\nu(F_1, F_2; z_0) = \nu(W_1, W_2; \lambda_0). \tag{2.27}$$

Proof. Since F_1 and F_2 are both non-singular at zero, we have $z_0 \neq 0$. Furthermore,

$$\operatorname{Ker} F_1(z_0) \cap \operatorname{Ker} F_2(z_0) = \operatorname{Ker} W_1(\lambda_0) \cap \operatorname{Ker} W_2(\lambda_0). \tag{2.28}$$

Thus z_0 is a common eigenvalue of F_1 and F_2 if and only if λ_0 is a common eigenvalue of W_1 and W_2. It remains to prove (2.27).

Let $y_0, \ldots, y_{r-1}$ be any common Jordan chain of F_1 and F_2 at z_0, and let

$$y(z) = y_0 + (z - z_0)y_1 + \cdots + (z - z_0)^{r-1}y_{r-1}.$$

In what follows we define $\widetilde{y}_0, \ldots, \widetilde{y}_{r-1}$ to be the first r Taylor coefficients of $\widetilde{y}(\lambda) = y(z)$ at λ_0, where $\lambda = z^{-1}$. Notice that $y_0 = \widetilde{y}_0$. We claim that $\widetilde{y}_0, \ldots, \widetilde{y}_{r-1}$ is a common Jordan chain of W_1 and W_2 at λ_0. To see this, let $i = 1, 2$, and consider $F_i(z)y(z)$. Since $y_0, \ldots, y_{r-1}$ is a Jordan chain of F_i at z_0, we have $F_i(z)y(z) = (z - z_0)^r \psi_i(z)$, with ψ_i being analytic at z_0. It follows that

$$W_i(\lambda)\widetilde{y}(\lambda) = F_i(z)y(z) = (z - z_0)^r \psi_i(z) = (\lambda - \lambda_0)^r \left(\frac{-1}{\lambda\lambda_0}\right)^r \psi_i\left(\frac{1}{\lambda}\right).$$

The function $(-\lambda\lambda_0)^{-r}\psi_i(\lambda^{-1})$ is analytic at λ_0. Thus $\widetilde{y}_0, \ldots, \widetilde{y}_{r-1}$ is a Jordan chain of W_i at λ_0.

Reversing the arguments used in the preceding paragraph, one proves that each common Jordan chain of W_1 and W_2 at λ_0 is of the form $\widetilde{y}_0, \ldots, \widetilde{y}_{r-1}$, where

$y_0, \ldots, y_{r-1}$ is some common Jordan chain of F_1 and F_2 at z_0 and $y_0 = \widetilde{y}_0$. We can then use (2.28) to show that the map

$$(y_0, \ldots, y_{r-1}) \mapsto (\widetilde{y}_0, \ldots, \widetilde{y}_{r-1}) \tag{2.29}$$

transforms a canonical system of common Jordan chains of F_1 and F_2 at z_0 into a canonical system of common Jordan chains of W_1 and W_2 at λ_0, which proves (2.27). $\qquad\square$

From Lemma 2.7 and the remark made in the last paragraph of the previous section we have the following result.

Corollary 2.8. *Let F_1 and F_2 be given by (2.23), and let $\mathcal{M}$ be the largest subspace contained in $\mathrm{Ker}\,(C_1 - C_2)$ that is invariant under $A_2^\times = A - BC_2$. Put $A_{\mathcal{M}}^\times = A_2^\times|_{\mathcal{M}}$. Then F_1 and F_2 have a finite number of common eigenvalues if and only if the non-zero part of the spectrum of $A_{\mathcal{M}}^\times$ is finite. Moreover, in that case $\nu(F_1, F_2)$ is equal to the rank of the Riesz projection corresponding to the non-zero part of the spectrum of $A_{\mathcal{M}}^\times$.*

Proof. Let $W_1(\lambda) = F_1(\lambda^{-1})$ and $W_2(\lambda) = F_2(\lambda^{-1})$. Using (2.23) we see that W_1 and W_2 are given by the realizations in (2.15), and hence we can apply the results of the previous section. Since the matrices $F_1(0)$ and $F_2(0)$ are non-singular, the common eigenvalues of F_1 and F_2 are all non-zero. Hence we can use Lemma 2.7 to show that F_1 and F_2 have a finite number of common eigenvalues if and only if W_1 and W_2 have a finite number of common eigenvalues in $\mathbb{C}\backslash\{0\}$. But then the remark made in the last paragraph of the previous section yields the first part of the corollary.

Assume now that F_1 and F_2 have a finite number of common eigenvalues, $z_1, \ldots, z_\ell$, say. For $j = 1, \ldots, \ell$ put $\lambda_j = z_j^{-1}$. Then $\lambda_1, \ldots, \lambda_\ell$ are the common eigenvalues of W_1 and W_2 in $\mathbb{C}\backslash\{0\}$. Using (2.27), this yields

$$\nu(F_1, F_2) = \sum_{j=1}^{\ell} \nu(F_1, F_2; z_j) = \sum_{j=1}^{\ell} \nu(W_1, W_2; \lambda_j) = \nu(W_1, W_2; \mathbb{C}\backslash\{0\}).$$

By the remark made in the last paragraph of the previous section the quantity $\nu(W_1, W_2; \mathbb{C}\backslash\{0\})$ is equal to the rank of the Riesz projection corresponding to the non-zero part of the spectrum of $A_{\mathcal{M}}^\times$, which completes the proof. $\qquad\square$

Proof of Theorem 2.6. The injectivity of $A_2^\times$ and the fact that $\mathcal{M}$ is invariant under $A_2^\times$ imply that $A_{\mathcal{M}}^\times = A_2^\times|_{\mathcal{M}}$ is injective too. By assumption, $\mathcal{M}$ is finite-dimensional. Hence the spectrum of $A_{\mathcal{M}}^\times$ is finite and consists of eigenvalues only. Since $A_{\mathcal{M}}^\times$ is injective, it follows that the point zero is not in the spectrum of $A_{\mathcal{M}}^\times$. Summarizing we see that the spectrum of $A_{\mathcal{M}}^\times$ is equal to the non-zero part of the spectrum of $A_{\mathcal{M}}^\times$ and is finite. In particular, $\mathcal{M}$ is equal to the range of the Riesz projection corresponding to the non-zero part of the spectrum of $A_{\mathcal{M}}^\times$. An application of Corollary 2.8 then yields (2.24).

Next we prove that the vectors in (2.26) form a basis of $\mathcal{M}$. Let $W_1(\lambda) = F_1(\lambda^{-1})$ and $W_2(\lambda) = F_2(\lambda^{-1})$. For $\nu = 1, \ldots, \ell$ put $\lambda_\nu = z_\nu^{-1}$. Using the map (2.29) with z_ν in place of z_0, we transform the canonical system (2.25) into

$$\widetilde{y}_{1,0}^{\nu}, \ldots, \widetilde{y}_{1,r_1^{(\nu)}-1}^{\nu}, \widetilde{y}_{2,0}^{\nu}, \ldots, \widetilde{y}_{2,r_2^{(\nu)}-1}^{\nu}, \ldots, \widetilde{y}_{p_\nu,0}^{\nu}, \ldots, \widetilde{y}_{p_\nu,r_{p_\nu}^{(\nu)}-1}^{\nu}. \tag{2.30}$$

From the proof of Lemma 2.7 we know that (2.30) forms a canonical system of common Jordan chains of W_1 and W_2 at $\lambda_\nu = z_\nu^{-1}$. But then we can use Corollary 2.5 to show that

$$\widetilde{y}_{j,k}^{\nu} = C_{\mathcal{M}} \widetilde{x}_{j,k}^{\nu}, \quad k = 0, \ldots, r_j^{(\nu)} - 1, \; j = 1, \ldots, p_\nu,$$

where

$$\widetilde{x}_{1,0}^{\nu}, \ldots, \widetilde{x}_{1,r_1^{(\nu)}-1}^{\nu}, \widetilde{x}_{2,0}^{\nu}, \ldots, \widetilde{x}_{2,r_2^{(\nu)}-1}^{\nu}, \ldots, \widetilde{x}_{p_\nu,0}^{\nu}, \ldots, \widetilde{x}_{p_\nu,r_{p_\nu}^{(\nu)}-1}^{\nu}$$

is a canonical system of Jordan chains of $A_{\mathcal{M}}^{\times} = A_2^{\times}|_{\mathcal{M}}$ at λ_ν. It follows that the set of vectors

$$\{\widetilde{x}_{j,k}^{\nu} \mid k = 0, \ldots, r_j^{(\nu)} - 1, \; j = 1, \ldots, p_\nu, \; \nu = 1, \ldots, \ell\} \tag{2.31}$$

forms a basis for $\mathcal{M}$.

We proceed by relating the vectors in the set (2.31) to the vectors $u_{j,k}^{\nu}$ in (2.26). From (2.13) we know that

$$\widetilde{x}_{j,k}^{\nu} = \sum_{\alpha=0}^{k} (A - \lambda_\nu I)^{-(\alpha+1)} B \widetilde{y}_{j,k-\alpha}^{\nu}, \quad k = 0, \ldots, r_j^{(\nu)} - 1.$$

Now put

$$\widetilde{x}_j^{\nu}(\lambda) = \widetilde{x}_{j,0}^{\nu} + (\lambda - \lambda_\nu)\widetilde{x}_{j,1}^{\nu} + \cdots + (\lambda - \lambda_\nu)^{r_j^{\nu}-1}\widetilde{x}_{j,r_j^{\nu}-1}^{\nu},$$

$$\widetilde{y}_j^{\nu}(\lambda) = \widetilde{y}_{j,0}^{\nu} + (\lambda - \lambda_\nu)\widetilde{y}_{j,1}^{\nu} + \cdots + (\lambda - \lambda_\nu)^{r_j^{\nu}-1}\widetilde{y}_{j,r_j^{\nu}-1}^{\nu}.$$

Then at λ_ν the function $\widetilde{x}_j^{\nu}(\lambda) + (\lambda I - A)^{-1} B \widetilde{y}_j^{\nu}(\lambda)$ has a zero of order at least $r_j^{(\nu)}$. Next, for $z = \lambda^{-1}$ put

$$x_j^{\nu}(z) = \widetilde{x}_j^{\nu}(\lambda), \quad y_j^{\nu}(z) = \widetilde{y}_j^{\nu}(\lambda).$$

Then we see that at z_ν the function $x_j^{\nu}(z) + z(I - zA)^{-1} B y_j^{\nu}(z)$ has a zero of order at least $r_j^{(\nu)}$ too. Let $x_{j,0}^{\nu}, \ldots, x_{j,r_j^{(\nu)}-1}^{\nu}$ be the first $r_j^{(\nu)}$ Taylor coefficients in the Taylor expansion of $x_j^{\nu}(z)$ at z_ν. By comparing the Taylor expansions of the functions $x_j^{\nu}(z)$ and $-z(I - zA)^{-1} B y_j^{\nu}(z)$ at z_ν we obtain

$$x_{j,0}^{\nu} = -z_\nu(I - z_\nu A)^{-1} B y_{j,0}^{\nu}, \tag{2.32}$$

$$x_{j,k}^{\nu} = -z_\nu(I - z_\nu A)^{-1} B y_{j,k}^{\nu} \tag{2.33}$$

$$- \sum_{\alpha=1}^{k} (I - z_\nu A)^{-(\alpha+1)} A^{\alpha-1} B y_{j,k-\alpha}^{\nu} \quad (k = 1, \ldots, r_j^{(\nu)} - 1).$$

To see this note that

$$z(I - zA)^{-1} = z\big((I - z_\nu A) - (z - z_\nu)A\big)^{-1}$$

$$= z \sum_{\alpha=0}^{\infty} (z - z_\nu)^\alpha (I - z_\nu A)^{-(\alpha+1)} A^\alpha$$

$$= \sum_{\alpha=0}^{\infty} (z - z_\nu)^{(\alpha+1)} (I - z_\nu A)^{-(\alpha+1)} A^\alpha + z_\nu (I - z_\nu A)^{-1}$$

$$+ \sum_{\alpha=1}^{\infty} (z - z_\nu)^\alpha (I - z_\nu A)^{-(\alpha+1)} (z_\nu A - I + I) A^{(\alpha-1)}$$

$$= \sum_{\alpha=0}^{\infty} (z - z_\nu)^{(\alpha+1)} (I - z_\nu A)^{-(\alpha+1)} A^\alpha + z_\nu (I - z_\nu A)^{-1}$$

$$- \sum_{\alpha=1}^{\infty} (z - z_\nu)^\alpha (I - z_\nu A)^{-\alpha} A^{(\alpha-1)}$$

$$+ \sum_{\alpha=1}^{\infty} (z - z_\nu)^\alpha (I - z_\nu A)^{-(\alpha+1)} A^{(\alpha-1)}$$

$$= z_\nu (I - z_\nu A)^{-1} + \sum_{\alpha=1}^{\infty} (z - z_\nu)^\alpha (I - z_\nu A)^{-(\alpha+1)} A^{(\alpha-1)}.$$

Thus

$$z(I - zA)^{-1} = z_\nu (I - z_\nu A)^{-1} + \sum_{\alpha=1}^{\infty} (z - z_\nu)^\alpha (I - z_\nu A)^{-(\alpha+1)} A^{(\alpha-1)}.$$

From the latter identity the formulas (2.32) and (2.33) are clear.

Finally, to complete the proof notice that for $\alpha \geq 1$ we have

$$z_\nu (I - z_\nu A)^{-(\alpha+1)} A^\alpha = (I - z_\nu A)^{-(\alpha+1)} (z_\nu A - I + I) A^{(\alpha-1)}$$

$$= -(I - z_\nu A)^{-\alpha} A^{(\alpha-1)} + (I - z_\nu A)^{-(\alpha+1)} A^{(\alpha-1)}.$$

Using this in (2.26) we obtain

$$z_\nu u_{j,0}^\nu = -x_{j,0}^\nu, \quad z_\nu u_{j,k}^\nu = -x_{j,k}^\nu - u_{j,k-1}^\nu \quad (k = 1, \ldots, r_j^{(\nu)} - 1).$$

Since the set (2.31) is a basis for $\mathcal{M}$ and $z_\nu \neq 0$, we conclude that vectors in (2.26) form a basis for $\mathcal{M}$ too. $\qquad\square$

In the next chapter we shall apply the results of this section to the entire matrix functions $\mathcal{B}$ and $\mathcal{D}$ appearing in (1.2) and (1.3).

3. The null space of the Bezout integral operator

In this chapter we prove Theorem 1.1. The proof will be based on Theorem 2.6. This requires to have appropriate co-realizations for the entire matrix functions

$\mathcal{A}, \mathcal{B}, \mathcal{C}, \mathcal{D}$. These co-realizations will be constructed in Section 3.2, using some preliminaries on convolution integral operators from Section 3.1. In Section 3.3 we use a result from [10] to restate the quasi commutativity property (1.1) in terms of convolution integral operators on $L_1^n[0, \omega]$. In Section 3.4 we establish intertwining relations between the operator $T = T\{\mathcal{A}, \mathcal{C}; \mathcal{B}, \mathcal{D}\}$ and the main operators of the inverses of the co-realizations in Section 3.2. We are then ready to give the proof in Section 3.5.

3.1. Preliminaries on convolution integral operators

Throughout this section V and W are the linear transformations defined by

$$(Vf)(t) = -i \int_0^t f(s)\, ds, \quad (Wf)(t) = i \int_t^\omega f(s)\, ds \quad (0 \le t \le \omega). \tag{3.1}$$

We view V and W as bounded linear operators on $L_1^n[0, \omega]$. We also need the following projection and embedding operators:

$$\pi : L_1^n[0, \omega] \to \mathbb{C}^n, \qquad \pi f = \int_0^\omega f(s)\, ds, \tag{3.2}$$

$$\tau : \mathbb{C}^n \to L_1^n[0, \omega], \quad (\tau x)(t) = x, \quad 0 \le t \le \omega. \tag{3.3}$$

Notice that $W - V = iL$ where $L = \tau\pi$. The operators V and W are *Volterra* operators, that is, the operators V and W are compact and their spectra consist of the number zero only.

Proposition 3.1. *Let $k \in L_1^{n \times n}[-\omega, \omega]$, and consider on $L_1^n[0, \omega]$ the integral operators*

$$(K\varphi)(t) = \int_0^\omega k(t - s)\varphi(s)\, ds, \qquad\qquad 0 \le t \le \omega,$$

$$(K_+\varphi)(t) = \int_0^t k(t - s)\varphi(s)\, ds, \qquad\qquad 0 \le t \le \omega,$$

$$(K_-\varphi)(t) = \int_t^\omega k(t - s)\varphi(s)\, ds, \qquad\qquad 0 \le t \le \omega.$$

Put $L = \tau\pi$, where π and τ are defined by (3.2) and (3.3). Then

$$VK - KV = iK_-L - iLK_-, \quad WK - KW = iLK_+ - iK_+L. \tag{3.4}$$

Moreover, K commutes with V if and only if k has its support on the positive half-line, and K commutes with W if and only if k has its support on the negative half-line.

Proof. We split the proof into three parts.

Part 1. In this part R is an arbitrary integral operator on $L_1^n[0, \omega]$,

$$(Rf)(t) = \int_0^\omega \rho(t, s)f(s)\, ds, \quad 0 \le t \le \omega.$$

We assume that the function $|\rho(t,s)f(s)|$ is integrable on $[0,\omega] \times [0,\omega]$. We claim that $WR - RV$ is an integral operator of which the kernel function γ_R is given by

$$\gamma_R(t,r) = i \int_r^\omega \rho(t,s)\,ds + i \int_t^\omega \rho(s,r)\,ds, \quad 0 \le t, r \le \omega. \tag{3.5}$$

Indeed, using Fubini's theorem, we have

$$\begin{aligned}
(WRf)(t) &= i \int_t^\omega (Rf)(s)\,ds = i \int_t^\omega \Big(\int_0^\omega \rho(s,r)f(r)\,dr \Big)\,ds \\
&= i \int_0^\omega \Big(\int_t^\omega \rho(s,r)\,ds \Big) f(r)\,dr,
\end{aligned}$$

and

$$\begin{aligned}
(RVf)(t) &= \int_0^\omega \rho(t,s)(Vf)(s)\,ds = -i \int_0^\omega \rho(t,s)\Big(\int_0^s f(r)\,dr \Big)\,ds \\
&= -i \int_0^\omega \Big(\int_r^\omega \rho(t,s)\,ds \Big) f(r)\,dr.
\end{aligned}$$

This shows that $WR - RV$ has (3.5) as its kernel function.

Part 2. In this part we apply the result of the previous part to $R = K$, and we show that

$$WK - KV = iLK_+ + iK_-L. \tag{3.6}$$

Indeed, when $R = K$, we have $\rho(t,s) = k(t-s)$, and the kernel function $\gamma_K(t,r)$ of $WK - KV$ is given by

$$\begin{aligned}
\gamma_K(t,r) &= i \int_r^\omega k(t-s)\,ds + i \int_t^\omega k(s-r)\,ds \\
&= i \int_{t-\omega}^{t-r} k(s)\,ds + i \int_{t-r}^{\omega-r} k(s)\,ds = i \int_{t-\omega}^{\omega-r} k(s)\,ds, \quad 0 \le t, r \le \omega.
\end{aligned}$$

It follows that

$$\begin{aligned}
((WK - KV)f)(t) &= \int_0^\omega i\Big(\int_{t-\omega}^{\omega-r} k(s)\,ds \Big) f(r)\,dr \\
&= i \int_0^\omega \Big(\int_0^{\omega-r} k(s)\,ds \Big) f(r)\,dr + i \int_0^\omega \Big(\int_{t-\omega}^0 k(s)\,ds \Big) f(r)\,dr \\
&= i \int_0^\omega \Big(\int_r^\omega k(s-r)\,ds \Big) f(r)\,dr + i \Big(\int_{t-\omega}^0 k(s)\,ds \Big) Lf \\
&= i \int_0^\omega \Big(\int_0^s k(s-r)f(r)\,dr \Big)\,ds + i \Big(\int_0^{\omega-t} k(-s)\,ds \Big) Lf \\
&= i(LK_+f)(t) + i \int_t^\omega k(t-s)(Lf)(s)\,ds \\
&= i(LK_+f)(t) + i(K_-Lf)(t),
\end{aligned}$$

which proves (3.6). Using that $W - V = iL$, it is straightforward to derive from (3.6) the two identities in (3.4).

Part 3. In this part we prove the final statements of the proposition. First note that the identities in (3.4) yield

$$VK_+ - K_+V = 0, \qquad WK_- - K_-W = 0. \tag{3.7}$$

If k has its support on the positive half-line, then $K = K_+$, and hence the first identity in (3.7) shows that K commutes with V. Since $K = K_+ + K_-$, to prove the reverse implication, it suffices to show that $VK_- - K_-V = 0$ implies $K_- = 0$. To do this, assume $VK_- - K_-V = 0$. Then the first identities in (3.7) and (3.4) yield that $K_-L = LK_-$. Put $k_- = k|_{[-\omega,0]}$. The identity $K_-L = LK_-$ implies that for each $x \in \mathbb{C}^n$ we have

$$\int_t^\omega k_-(t - s)x\,ds = (K_-L\tau x)(t) = (LK_-\tau x)(t),$$

and hence $\int_t^\omega k_-(t - s)x\,ds$ does not depend on t. It follows that

$$\int_t^\omega k_-(t - s)x\,ds = \int_{t-\omega}^0 k_-(s)x\,ds, \quad 0 \le t \le \omega,$$

does not depend on t, which implies that $k_- = 0$, and hence $K_- = 0$.

In a similar way one proves that K commutes with W if and only if k has its support on the negative half-line. $\qquad\square$

Let K, K_+ and K_- be as in the above proposition. We say that $K \in \mathcal{P}$ if $K = K_+$, and $K \in \mathcal{N}$ if $K = K_-$. In other words, $K \in \mathcal{P}$ if and only if k has its support on the positive half-line, and $K \in \mathcal{N}$ if and only if k has its support on the negative half-line. Using this terminology, the final part of Proposition 3.1 can be summarized as follows: $K \in \mathcal{P}$ if and only if K commutes with V, and $K \in \mathcal{N}$ if and only if K commutes with W.

3.2. Co-realizations for the functions $\mathcal{A}, \mathcal{B}, \mathcal{C}, \mathcal{D}$

In this section we show that the entire matrix functions $\mathcal{A}, \mathcal{B}, \mathcal{C}, \mathcal{D}$ defined by (1.2), (1.3) admit the following co-realizations:

$$\mathcal{A}(\lambda) = \mathcal{A}(0) + i\lambda\pi(I - \lambda W)^{-1}Y_\mathcal{A}, \tag{3.8}$$

$$\mathcal{B}(\lambda) = \mathcal{B}(0) + i\lambda Z_\mathcal{B}(I - \lambda V)^{-1}\tau, \tag{3.9}$$

$$\mathcal{C}(\lambda) = e^{-i\lambda\omega}\{\mathcal{C}(0) + i\lambda\pi(I - \lambda W)^{-1}Y_\mathcal{C}\}, \tag{3.10}$$

$$\mathcal{D}(\lambda) = e^{i\lambda\omega}\{\mathcal{D}(0) + i\lambda Z_\mathcal{D}(I - \lambda V)^{-1}\tau\}. \tag{3.11}$$

Here V and W are the operators on $L_1^n[0, \omega]$ defined by (3.1), the operators π and τ are given by (3.2) and (3.3), respectively, and the operators $Y_\mathcal{A}, Y_\mathcal{C}$ from $\mathbb{C}^n$ into

$L_1^n[0,\omega]$, and $Z_\mathcal{B}$, $Z_\mathcal{D}$ from $L_1^n[0,\omega]$ into $\mathbb{C}^n$ are given by

$$Y_\mathcal{A} = A_1\tau, \quad (A_1 f)(t) = \int_t^\omega a(t+\omega-s)f(s)\,ds, \quad 0 \le t \le \omega, \quad (3.12)$$

$$Y_\mathcal{C} = (I+C_0)\tau, \quad (C_0 f)(t) = \int_t^\omega c(t-s)f(s)\,ds, \quad 0 \le t \le \omega, \quad (3.13)$$

$$Z_\mathcal{B} = -\pi B_{-1}, \quad (B_{-1} f)(t) = \int_0^t b(t-\omega-s)f(s)\,ds, \quad 0 \le t \le \omega, \quad (3.14)$$

$$Z_\mathcal{D} = -\pi(I+D_0), \quad (D_0 f)(t) = \int_0^t d(t-s)f(s)\,ds, \quad 0 \le t \le \omega. \quad (3.15)$$

Here π and τ are defined by (3.2) and (3.3), respectively. To derive formulas (3.8)–(3.11) we need some auxiliary results.

Recall that the spectra of the operators V and W consist of the point zero only. Hence $(I-\lambda V)^{-1}$ and $(I-\lambda W)^{-1}$ are well defined for each $\lambda \in \mathbb{C}$. In fact, elementary calculations show that for each $\lambda \in \mathbb{C}$ we have

$$\big((I-\lambda V)^{-1}f\big)(t) = f(t) - i\lambda \int_0^t e^{i\lambda(r-t)}f(r)\,dr, \; 0 \le t \le \omega, \quad (3.16)$$

$$\big((I-\lambda W)^{-1}f\big)(t) = f(t) + i\lambda \int_t^\omega e^{i\lambda(r-t)}f(r)\,dr, \; 0 \le t \le \omega. \quad (3.17)$$

From (3.16) and (3.17) it is straightforward to derive the following equalities which will be useful later:

$$\big((I-\lambda V)^{-1}\tau x\big)(t) \;=\; e^{-i\lambda t}x, \quad 0 \le t \le \omega, \quad x \in \mathbb{C}^n, \quad (3.18)$$

$$\big((I-\lambda W)^{-1}\tau x\big)(t) \;=\; e^{i\lambda(\omega-t)}x, \quad 0 \le t \le \omega, \quad x \in \mathbb{C}^n, \quad (3.19)$$

$$i\lambda\pi(I-\lambda V)^{-1}\tau \;=\; (1-e^{-i\lambda\omega})I_n, \quad \lambda \in \mathbb{C}, \quad (3.20)$$

$$i\lambda\pi(I-\lambda W)^{-1}\tau \;=\; (e^{i\lambda\omega}-1)I_n, \quad \lambda \in \mathbb{C}. \quad (3.21)$$

Here I_n denotes the $n \times n$ identity matrix.

To derive the co-realizations (3.8)–(3.11) the following two propositions will be useful.

Proposition 3.2. *Let $m \in L_1^{n\times n}[0,\omega]$, and let M_0 and M_1 be the operators on $L_1^n[0,\omega]$ defined by*

$$(M_0 f)(t) \;=\; \int_0^t m(t-s)f(s)\,ds, \quad 0 \le t \le \omega,$$

$$(M_1 f)(t) \;=\; \int_t^\omega m(t+\omega-s)f(s)\,ds, \quad 0 \le t \le \omega.$$

Then the Fourier transform $\hat{m}$ of m admits the following representations:

$$\hat{m}(\lambda) \;=\; e^{i\lambda\omega}\{\hat{m}(0) - i\lambda\pi M_0(I - \lambda V)^{-1}\tau\}, \tag{3.22}$$

$$\hat{m}(\lambda) \;=\; \hat{m}(0) + i\lambda\pi M_1(I - \lambda W)^{-1}\tau. \tag{3.23}$$

Proof. First notice that

$$\hat{m}(\lambda) = \int_0^\omega e^{i\lambda r} m(r)\,dr = \int_0^\omega e^{i\lambda(\omega - r)} m(\omega - r)\,dr. \tag{3.24}$$

Using (3.18), we see that

$$\hat{m}(\lambda) = e^{i\lambda\omega} M(I - \lambda V)^{-1}\tau = e^{i\lambda\omega}\big[M\tau + \lambda MV(I - \lambda V)^{-1}\tau\big],$$

where M is the operator from $L_1^n[0,\omega]$ into $\mathbb{C}^n$ given by

$$Mf = \int_0^\omega m(\omega - r) f(r)\,dr. \tag{3.25}$$

Obviously, $M\tau = \hat{m}(0)$. Notice that $Mf = (M_0 f)(\omega)$. The latter identity, together with the fact M_0 and V commute (see the final paragraph of the previous section), yields

$$MVf = (M_0 Vf)(\omega) = (VM_0 f)(\omega) = -i\int_0^\omega (M_0 f)(t)\,dt = -i\pi M_0 f,$$

which proves (3.22).

Formula (3.23) is proved in a similar way. Indeed, using (3.24) and (3.19), we have

$$\hat{m}(\lambda) = M(I - \lambda W)^{-1}\tau = M\tau + \lambda MW(I - \lambda W)^{-1}\tau.$$

Since $Mf = (M_1 f)(0)$ and M_1 and W commute (see the final paragraph of the previous section), we have

$$MWf = (M_1 Wf)(0) = (WM_1 f)(0) = i\int_0^\omega M_1 f(t)\,dt = i\pi M_1 f,$$

which proves (3.23). $\qquad\qquad\qquad\qquad\qquad\qquad\qquad\qquad\square$

Proposition 3.3. *Let $\ell \in L_1^{n\times n}[-\omega, 0]$, and let L_0 and L_{-1} be the operators on $L_1^n[0,\omega]$ defined by*

$$(L_0 f)(t) \;=\; \int_t^\omega \ell(t - s) f(s)\,ds, \quad 0 \le t \le \omega,$$

$$(L_{-1} f)(t) \;=\; \int_0^t \ell(t - \omega - s) f(s)\,ds, \quad 0 \le t \le \omega.$$

Then the Fourier transform $\hat{\ell}$ of ℓ admits the following representations:

$$\hat{\ell}(\lambda) \;=\; \hat{\ell}(0) - i\lambda\pi L_{-1}(I - \lambda V)^{-1}\tau, \tag{3.26}$$

$$\hat{\ell}(\lambda) \;=\; e^{-i\lambda\omega}\{\hat{\ell}(0) + i\lambda\pi L_0(I - \lambda W)^{-1}\tau\}. \tag{3.27}$$

Proof. We obtain this proposition as a corollary of the previous one. Indeed, define $m(t) = \ell(t - \omega)$ for $0 \le t \le \omega$. Then $m \in L_1^{n \times n}[0, \omega]$, and

$$\hat{m}(\lambda) = \int_0^\omega e^{i\lambda t} m(t)\, dt = \int_0^\omega e^{i\lambda t} \ell(t - \omega)\, dt$$

$$= e^{i\lambda\omega} \left(\int_0^\omega e^{i\lambda(t-\omega)} \ell(t - \omega)\, dt \right)$$

$$= e^{i\lambda\omega} \left(\int_{-\omega}^0 e^{i\lambda t} \ell(t)\, dt \right) = e^{i\lambda\omega} \hat{\ell}(\lambda).$$

Now apply Proposition 3.2 to this m. Notice that $M_0 = L_{-1}$ and $M_1 = L_0$. Since $\hat{m}(0) = \hat{\ell}(0)$, and $\hat{m}(\lambda) = e^{i\lambda\omega}\hat{\ell}(\lambda)$, we see that formula (3.22) yields (3.26), and (3.23) yields (3.27). $\qquad\square$

Let us now derive formulas (3.8)–(3.11) by applying the above two propositions.

Proof of (3.8). We apply Proposition 3.2 with $m = a$. In this case the operator $M_1 = A_1$, where A_1 is given by (3.12). Hence (3.23), together with the fact that A_1 and W commute, yields

$$\hat{a}(\lambda) = \hat{a}(0) + i\lambda\pi(I - \lambda W)^{-1} A_1 \tau. \tag{3.28}$$

Recall that $Y_{\mathcal{A}} = A_1\tau$. Since $\mathcal{A}(\lambda) = I_n + \hat{a}(\lambda)$, we have $\mathcal{A}(0) = I_n + \hat{a}(0)$. Thus (3.28) yields (3.8). $\qquad\square$

Proof of (3.9). We apply Proposition 3.3 with $\ell = b$. In this case $L_{-1} = B_{-1}$ and $\hat{\ell}(\lambda) = \hat{b}(\lambda)$. Thus (3.26) yields

$$\hat{b}(\lambda) = \hat{b}(0) - i\lambda\pi B_{-1}(I - \lambda V)^{-1}\tau.$$

Since $Z_{\mathcal{B}} = -\pi B_{-1}$ and $\mathcal{B}(\lambda) = I_n + \hat{b}(\lambda)$, we see that (3.9) holds. $\qquad\square$

Proof of (3.10). We apply Proposition 3.3 with $\ell = c$. In this case $L_0 = C_0$ and $\hat{\ell}(\lambda) = \hat{c}(\lambda)$. The operator C_0 commutes with W. Thus (3.27) yields

$$e^{i\lambda\omega}\hat{c}(\lambda) = \hat{c}(0) - i\lambda\pi(I - \lambda W)^{-1} C_0 \tau$$
$$= \hat{c}(0) - i\lambda\pi(I - \lambda W)^{-1}(I + C_0)\tau - i\lambda\pi(I - \lambda W)^{-1}\tau.$$

According to (3.21) the last term is equal to $e^{i\lambda\omega} I_n - I_n$. Recall that $Y_{\mathcal{C}} = (I + C_0)\tau$, and $\mathcal{C}(\lambda) = I_n + \hat{c}(\lambda)$. It follows that

$$e^{i\lambda\omega}\mathcal{C}(\lambda) = e^{i\lambda\omega} I_n + e^{i\lambda\omega}\hat{c}(\lambda)$$
$$= e^{i\lambda\omega} I_n + \hat{c}(0) + i\lambda\pi(I - \lambda W)^{-1} Y_{\mathcal{C}} - e^{i\lambda\omega} I_n + I_n$$
$$= \mathcal{C}(0) + i\lambda\pi(I - \lambda W)^{-1} Y_{\mathcal{C}},$$

and (3.10) is proved. $\qquad\square$

Proof of (3.11). We apply Proposition 3.2 with $m = d$. In this case $M_0 = D_0$ and $\hat{m}(\lambda) = \hat{d}(\lambda)$. Thus (3.22) yields

$$e^{-i\lambda\omega}\hat{d}(\lambda) = \hat{d}(0) - i\lambda\pi D_0(I - \lambda V)^{-1}\tau.$$

Recall that $Z_{\mathcal{D}} = -\pi(I + D_0)$. Thus

$$e^{-i\lambda\omega}\hat{d}(\lambda) = \hat{d}(0) + i\lambda Z_{\mathcal{D}}(I - \lambda V)^{-1}\tau + i\lambda\pi(I - \lambda V)^{-1}\tau.$$

According to (3.20) the last term is equal to $I_n - e^{-i\lambda\omega}I_n$. Since $\mathcal{D}(\lambda) = I_n + \hat{d}(\lambda)$, we conclude that

$$\begin{aligned}
e^{-i\lambda\omega}\mathcal{D}(\lambda) &= e^{-i\lambda\omega}I_n + e^{-i\lambda\omega}\hat{d}(\lambda) \\
&= I_n + \hat{d}(0) + i\lambda Z_{\mathcal{D}}(I - \lambda V)^{-1}\tau \\
&= \mathcal{D}(0) + i\lambda Z_{\mathcal{D}}(I - \lambda V)^{-1}\tau,
\end{aligned}$$

which completes the proof of (3.11). $\square$

3.3. Quasi commutativity in operator form

In this section we recall Proposition 3.2 from [10]. This proposition restates the quasi commutativity property (1.1) in operator form. To state the precise result (see Proposition 3.4 below) we need some preliminaries.

First some additional notation and terminology. If a lower case letter f denotes a function in $L_1^{n\times n}(\mathbb{R})$, then we let the calligraphic letter $\mathcal{F}$ denote the function defined by

$$\mathcal{F}(\lambda) = I_n + \int_{-\infty}^{\infty} e^{i\lambda s} f(s)\,ds. \tag{3.29}$$

Furthermore, for each $\nu \in \mathbb{Z}$ we let capital F_ν be the convolution operator on $L_1^n[0,\omega]$ given by

$$(F_\nu\varphi)(t) = \int_0^{\omega} f(t - s + \nu\omega)\varphi(s)\,ds, \quad 0 \le t \le \omega. \tag{3.30}$$

We call $\mathcal{F}$ the *entire matrix function defined by* f, and we shall refer to the operators F_ν, $\nu \in \mathbb{Z}$, as the *convolution operators corresponding to* $\mathcal{F}$.

Now, let a, b, c, and d be the functions appearing in (1.2) and (1.3). We shall view a, b, c, d as functions in $L_1^{n\times n}(\mathbb{R})$, with a and d having their support in $[0,\omega]$, while the support of b and c is in $[-\omega, 0]$. Using the terminology introduced in the previous paragraph, the functions $\mathcal{A}$, $\mathcal{B}$, $\mathcal{C}$, and $\mathcal{D}$ in (1.2), (1.3) are the entire matrix functions defined by the functions a, b, c, and d, respectively.

Next, we consider the convolution operators corresponding to $\mathcal{A}$, $\mathcal{B}$, $\mathcal{C}$, and $\mathcal{D}$. Since a has its support in $[0,\omega]$, the convolution operators A_ν, $\nu \in \mathbb{Z}$, corresponding to $\mathcal{A}$ have the following properties:

(i) $(A_0\varphi)(t) = \displaystyle\int_0^t a(t - s)\varphi(s)\,ds, \quad 0 \le t \le \omega,$

(ii) $(A_1\varphi)(t) = \displaystyle\int_t^{\omega} a(t + \omega - s)\varphi(s)\,ds, \quad 0 \le t \le \omega,$

(iii) $A_\nu = 0$ for $\nu \neq 0, \nu \neq 1$.

Similarly, since b has its support in $[-\omega, 0]$, the convolution operators B_ν, $\nu \in \mathbb{Z}$, corresponding to $\mathcal{B}$ have the following properties:

(j) $(B_0\varphi)(t) = \displaystyle\int_t^\omega b(t-s)\varphi(s)\,ds, \quad 0 \le t \le \omega,$

(jj) $(B_{-1}\varphi)(t) = \displaystyle\int_0^t b(t-s-\omega)\varphi(s)\,ds, \quad 0 \le t \le \omega,$

(jjj) $B_\nu = 0$ for $\nu \ne 0, \nu \ne -1$.

Analogous results hold for the convolution operators C_ν and D_ν, $\nu \in \mathbb{Z}$, corresponding to $\mathcal{C}$ and $\mathcal{D}$, respectively.

Notice that the notations introduced in the previous paragraph are consistent with the notations used in (3.12)–(3.15).

We are now ready to restate the quasi commutativity property in operator form. For the sake of completeness we repeat the proof given in [10].

Proposition 3.4. *The quasi commutativity property* (1.1) *is equivalent to the following two conditions:*

$$(I + A_0)B_{-1} = C_{-1}(I + D_0), \quad (I + C_0)D_1 = A_1(I + B_0). \tag{3.31}$$

Moreover, the identities in (3.31) *imply*

$$A_0 + B_0 + A_0 B_0 + A_1 B_{-1} = C_0 + D_0 + C_0 D_0 + C_{-1} D_1. \tag{3.32}$$

Proof. We begin with some additional notation. Given f in $L_1^{n \times n}(\mathbb{R})$, we denote by the bold face capital letter $\mathbf{F}$ the convolution operator on $L_1^n(\mathbb{R})$ defined by

$$(\mathbf{F}\varphi)(t) = \int_{-\infty}^\infty f(t-s)\varphi(s)\,ds, \quad -\infty < t < \infty. \tag{3.33}$$

Notice that the function $\mathcal{F}$ given by (3.29) is the symbol of the operator $I + \mathbf{F}$. With the convolution operator $\mathbf{F}$ we associate the block Laurent operator

$$L_{\mathbf{F}} = \begin{bmatrix} \ddots & & & & \\ & F_0 & F_{-1} & F_{-2} & \\ & F_1 & \boxed{F_0} & F_{-1} & \\ & F_2 & F_1 & F_0 & \\ & & & & \ddots \end{bmatrix}.$$

Here F_ν is the νth convolution operators corresponding to $\mathcal{F}$, see (3.30). We consider $L_{\mathbf{F}}$ as a bounded linear operator on the space $\ell_{1,\mathbb{Z}}(L_1^n[0,\omega])$. The latter space consists of all doubly infinite sequences $\varphi = (\varphi_j)_{j \in \mathbb{Z}}$ with $\varphi_j \in L_1^n[0,\omega]$ such that

$$\|\varphi\|_{\ell_{1,\mathbb{Z}}(L_1^n[0,\omega])} := \sum_{j=-\infty}^\infty \|\varphi_j\|_{L_1^n[0,\omega]} < \infty.$$

The spaces $L_1^n(R)$ and $\ell_{1,\mathbb{Z}}(L_1^n[0,\omega])$ are isometrically equivalent, and for f and g in $L_1^{n \times n}(\mathbb{R})$ we have

$$L_{\mathbf{FG}} = L_{\mathbf{F}} L_{\mathbf{G}}. \tag{3.34}$$

250   I. Gohberg, I. Haimovici, M.A. Kaashoek and L. Lerer

Now let us consider the functions $\mathcal{A}$, $\mathcal{B}$, $\mathcal{C}$, and $\mathcal{D}$ given by (1.2) and (1.3). Notice that $\mathcal{A}$, $\mathcal{B}$, $\mathcal{C}$, and $\mathcal{D}$ are the symbols of the convolution operators $I + \mathbf{A}$, $I + \mathbf{B}$, $I + \mathbf{C}$, and $I + \mathbf{D}$, respectively. It follows that condition (1.1) is equivalent to

$$(I + \mathbf{A})(I + \mathbf{B}) = (I + \mathbf{C})(I + \mathbf{D}), \tag{3.35}$$

which according to (3.34) can be rewritten as

$$L_{\mathbf{A}} + L_{\mathbf{B}} + L_{\mathbf{A}}L_{\mathbf{B}} = L_{\mathbf{C}} + L_{\mathbf{D}} + L_{\mathbf{C}}L_{\mathbf{D}}. \tag{3.36}$$

Now recall the properties (i)–(iii) for the operator A_ν, the properties (j)–(jjj) for the operators B_ν, and the analogous properties for the operators C_ν and D_ν ($\nu \in \mathbb{Z}$). By comparing the entries in the infinite operator matrices determined by the left- and right-hand sides of (3.36) we see that (1.1) is equivalent to

(α) $B_{-1} + A_0 B_{-1} = C_{-1} + C_{-1} D_0$,
(β) $A_0 + B_0 + A_0 B_0 + A_1 B_{-1} = C_0 + D_0 + C_0 D_0 + C_{-1} D_1$
(γ) $A_1 + A_1 B_0 = D_1 + C_0 D_1$.

Obviously, (α) is the same as the first part of (3.31), and (γ) is the same as the second part of (3.31). Thus to complete the proof we have to show that (3.31) implies condition (β).

Consider the functions $f = a + b + a * b$ and $g = c + d + c * d$, where $*$ denotes the convolution product in $L_1^{n \times n}(\mathbb{R})$. Then $L_{\mathbf{F}}$ is equal to the left-hand side of (3.36), and $L_{\mathbf{G}}$ to the right-hand side of (3.36). The first part of (3.31) yields $F_{-1} = G_{-1}$, and the second part of (3.31) implies $F_1 = G_1$. Now notice that $F_{-1} = G_{-1}$ is equivalent to $f(t) = g(t)$ for each $-2\omega \le t \le 0$, and $F_1 = G_1$ is equivalent to $f(t) = g(t)$ for each $0 \le t \le 2\omega$, i.e.,

$$F_{-1} = G_{-1} \iff f|_{[-2\omega,\,0]} = g|_{[-2\omega,\,0]},$$
$$F_1 = G_1 \iff f|_{[0,\,2\omega]} = g|_{[0,\,2\omega]}.$$

In particular, if (3.31) holds, then $f|_{[-\omega,\,\omega]} = g|_{[-\omega,\,\omega]}$, which is equivalent to $F_0 = G_0$. But $F_0 = G_0$ is equivalent to (β). This completes the proof. $\qquad\square$

For latter purposes we present the following lemma.

Lemma 3.5. *Let $\mathcal{F}$ be the entire matrix function defined by $f \in L_1^{n \times n}(\mathbb{R})$. Then*

$$\pi\varphi + \sum_{\nu=-\infty}^{\infty} \pi F_\nu \varphi = \mathcal{F}(0)\pi\varphi, \qquad \varphi \in L_1^n[0, \omega], \tag{3.37}$$

$$\tau x + \sum_{\nu=-\infty}^{\infty} F_\nu \tau x = \tau \mathcal{F}(0)x, \qquad x \in \mathbb{C}^n. \tag{3.38}$$

Here π and τ are the operators defined by (3.2) and (3.3), respectively.

Proof. For $\varphi \in L_1^n[0, \omega]$ we have

$$
\begin{aligned}
\sum_{\nu=-\infty}^{\infty} \pi F_\nu \varphi &= \sum_{\nu=-\infty}^{\infty} \int_0^\omega \left(\int_0^\omega f(t - s + \nu\omega)\varphi(s)\, ds \right) dt \\
&= \sum_{\nu=-\infty}^{\infty} \int_{\nu\omega}^{(\nu+1)\omega} \left(\int_0^\omega f(t - s)\varphi(s)\, ds \right) dt \\
&= \int_{-\infty}^{\infty} \left(\int_0^\omega f(t - s)\varphi(s)\, ds \right) dt \\
&= \int_0^\omega \left(\int_{-\infty}^{\infty} f(t - s)\, dt \right)\varphi(s)\, ds \\
&= \left(\int_{-\infty}^{\infty} f(t)\, dt \right) \int_0^\omega \varphi(s)\, ds \\
&= \mathcal{F}(0)\pi\varphi - \pi\varphi,
\end{aligned}
$$

which proves (3.37). The proof of (3.38) is similar. $\square$

3.4. Intertwining properties

In this section we prove two propositions about intertwining relations between T and the data from the co-realizations (3.8)–(3.11).

Proposition 3.6. *Assume the quadruple $\{\mathcal{A}, \mathcal{C}; \mathcal{B}, \mathcal{D}\}$ satisfies the quasi commutativity property. Then the Bezout integral operator T associated with $\{\mathcal{A}, \mathcal{C}; \mathcal{B}, \mathcal{D}\}$ satisfies the equation*

$$
WT - TV = iY_{\mathcal{A}}Z_{\mathcal{B}} - iY_{\mathcal{C}}Z_{\mathcal{D}}, \tag{3.39}
$$

where $Y_{\mathcal{A}}$, $Y_{\mathcal{C}}$, $Z_{\mathcal{B}}$, $Z_{\mathcal{D}}$ are the operators on $L_1^n[0, \omega]$ defined by (3.12)–(3.15). Furthermore,

$$
\pi T = \mathcal{A}(0)Z_{\mathcal{B}} - \mathcal{C}(0)Z_{\mathcal{D}}, \tag{3.40}
$$

$$
T\tau = Y_{\mathcal{C}}\mathcal{D}(0) - Y_{\mathcal{A}}\mathcal{B}(0). \tag{3.41}
$$

Proof. Recall that the Bezout integral operator T is given by (1.4) and (1.5). Using these formulas and the notations introduced in the previous section we see that

$$
T = (I + A_0)(I + B_0) - C_{-1}D_1. \tag{3.42}
$$

On the other hand, the quasi commutativity property implies that

$$
A_0 + B_0 + A_0 B_0 + A_1 B_{-1} = C_0 + D_0 + C_0 D_0 + C_{-1} D_1.
$$

It follows that the Bezout integral operator is also given by

$$
T = (I + C_0)(I + D_0) - A_1 B_{-1}. \tag{3.43}
$$

Next, notice that the operator A_1 and C_0 belong to the class $\mathcal{N}$, and the operators B_{-1} and D_0 to the class $\mathcal{P}$. Thus (see the final paragraph of Section 3.1)

252 I. Gohberg, I. Haimovici, M.A. Kaashoek and L. Lerer

the operators A_{-1} and C_0 commute with W and the operators B_{-1} and D_0 with V. Using (3.43) this yields

$$
\begin{aligned}
WT &= (I + C_0)W(I + D_0) - A_1 W B_{-1}, \\
TV &= (I + C_0)V(I + D_0) - A_1 V B_{-1}.
\end{aligned}
$$

Since $W - V = i\tau\pi$ we see that

$$
WT - VT = i(I + C_0)\tau\pi(I + D_0) - iA_1\tau\pi B_{-1}. \tag{3.44}
$$

Now, recall that

$$
Y_{\mathcal{A}} = A_1\tau, \quad Y_{\mathcal{C}} = (I + C_0)\tau, \quad Z_{\mathcal{B}} = -\pi B_{-1}, \quad Z_{\mathcal{D}} = -\pi(I + D_0). \tag{3.45}
$$

By using these identities in (3.44) we obtain (3.39).

To prove (3.40) we first note that (3.37) applied to the matrix functions a and c yields the following two identities:

$$
\pi(I + A_0) + \pi A_1 = \mathcal{A}(0)\pi, \quad \pi(I + C_0) + \pi C_{-1} = \mathcal{C}(0)\pi.
$$

Thus

$$
\begin{aligned}
\pi T &= \pi(I + C_0)(I + D_0) - \pi A_1 B_{-1} \\
&= \mathcal{C}(0)\pi(I + D_0) - \pi C_{-1}(I + D_0) - \mathcal{A}(0)\pi B_{-1} + \pi(I + A_0)B_{-1}.
\end{aligned}
$$

Since the quadruple $\{\mathcal{A}, \mathcal{C}; \mathcal{B}, \mathcal{D}\}$ satisfies the quasi commutativity property, the first identity in (3.31) holds true. This yields that

$$
\pi T = \mathcal{C}(0)\pi(I + D_0) - \mathcal{A}(0)\pi B_{-1}.
$$

But then we can use the third and fourth identity in (3.45) to show that (3.40) holds.

To prove (3.41) we apply (3.38) to the functions b and d. This yields

$$
(I + B_0)\tau + B_{-1}\tau = \tau\mathcal{B}(0), \quad (I + D_0)\tau + D_1\tau = \tau\mathcal{D}(0).
$$

By using this together with the second identity in (3.31) we obtain

$$
\begin{aligned}
T\tau &= (I + C_0)(I + D_0)\tau - A_1 B_{-1}\tau \\
&= (I + C_0)\tau\mathcal{D}(0) - (I + C_0)D_1\tau - A_1\tau\mathcal{B}(0) + A_1(I + B_0)\tau \\
&= (I + C_0)\tau\mathcal{D}(0) - A_1\tau\mathcal{B}(0).
\end{aligned}
$$

But then we can use the first two identities in (3.45) to show that (3.41) holds. $\square$

Next, we assume that the matrices $\mathcal{A}(0)$ and $\mathcal{D}(0)$ are non-singular. This allows us to introduce the operators

$$
W_{\mathcal{A}}^{\times} = W - iY_{\mathcal{A}}\mathcal{A}(0)^{-1}\pi, \quad V_{\mathcal{D}}^{\times} = V - i\tau\mathcal{D}(0)^{-1}Z_{\mathcal{D}}. \tag{3.46}
$$

Notice that $W_{\mathcal{A}}^{\times}$ and $V_{\mathcal{D}}^{\times}$ are finite rank perturbations of W and V, respectively. Thus both $W_{\mathcal{A}}^{\times}$ and $V_{\mathcal{D}}^{\times}$ are compact operators. Using the operators $W_{\mathcal{A}}^{\times}$ and $V_{\mathcal{D}}^{\times}$, we can now give a first description of the kernel of the Bezout integral operator T.

Proposition 3.7. *Assume that the quadruple $\{\mathcal{A}, \mathcal{C}; \mathcal{B}, \mathcal{D}\}$ satisfies the quasi commutativity property, and let the matrices $\mathcal{A}(0)$ and $\mathcal{D}(0)$ be non-singular. Then the Bezout integral operator T associated with $\{\mathcal{A}, \mathcal{C}; \mathcal{B}, \mathcal{D}\}$ satisfies the intertwining relation*

$$W_{\mathcal{A}}^{\times} T = T V_{\mathcal{D}}^{\times}, \tag{3.47}$$

and the null space of T is equal to the maximal $V_{\mathcal{D}}^{\times}$-invariant subspace contained in $\operatorname{Ker} \pi T$. Here $W_{\mathcal{A}}^{\times}$ and $V_{\mathcal{D}}^{\times}$ are the compact operators defined by (3.46).

Proof. From (3.40) it follows

$$W_{\mathcal{A}}^{\times} T = W T - i Y_{\mathcal{A}}(0)^{-1} \pi T = W T - i Y_{\mathcal{A}} Z_{\mathcal{B}} + i Y_{\mathcal{A}} \mathcal{A}(0)^{-1} \mathcal{C}(0) Z_{\mathcal{D}}.$$

Similarly, (3.41) yields

$$T V_{\mathcal{D}}^{\times} = T V - i T \tau \mathcal{D}(0)^{-1} Z_{\mathcal{D}} = T V - i Y_{\mathcal{C}} Z_{\mathcal{D}} + i Y_{\mathcal{A}} \mathcal{B}(0) \mathcal{D}(0)^{-1} Z_{\mathcal{D}}.$$

Since $\{\mathcal{A}, \mathcal{C}; \mathcal{B}, \mathcal{D}\}$ satisfies the quasi commutativity property, we have $\mathcal{A}(0)\mathcal{B}(0) = \mathcal{C}(0)\mathcal{D}(0)$, and hence $\mathcal{A}(0)^{-1}\mathcal{C}(0) = \mathcal{B}(0)\mathcal{D}(0)^{-1}$. Thus

$$W_{\mathcal{A}}^{\times} T - T V_{\mathcal{D}}^{\times} = W T - T V - i Y_{\mathcal{A}} Z_{\mathcal{B}} + i Y_{\mathcal{C}} Z_{\mathcal{D}} = 0,$$

because of (3.39). Thus (3.47) holds.

Next, we prove the statement about the null space of T. From (3.47) it follows that $\operatorname{Ker} T$ is invariant under $V_{\mathcal{D}}^{\times}$. Obviously, $\operatorname{Ker} T \subset \operatorname{Ker} \pi T$. Now, let $\mathcal{M}$ be the maximal $V_{\mathcal{D}}^{\times}$-invariant subspace in $\operatorname{Ker} \pi T$. Since $\mathcal{M}$ is maximal, it suffices to show that $\mathcal{M} \subset \operatorname{Ker} T$.

Take $f \in \mathcal{M}$. Then $\pi T f = 0$, and hence

$$W_{\mathcal{A}}^{\times} T f = (W - i Y_{\mathcal{A}} \mathcal{A}(0)^{-1} \pi) T f = W T f. \tag{3.48}$$

Since $\mathcal{M}$ is invariant under $V_{\mathcal{D}}^{\times}$, we have $(V_{\mathcal{D}}^{\times})^{k} f \in \mathcal{M}$ for each k. It follows

$$T (V_{\mathcal{D}}^{\times})^{k} f = W^{k} T f, \quad k = 0, 1, 2, \ldots. \tag{3.49}$$

Indeed, using (3.48) with $(V_{\mathcal{D}}^{\times})^{k-1} f$ in place of f, we obtain

$$T(V_{\mathcal{D}}^{\times})^{k} f = T V_{\mathcal{D}}^{\times} \big((V_{\mathcal{D}}^{\times})^{k-1} f \big) = W_{\mathcal{A}}^{\times} T \big((V_{\mathcal{D}}^{\times})^{k-1} f \big) = W T \big((V_{\mathcal{D}}^{\times})^{k-1} f \big).$$

Since $T V_{\mathcal{D}}^{\times} f = W_{\mathcal{A}}^{\times} T f = W T f$ by (3.48), the above calculation shows that we can prove (3.49) by induction.

From (3.49) we see that

$$\pi W^{k} T f = \pi T \big((V_{\mathcal{D}}^{\times})^{k} f \big) = 0, \quad k = 0, 1, 2, \ldots.$$

But then $\pi (I - \lambda W)^{-1} T f = 0$ for each $\lambda \in \mathbb{C}$. Using (3.17) it is straightforward to show that

$$\pi (I - \lambda W)^{-1} g = \hat{g}(\lambda), \quad \lambda \in \mathbb{C},$$

where $\hat{g}$ is the Fourier transform of $g \in L_1^n[0, \omega]$. Thus $\pi (I - \lambda W)^{-1} T f = 0$ implies that $\widehat{T f}(\lambda) = 0$ for each $\lambda \in \mathbb{C}$. Hence $T f = 0$. This proves that $\mathcal{M} \subset \operatorname{Ker} T$. $\square$

254 I. Gohberg, I. Haimovici, M.A. Kaashoek and L. Lerer

3.5. Proof of the first main theorem on the Bezout integral operator

In this section we shall prove Theorem 1.1. We split the proof into two parts. In the first part we show that without loss of generality we can assume that the matrices $\mathcal{A}(0)$, $\mathcal{B}(0)$, $\mathcal{C}(0)$, $\mathcal{D}(0)$ are non-singular. In the second part we assume that this non-singularity condition is satisfied, and we apply Theorem 2.6 and Proposition 3.7 to complete the proof.

Part 1. For $\alpha \in \mathbb{R}$ let $\mathcal{A}_\alpha$, $\mathcal{B}_\alpha$, $\mathcal{C}_\alpha$, $\mathcal{D}_\alpha$ be the $n \times n$ matrix functions which one obtains when in formulas (1.2)–(1.3) the functions $a(s)$, $b(s)$, $c(s)$, $d(s)$ are replaced by $e^{i\alpha s}a(s)$, $e^{i\alpha s}b(s)$, $e^{i\alpha s}c(s)$, $e^{i\alpha s}d(s)$, respectively. In other words

$$\mathcal{A}_\alpha(\lambda) = \mathcal{A}(\alpha + \lambda), \qquad \mathcal{B}_\alpha(\lambda) = \mathcal{B}(\alpha + \lambda), \tag{3.50}$$

$$\mathcal{C}_\alpha(\lambda) = \mathcal{C}(\alpha + \lambda), \qquad \mathcal{D}_\alpha(\lambda) = \mathcal{D}(\alpha + \lambda). \tag{3.51}$$

Let T_α be the Bezout integral operator associated with $\{\mathcal{A}_\alpha, \mathcal{C}_\alpha; \mathcal{B}_\alpha, \mathcal{D}_\alpha\}$. Then formulas (1.4), (1.5) show that $T_\alpha = M_\alpha T M_{-\alpha}$, where M_α is the operator on $L_1^n[0, \omega]$ defined by

$$(M_\alpha f)(t) = e^{i\alpha t} f(t), \quad 0 \le t \le \omega. \tag{3.52}$$

Since $\alpha \in \mathbb{R}$, we have $|e^{i\alpha t}| = 1$ for each t. It follows that M_α is an invertible bounded linear operator on $L_1^n[0, \omega]$ and $M_\alpha^{-1} = M_{-\alpha}$. Thus T and T_α are similar operators, and hence $\dim \operatorname{Ker} T = \dim \operatorname{Ker} T_\alpha$. From the identities in the right-hand sides of (3.50) and (3.51) it is clear that the total common multiplicity of $\mathcal{B}$ and $\mathcal{D}$ is equal to the total common multiplicity of $\mathcal{B}_\alpha$ and $\mathcal{D}_\alpha$. In other words, $\nu(\mathcal{B}, \mathcal{D}) = \nu(\mathcal{B}_\alpha, \mathcal{D}_\alpha)$. Thus

$$\dim \operatorname{Ker} T = \nu(\mathcal{B}, \mathcal{D}) \iff \dim \operatorname{Ker} T_\alpha = \nu(\mathcal{B}_\alpha, \mathcal{D}_\alpha).$$

From the identities in (3.50) and (3.51) it is also clear that $\{\mathcal{A}, \mathcal{C}; \mathcal{B}, \mathcal{D}\}$ has the quasi commutativity property if and only if this property is satisfied for the quadruple $\{\mathcal{A}_\alpha, \mathcal{C}_\alpha; \mathcal{B}_\alpha, \mathcal{D}_\alpha\}$.

The above results show that it suffices to prove Theorem 1.1 for some quadruple $\{\mathcal{A}_\alpha, \mathcal{C}_\alpha; \mathcal{B}_\alpha, \mathcal{D}_\alpha\}$ in place of $\{\mathcal{A}, \mathcal{C}; \mathcal{B}, \mathcal{D}\}$. We claim that we can choose α in such a way that the values of $\mathcal{A}_\alpha$, $\mathcal{B}_\alpha$, $\mathcal{C}_\alpha$, $\mathcal{D}_\alpha$ at zero are non-singular matrices. To see this, we first note that

$$\mathcal{A}_\alpha(0) = \mathcal{A}(\alpha), \quad \mathcal{B}_\alpha(0) = \mathcal{B}(\alpha), \quad \mathcal{C}_\alpha(0) = \mathcal{C}(\alpha), \quad \mathcal{D}_\alpha(0) = \mathcal{D}(\alpha).$$

Next, since the functions a, b, c, d have their support in a finite interval, the Riemann-Lebesgue lemma shows that for $\alpha \in \mathbb{R}$, $\alpha \to \infty$, the values $\mathcal{A}_\alpha$, $\mathcal{B}_\alpha$, $\mathcal{C}_\alpha$, $\mathcal{D}_\alpha$ tend to the $n \times n$ identity matrix. Thus $\mathcal{A}_\alpha(0)$, $\mathcal{B}_\alpha(0)$, $\mathcal{C}_\alpha(0)$, $\mathcal{D}_\alpha(0)$ are all non-singular for $\alpha \in \mathbb{R}$ and α sufficiently large.

Part 2. In this part we assume that the values of $\mathcal{A}$, $\mathcal{B}$, $\mathcal{C}$, $\mathcal{D}$ at zero are non-singular. This allows us to introduce the functions $F_\mathcal{B}(\lambda) = \mathcal{B}(0)^{-1}\mathcal{B}(\lambda)$ and $F_\mathcal{D}(\lambda) = e^{-i\lambda\omega}\mathcal{D}(0)^{-1}\mathcal{D}(\lambda)$. In other words, using the representations (3.9) and

(3.11) we have

$$F_{\mathcal{B}}(\lambda) = I_n + i\lambda\mathcal{B}(0)^{-1}Z_{\mathcal{B}}(I - \lambda V)^{-1}\tau,$$
$$F_{\mathcal{D}}(\lambda) = I_n + i\lambda\mathcal{D}(0)^{-1}Z_{\mathcal{D}}(I - \lambda V)^{-1}\tau.$$

Since $\mathcal{B}(\lambda) = \mathcal{B}(0)F_{\mathcal{B}}(\lambda)$ and $\mathcal{D}(\lambda) = e^{i\lambda\omega}\mathcal{D}(0)F_{\alpha}(\lambda)$, the common eigenvalues of $\mathcal{B}$ and $\mathcal{D}$ are the same as those of $F_{\mathcal{B}}$ and $F_{\mathcal{D}}$. Furthermore, if $x_0, \ldots, x_{r-1}$ is a common Jordan chain of $\mathcal{B}$ and $\mathcal{D}$, then it is a common Jordan chain of $F_{\mathcal{B}}$ and $F_{\mathcal{D}}$, and conversely. It follows that

$$\nu(\mathcal{B}, \mathcal{D}) = \nu(F_{\mathcal{B}}, F_{\mathcal{D}}).$$

To compute $\nu(F_{\mathcal{B}}, F_{\mathcal{D}})$ we apply Theorem 2.6. This requires to determine the largest $V_{\mathcal{D}}^{\times}$-invariant subspace contained in $\mathrm{Ker}\,\left(i\mathcal{B}(0)^{-1}Z_{\mathcal{B}} - i\mathcal{D}(0)^{-1}Z_{\mathcal{D}}\right)$. To do this, we first show that

$$\mathrm{Ker}\,\left(i\mathcal{B}(0)^{-1}Z_{\mathcal{B}} - i\mathcal{D}(0)^{-1}Z_{\mathcal{D}}\right) = \mathrm{Ker}\,\pi T. \tag{3.53}$$

Indeed, using the quasi commutativity property and (3.40) we have

$$\begin{aligned}
i\mathcal{B}(0)^{-1}Z_{\mathcal{B}} - i\mathcal{D}(0)^{-1}Z_{\mathcal{D}} &= i\mathcal{D}(0)^{-1}\left(\mathcal{D}(0)\mathcal{B}(0)^{-1}Z_{\mathcal{B}} - Z_{\mathcal{D}}\right) \\
&= i\mathcal{D}(0)^{-1}\left(\mathcal{A}(0)^{-1}\mathcal{C}(0)Z_{\mathcal{B}} - Z_{\mathcal{D}}\right) \\
&= i\mathcal{D}(0)^{-1}\mathcal{A}(0)^{-1}\left(\mathcal{C}(0)Z_{\mathcal{B}} - \mathcal{A}(0)Z_{\mathcal{D}}\right) \\
&= -i\mathcal{D}(0)^{-1}\mathcal{A}(0)^{-1}\pi T,
\end{aligned}$$

which proves (3.53)

Let $\mathcal{M}$ be the largest $V_{\mathcal{D}}^{\times}$-invariant subspace contained in the null space of $i\mathcal{B}(0)^{-1}Z_{\mathcal{B}} - i\mathcal{D}(0)^{-1}Z_{\mathcal{D}}$. Using (3.53) and Proposition 3.7, we see that $\mathcal{M} = \mathrm{Ker}\,T$. Since T is of the form $I + \Gamma$, with Γ a compact operator, $\dim \mathrm{Ker}\,T < \infty$, and hence $\dim \mathcal{M} < \infty$. Thus by Theorem 2.6,

$$\dim \mathrm{Ker}\,T = \dim \mathcal{M} = \nu(F_{\mathcal{B}}, F_{\mathcal{D}}) = \nu(\mathcal{B}, \mathcal{D}),$$

provided $V_{\mathcal{D}}^{\times}$ is injective.

Thus to complete the proof it remains to show that $V_{\mathcal{D}}^{\times} f = 0$ implies $f = 0$. To do this, recall that $V_{\mathcal{D}}^{\times} = V - i\tau\mathcal{D}(0)^{-1}Z_{\mathcal{D}}$. Hence, using (3.15), the hypotheses $V_{\mathcal{D}}^{\times} f = 0$ implies that

$$\int_0^t f(s)\,ds = \mathcal{D}(0)^{-1}\pi(I + D_0)f, \quad 0 \le t \le \omega.$$

The right-hand side in the previous identity does not depend on t. Hence $f(s) = 0$ a.e. on $[0, \omega]$, and therefore $f = 0$. Thus $V_{\mathcal{D}}^{\times}$ is injective. This completes the proof of Theorem 1.1. $\qquad\square$

At this stage, using the second part of Theorem 2.6 and the arguments used in the above proof, we could also prove Theorem 4.6. However, we prefer first to clarify the general scheme underlying the definition of the Bezout integral operator.

Example. We conclude with an example showing that Theorem 1.1 does not remain true when the quasi commutativity property is not fulfilled. For this purpose, take $n = 1$, and let $a, d \in L_1[0, 1]$, and $b, c \in L_1[-1, 0]$ be given by

$$a(t) = 0, \ b(-t) = -1, \ c(-t) = 0, \ d(t) = -1, \quad 0 \le t \le 1. \tag{3.54}$$

With this choice of a, b, c, d, and $\omega = 1$, we let T be the operator on $L_1[0, 1]$ defined by (1.4) and (1.5). One computes that the action of T is given by

$$(T\varphi)(t) = \varphi(t) - \int_t^1 \varphi(s)\, ds, \quad 0 \le t \le 1.$$

Hence T is an invertible operator on $L_1[0, 1]$.

Next consider the functions $\mathcal{A}, \mathcal{B}, \mathcal{C}, \mathcal{D}$ defined by (1.2) and (1.3), where a, b, c, d are as above (and $\omega = 1$). It follows that for each $\lambda \in \mathbb{C}$ we have

$$\mathcal{A}(\lambda) = 1, \ \mathcal{B}(\lambda) = 1 - \frac{1 - e^{-i\lambda}}{i\lambda}, \ \mathcal{C}(\lambda) = 1, \ \mathcal{D}(\lambda) = 1 - \frac{e^{i\lambda} - 1}{i\lambda}.$$

Obviously, in this case the quasi commutativity property (1.1) is not satisfied.

Now let us show that in this case the conclusion of Theorem 1.1 does not hold. Since the derivative of e^z at zero is equal to one, we have $\mathcal{D}(0) = 1 - 1 = 0$. Similarly, $\mathcal{B}(0) = 1 - 1 = 0$. Thus 0 is a common zero of $\mathcal{B}$ and $\mathcal{D}$. Hence the total multiplicity $\nu(\mathcal{B}, \mathcal{D})$ is positive, and therefore strictly larger than $\dim \operatorname{Ker} T$, which is equal to zero. We conclude that in this case the result of Theorem 1.1 does not hold.

Note that in this scalar case we can use the result of [9] to express $\nu(\mathcal{B}, \mathcal{D})$ as the dimension of the null space of a suitable Bezout integral operator. In fact, this would mean to keep b and d as in (3.54), and to take $a = d$ and $c = b$. With this choice of a, b, c, d the quasi commutativity property is trivially satisfied. The corresponding Bezout integral operator T is now given by

$$(T\varphi)(t) = \varphi(t) - \int_0^1 \varphi(s)\, ds, \quad 0 \le t \le 1.$$

For this operator T we have $\dim \operatorname{Ker} T = \nu(\mathcal{B}, \mathcal{D})$, and this number is equal to one.

4. A general scheme for defining Bezout operators

In this chapter we present a refinement of the general scheme from [17] for defining Bezout operators. The first section has a preliminary character. The main result of this chapter is Theorem 4.3 in Section 4.2. This theorem includes a description of the null space of an abstract Bezout operator in terms of a certain invariant subspace. In the third section we show that the abstract Bezout operator defined in Theorem 4.3 is the closure of a Bezout operator in the sense of [17]. In the final section we show that the Bezout integral operator defined in Chapter 1 is also a Bezout operator according to the general scheme.

4.1. A preliminary proposition

In this section we prove a proposition which can be viewed as a generalization of the classical state space similarity theorem from mathematical system theory. In the next section the proposition will be used to define an abstract Bezout operator.

Consider two operator-valued functions H_1 and H_2 given in the following form:

$$H_1(\lambda) = I_{\mathcal{U}} + \lambda C_1(I_{\mathcal{X}_1} - \lambda A_1)^{-1}B_1, \quad H_2(\lambda) = I_{\mathcal{U}} + \lambda C_2(I_{\mathcal{X}_2} - \lambda A_2)^{-1}B_2. \quad (4.1)$$

Here $\mathcal{U}$, $\mathcal{X}_1$, $\mathcal{X}_2$ are complex Banach spaces, $I_{\mathcal{U}}$, $I_{\mathcal{X}_1}$, $I_{\mathcal{X}_2}$ are the identity operators on the corresponding spaces, and

$$A_j : \mathcal{X}_j \to \mathcal{X}_j, \quad B_j : \mathcal{U} \to \mathcal{X}_j, \quad C_j : \mathcal{X}_j \to \mathcal{U} \quad (j = 1, 2)$$

are bounded linear operators. We refer to the representations in (4.1) of H_1 and H_2 as *co-realizations*. If in (4.1) the variable λ is replaced by λ^{-1}, then the formulas in (4.1) yield the usual realizations which are known from mathematical system theory (see, e.g., the books [5, 7]).

Let $T(\mathcal{X}_1 \to \mathcal{X}_2)$ be a closed linear operator with domain $\mathcal{D}(T)$ in $\mathcal{X}_1$ and range in $\mathcal{X}_2$, and let $\mathcal{L}$ be a linear submanifold of $\mathcal{X}_1$. We say $\mathcal{L}$ is a *core* for T if $\mathcal{L}$ is contained in $\mathcal{D}(T)$ and T is equal to the closure of the restriction $T|\mathcal{L}$. In what follows, for $j = 1, 2$,

$$\mathrm{Im}\,(A_j|B_j) = \mathrm{span}\{A_j^n B_j \mathcal{U} \mid n \geq 0\}, \quad \mathrm{Ker}\,(C_j|A_j) = \bigcap_{n=0}^{\infty} \mathrm{Ker}\, C_j A_j^n.$$

We are now ready to state the proposition.

Proposition 4.1. *Put $\mathcal{R} = \mathrm{Im}\,(A_1|B_1)$, and assume $\mathrm{Ker}\,(C_2|A_2) = \{0\}$. Then H_1 and H_2 coincide in a neighborhood of zero if and only if there exists a closed linear operator $T(\mathcal{X}_1 \to \mathcal{X}_2)$ such that $\mathcal{R}$ is a core for T,*

$$TB_1 = B_2, \quad TA_1 x = A_2 T x, \quad C_2 T x = C_1 x \quad (x \in \mathcal{D}(T)). \quad (4.2)$$

Moreover, in that case T is uniquely determined, and $\mathrm{Ker}\, T$ is the maximal A_1-invariant subspace of $\mathcal{X}_1$ contained in $\mathcal{D}(T) \cap \mathrm{Ker}\, C_1$. Finally, the second identity in (4.2) includes the statement that $\mathcal{D}(T)$ is invariant under the operator A_1.

If we assume that the co-realization for H_1 and H_2 are minimal, that is, if for $j = 1, 2$ the space $\mathrm{Im}\,(A_j|B_j)$ is dense in $\mathcal{X}_j$ and $\mathrm{Ker}\,(C_j|A_j) = \{0\}$, then Proposition 4.1 reduces to the state space similarity theorem for (possibly infinite-dimensional) systems with bounded coefficients, and in this case the operator T is known as a pseudo-similarity (see [19], Theorem 3b.1, [4], Theorem 3.2, and [2], Proposition 6). The proof of Proposition 4.1 given below follows that of the state space similarity theorem.

In general, without the core condition, the operator T in Proposition 4.1 is not unique; this follows from Section 3.3 in [3].

Proof of Proposition 4.1. Let $T(\mathcal{X}_1 \to \mathcal{X}_2)$ be a closed linear operator such that $\mathcal{R}$ is contained in $\mathcal{D}(T)$, and assume (4.2) holds. We show that H_1 and H_2 coincide in a neighborhood of zero. Since $\mathcal{R}$ is invariant under A_1 and $\mathcal{R} \subset \mathcal{D}(T)$, the second identity in (4.2) implies that

$$T A_1^j x = A_2^j T x \quad (x \in \mathcal{R}). \tag{4.3}$$

Now, take $u \in \mathcal{U}$. Then, using (4.2) and (4.3), we have

$$C_2 A_2^n B_2 u = C_2 A_2^n T B_1 u = C_2 T A_1^n B_1 u = C_1 A_1^n B_1 u, \quad n = 0, 1, 2, \ldots.$$

Thus at zero the functions H_1 and H_2 have the same Taylor coefficients, and therefore H_1 and H_2 coincide in a neighborhood of zero.

Next, let $T'(\mathcal{X}_1 \to \mathcal{X}_2)$ be another closed linear operator such that $\mathcal{R}$ is contained in $\mathcal{D}(T')$ and (4.2) holds with T' in place of T. Then (4.3) holds with T' in place of T, and hence

$$T A_1^n B_1 = A_2^n T B_1 = A_2^n B_2 = A_2^n T' B_1 = T' A_1^n B_1, \quad n = 0, 1, 2, \ldots.$$

It follows that T and T' coincide on $\mathcal{R}$. But then the closures of $T|\mathcal{R}$ and $T'|\mathcal{R}$ are equal too. This proves the uniqueness statement.

In the remaining part we assume that H_1 and H_2 coincide in a neighborhood of zero. This assumption is equivalent to the requirement that

$$C_1 A_1^n B_1 = C_2 A_2^n B_2, \quad n = 0, 1, 2, \ldots. \tag{4.4}$$

We shall use (4.4) to construct T and to describe its kernel.

Let $\mathcal{X}_1 \oplus \mathcal{X}_2$ be the Banach space consisting of all pairs $\begin{pmatrix} x_1 \\ x_2 \end{pmatrix}$, $x_1 \in \mathcal{X}_1$ and $x_2 \in \mathcal{X}_2$, with the norm being given by

$$\left\| \begin{pmatrix} x_1 \\ x_2 \end{pmatrix} \right\| = \|x_1\| + \|x_2\|.$$

Consider the operators

$$A : \mathcal{X}_1 \oplus \mathcal{X}_2 \to \mathcal{X}_1 \oplus \mathcal{X}_2, \quad A \begin{pmatrix} x_1 \\ x_2 \end{pmatrix} = \begin{pmatrix} A_1 x_1 \\ A_2 x_2 \end{pmatrix},$$

$$B : \mathcal{U} \to \mathcal{X}_1 \oplus \mathcal{X}_2, \qquad Bu = \begin{pmatrix} B_1 u \\ B_2 u \end{pmatrix},$$

$$C : \mathcal{X}_1 \oplus \mathcal{X}_2 \to \mathcal{U}, \qquad C \begin{pmatrix} x_1 \\ x_2 \end{pmatrix} = C_1 x_1 - C_2 x_2.$$

The operators A, B, and C are bounded linear operators. Introduce the space $\widehat{G} = \bigcap_{n \geq 0} \operatorname{Ker} C A^n$. Obviously, $\widehat{G}$ is a closed linear submanifold of $\mathcal{X}_1 \oplus \mathcal{X}_2$. The fact that $\operatorname{Ker}(C_2|A_2) = \{0\}$ implies that $\widehat{G}$ is a graph space, that is,

$$\begin{pmatrix} 0 \\ x_2 \end{pmatrix} \in \widehat{G} \implies x_2 = 0. \tag{4.5}$$

Next, consider the space

$$G_0 = \operatorname{Im}(A|B) = \operatorname{span}\{A^n B u \mid n = 0, 1, 2, \ldots, u \in \mathcal{U}\}.$$

Condition (4.4) is equivalent to the statement that $CA^nB = 0$ for $n = 0, 1, 2, \ldots$. Hence (4.4) implies that $G_0 \subset \widehat{G}$. Let G be the closure of G_0 in $\mathcal{X}_1 \oplus \mathcal{X}_2$. Then $G \subset \widehat{G}$. Hence G is a closed linear submanifold of $\mathcal{X}_1 \oplus \mathcal{X}_2$ and a graph space by (4.5). Thus there exists a closed linear operator T with domain $\mathcal{D}(T)$ in $\mathcal{X}_1$ and range in $\mathcal{X}_2$ such that

$$G = G(T) = \left\{ \begin{pmatrix} x \\ Tx \end{pmatrix} \mid x \in \mathcal{D}(T) \right\}.$$

We claim that T has the desired properties. Notice that $\operatorname{Im} B \subset G_0 \subset G$. Hence $B_1 = TB_2$. From the definition of G_0 it follows that G_0 is invariant under A. But A is bounded, and thus, by continuity, the space G is also invariant under A. This shows that $A_1\mathcal{D}(T) \subset \mathcal{D}(T)$ and $TA_1x = A_2Tx$ for each $x \in \mathcal{D}(T)$. Since $G \subset \widehat{G}$, we have $G \subset \operatorname{Ker} C$, and thus $C_1x = C_2Tx$ for each $x \in \mathcal{D}(T)$. Finally, note that

$$G_0 = \left\{ \begin{pmatrix} m \\ Tm \end{pmatrix} \mid m \in \operatorname{Im}(A_1|B_1) \right\}.$$

Since $G = G(T)$ is the closure of G_0, this shows that $\mathcal{R}$ is a core for T.

We conclude with the description of $\operatorname{Ker} T$. Obviously, $\operatorname{Ker} T$ is closed and contained in $\mathcal{D}(T)$. The third identity in (4.2) implies that $\operatorname{Ker} T \subset \operatorname{Ker} C_1$, and from the second identity in (4.2) we conclude that $\operatorname{Ker} T$ is invariant under A_1. Thus $\operatorname{Ker} T$ is an A_1-invariant subspace contained in $\mathcal{D}(T) \cap \operatorname{Ker} C_1$.

Let $\mathcal{N}$ be an arbitrary A_1-invariant subspace of $\mathcal{X}$, contained in the intersection $\mathcal{D}(T) \cap \operatorname{Ker} C_1$. To finish the proof it suffices to show that $\mathcal{N} \subset \operatorname{Ker} T$. Take $x \in \mathcal{N}$. Then $x \in \mathcal{D}(T)$, and we see from (4.2) and (4.3) that

$$C_2A_2^kTx = C_2TA_1^kx = C_1A_1^kx, \quad k = 0, 1, 2, \ldots.$$

On the other hand, $\mathcal{N}$ is A_1-invariant. Thus $A_1^kx \in \mathcal{N} \subset \operatorname{Ker} C_1$ for each k, that is, $C_1A_1^kx = 0$ for $k = 0, 1, 2, \ldots$. Thus $C_2A_2^kTx = 0$ for $k = 0, 1, 2, \ldots$. But $\operatorname{Ker}(C_2|A_2) = \{0\}$, and hence $Tx = 0$. We proved that $x \in \operatorname{Ker} T$, and therefore $\mathcal{N} \subset \operatorname{Ker} T$. $\qquad\square$

Corollary 4.2. *Put* $\mathcal{R} = \operatorname{Im}(A_1|B_1)$, *and assume that* $\operatorname{Ker}(C_2|A_2) = \{0\}$. *Let* $T(\mathcal{X}_1 \to \mathcal{X}_2)$ *be a closed linear operator such that* $\mathcal{R} \subset \mathcal{D}(T)$ *and* (4.2) *holds. Then for* $|\lambda|$ *and* $|\mu|$ *sufficiently small we have*

$$H_2(\lambda) - H_1(\mu) = (\lambda - \mu)C_2(I_{\mathcal{X}_2} - \lambda A_2)^{-1}T(I_{\mathcal{X}_1} - \mu A_1)^{-1}B_1. \qquad (4.6)$$

The previous identity includes the statement that for $|\mu|$ *sufficiently small the set* $(I_{\mathcal{X}_1} - \mu A_1)^{-1}B_1\mathcal{U}$ *is contained in* $\mathcal{D}(T)$.

Proof. First we use the second identity in (4.2) to prove that for $|\mu|$ sufficiently small we have

$$(I_{\mathcal{X}_1} - \mu A_1)^{-1}\mathcal{D}(T) \subset \mathcal{D}(T), \quad \text{and} \qquad (4.7)$$

$$T(I_{\mathcal{X}_1} - \mu A_1)^{-1}x = (I_{\mathcal{X}_2} - \mu A_2)^{-1}Tx \quad (x \in \mathcal{D}(T)).$$

To see this, let $x \in \mathcal{D}(T)$, and take $|\mu| < (\|A_1\| + \|A_2\|)^{-1}$. Then

$$\lim_{N\to\infty} \sum_{n=0}^{N} \mu^n A_j^n x = (I_{\mathcal{X}_j} - \mu A_j)^{-1} x, \quad j = 1, 2.$$

Since $x \in \mathcal{D}(T)$, the second identity in (4.2) yields

$$\lim_{N\to\infty} T\left(\sum_{n=0}^{N} \mu^n A_1^n x\right) = \lim_{N\to\infty} \sum_{n=0}^{N} \mu^n A_2^n T x = (I_{\mathcal{X}_2} - \mu A_2)^{-1} T x.$$

But T is closed. Thus the above formulas prove (4.7). Recall that $B_1 \mathcal{U} \subset \mathcal{R} \subset \mathcal{D}(T)$. Thus, as a corollary of (4.7), we have

$$(I_{\mathcal{X}_1} - \mu A_1)^{-1} B_1 \mathcal{U} \subset \mathcal{D}(T), \quad T(I_{\mathcal{X}_1} - \mu A_1)^{-1} B_1 = (I_{\mathcal{X}_2} - \mu A_2)^{-1} B_2. \quad (4.8)$$

for $|\mu|$ sufficiently small.

Next, we consider $H_1(\lambda) - H_2(\mu)$. By Proposition 4.1 the functions $H_1(\lambda)$ and $H_2(\lambda)$ coincide in a neighborhood of zero. Hence $C_1(I_{\mathcal{X}_1} - \lambda A_1)^{-1} B_1$ is equal to $C_2(I_{\mathcal{X}_2} - \lambda A_2)^{-1} B_2$ for $|\lambda|$ sufficiently small. This, together with the resolvent formula, yields

$$\begin{aligned}
H_2(\lambda) - H_1(\mu) &= \lambda C_2(I_{\mathcal{X}_2} - \lambda A_2)^{-1} B_2 - \mu C_1(I_{\mathcal{X}_1} - \mu A_1)^{-1} B_1 \\
&= \lambda C_2(I_{\mathcal{X}_2} - \lambda A_2)^{-1} B_2 - \mu C_2(I_{\mathcal{X}_2} - \mu A_2)^{-1} B_2 \\
&= (\lambda - \mu) C_2(I_{\mathcal{X}_2} - \lambda A_2)^{-1}(I_{\mathcal{X}_2} - \mu A_2)^{-1} B_2
\end{aligned}$$

for $|\lambda|$ and $|\mu|$ sufficiently small. Using the second equality in (4.8), we obtain (4.6). $\qquad\square$

For later purposes we note that the second identity in (4.2) also shows that

$$(I_{\mathcal{X}_1} - \mu A_1)^{-k} A_1^j B_1 \mathcal{U} \subset \mathcal{D}(T), \quad \text{and} \qquad (4.9)$$

$$T(I_{\mathcal{X}_1} - \mu A_1)^{-k} A_1^j B_1 = (I_{\mathcal{X}_2} - \mu A_2)^{-k} A_2^j B_2, \quad j, k = 0, 1, 2, \dots.$$

Here $|\mu|$ is assumed to be sufficiently small. Indeed, this follows directly from (4.7) using the fact that $A_1^j B_1 \mathcal{U}$ is contained in $\mathcal{D}(T)$ and $T A_1^j B_1 u = A_2^j B_2 u$ for each $u \in \mathcal{U}$ and each j.

4.2. Definition of an abstract Bezout operator

In this section we state and prove the theorem that we shall use to define an abstract Bezout operator.

Consider the following four operator-valued functions:

$$F_j(\lambda) = I_{\mathcal{U}} + \lambda C_j(I_{\mathcal{X}} - \lambda A)^{-1} B, \quad j = 1, 2, \qquad (4.10)$$

$$G_j(\lambda) = I_{\mathcal{U}} + \lambda C(I_{\widetilde{\mathcal{X}}} - \lambda \widetilde{A})^{-1} B_j, \quad j = 1, 2. \qquad (4.11)$$

Here $\mathcal{U}$, $\mathcal{X}$ and $\widetilde{\mathcal{X}}$ are Banach spaces, $I_{\mathcal{U}}$, $I_{\mathcal{X}}$ and $I_{\widetilde{\mathcal{X}}}$ are the identity operators on $\mathcal{U}$, $\mathcal{X}$ and $\widetilde{\mathcal{X}}$, respectively, and

$$A: \mathcal{X} \to \mathcal{X}, \quad B: \mathcal{U} \to \mathcal{X}, \qquad C_1, C_2 : \mathcal{X} \to \mathcal{U},$$
$$\widetilde{A}: \widetilde{\mathcal{X}} \to \widetilde{\mathcal{X}}, \quad B_1, B_2 : \mathcal{U} \to \widetilde{\mathcal{X}}, \quad C : \widetilde{\mathcal{X}} \to \mathcal{U}$$

are bounded linear operators. We shall prove the following theorem.

Theorem 4.3. *Put* $\mathcal{R} = \operatorname{Im}(A|B)$, *and assume* $\operatorname{Ker}(C|\widetilde{A}) = \{0\}$. *Then the functions* $G_1(\cdot)F_1(\cdot)$ *and* $G_2(\cdot)F_2(\cdot)$ *coincide in a neighborhood of zero if and only if there exists a closed linear operator* $T(\mathcal{X} \to \widetilde{\mathcal{X}})$ *such that* $\mathcal{R}$ *is a core for* T *and*

(i) $TAx - \widetilde{A}Tx = B_2C_2x - B_1C_1x, \quad x \in \mathcal{D}(T),$
(ii) $TB = B_2 - B_1$
(iii) $CTx = C_1x - C_2x, \quad x \in \mathcal{D}(T).$

Moreover, in that case the operator T *is uniquely determined and* $\operatorname{Ker}T$ *is the maximal* $(A - BC_1)$*-invariant subspace contained in* $\mathcal{D}(T) \cap \operatorname{Ker}(C_1 - C_2)$. *Finally, item* (i) *includes the statement that* $\mathcal{D}(T)$ *is invariant under the operator* A.

Notice that Theorem 4.3 contains Proposition 4.1 as a special case. Indeed, Proposition 4.1 appears when in Theorem 4.3 we take $F_1 = H_1$, $G_2 = H_2$, $C_2 = 0$, and $B_1 = 0$. Then $G_1(\lambda)F_1(\lambda) = H_1(\lambda)$ and $G_2(\lambda)F_2(\lambda) = H_2(\lambda)$, and the statements (i)–(iii) reduce to (4.2). On the other hand, as we shall see below, Theorem 4.3 is an immediate corollary of Proposition 4.1.

We shall refer to the operator T defined by Theorem 4.3 as the *abstract Bezout operator* associated to the operator-valued functions $\{G_1, G_2; F_1, F_2\}$ and the co-realizations (4.10) and (4.11). The use of this terminology will be justified in the next section. First we prove Theorem 4.3.

Proof of Theorem 4.3. For λ in an appropriate neighborhood of zero, put

$$H_1(\lambda) = F_1(\lambda)F_2(\lambda)^{-1}, \quad H_2(\lambda) = G_1(\lambda)^{-1}G_2(\lambda).$$

Then $G_1(\cdot)F_1(\cdot)$ and $G_2(\cdot)F_2(\cdot)$ coincide in a neighborhood of zero if and only if H_1 and H_2 coincide in a neighborhood of zero. Using the co-realizations for F_2 and G_1, we have

$$F_2(\lambda)^{-1} = I_{\mathcal{U}} - \lambda C_2(I_{\mathcal{X}} - \lambda A_2^{\times})^{-1}B, \quad A_2^{\times} = A - BC_2,$$
$$G_1(\lambda)^{-1} = I_{\mathcal{U}} - \lambda C(I_{\widetilde{\mathcal{X}}} - \lambda \widetilde{A}_1^{\times})B_1, \quad \widetilde{A}_1^{\times} = \widetilde{A} - B_1C.$$

A straightforward calculation then yields

$$H_1(\lambda) = I_{\mathcal{U}} + \lambda(C_1 - C_2)(I_{\mathcal{X}} - \lambda A_2^{\times})^{-1}B, \tag{4.12}$$
$$H_2(\lambda) = I_{\mathcal{U}} + \lambda C(I_{\widetilde{\mathcal{X}}} - \lambda \widetilde{A}_1^{\times})^{-1}(B_2 - B_1). \tag{4.13}$$

Next, using a standard feedback argument, we have

$$\operatorname{Im}(A_2^{\times}|B) = \operatorname{Im}(A|B), \quad \operatorname{Ker}(C|\widetilde{A}_1^{\times}) = \operatorname{Ker}(C|\widetilde{A}).$$

262   I. Gohberg, I. Haimovici, M.A. Kaashoek and L. Lerer

It follows that $\mathcal{R} = \operatorname{Im}(A_2^\times | B)$ and $\operatorname{Ker}(C | \widetilde{A}_1^\times) = \{0\}$. Thus we can apply Proposition 4.1 to show that $G_1(\cdot)F_1(\cdot)$ and $G_2(\cdot)F_2(\cdot)$ coincide in a neighborhood of zero if and only if there exists a closed linear operator $T(\mathcal{X} \to \widetilde{\mathcal{X}})$ such that $\mathcal{R}$ is a core for T and

  (a) $\ TA_2^\times x = \widetilde{A}_1^\times Tx, \quad x \in \mathcal{D}(T),$
  (b) $\ TB = B_2 - B_1,$
  (c) $\ CTx = C_1 x - C_2 x, \quad x \in \mathcal{D}(T).$

Moreover there is only one such T, and its null space is the maximal $A_2^\times$-invariant subspace of $\mathcal{X}$ contained in $\mathcal{D}(T) \cap \operatorname{Ker}(C_1 - C_2)$. Here $A_2^\times = A - BC_2$.

Note that (b) and (c) coincide with (ii) and (iii), respectively. Hence to complete the proof it remains to show that (a) and (i) are equivalent whenever (b),(c) or (ii), (iii) are satisfied. To do this, notice that $\operatorname{Im} B \subset \mathcal{R} \subset \mathcal{D}(T)$, and hence the identity $A_2^\times = A - BC_2$ implies that

$$A_2^\times \mathcal{D}(T) \subset \mathcal{D}(T) \quad \Longleftrightarrow \quad A\mathcal{D}(T) \subset \mathcal{D}(T)$$

In particular, $TA_2^\times x$ is well defined for each $x \in \mathcal{D}(T)$ if and only if TAx is well defined for each $x \in \mathcal{D}(T)$, and in that case $TA_2^\times x = TAx - TBC_2 x$ for each $x \in \mathcal{D}(T)$. From $\widetilde{A}_1^\times = \widetilde{A} - B_1 C$ it follows that $\widetilde{A}_1^\times Tx = \widetilde{A}Tx - B_1 CTx$ for each $x \in \mathcal{D}(T)$. Finally, using (b),(c) or, equivalently (ii), (iii), we obtain

$$\begin{aligned} B_2 C_2 x - B_1 C_1 x &= (B_2 - B_1)C_2 x - B_1(C_1 - C_2)x \\ &= TBC_2 x - B_1 CTx, \quad x \in \mathcal{D}(T). \end{aligned}$$

With these remarks the equivalence of (a) and (i) is clear. $\square$

4.3. The Haimovici-Lerer scheme for defining an abstract Bezout operator

In this section we show that the operator T appearing in Theorem 4.3 is the closure of a Bezout operator in the sense of [17]. To do this we first briefly describe the general set-up from the latter paper.

In [17] the construction of a Bezout operator requires that two basic assumptions are satisfied. The first is that along with a pair of given analytic operator-valued functions F_1 and F_2 one has two other analytic operator-valued functions G_1 and G_2 such that

$$G_2(\lambda)F_2(\lambda) - G_1(\lambda)F_1(\lambda) = 0 \tag{4.14}$$

in some open domain Ω. In our setting Ω is a sufficiently small neighborhood of zero, and (4.14) is the condition that $G_1(\cdot)F_1(\cdot)$ and $G_2(\cdot)F_2(\cdot)$ coincide in a neighborhood of zero. The second assumption is that we have co-realizations of F_1 and F_2 as in (4.10), and co-realizations of G_1 and G_2 as in (4.11).

These two basic assumptions are not artificial and, in general, it is quite straightforward to satisfy these assumptions. For instance, as has been shown in Section 1.2 in [17], given a pair of analytic operator-valued functions F_1 and F_2 one can always construct another pair G_1 and G_2 such that (4.14) holds. Furthermore, as we mentioned in the last but one paragraph of the introduction, in many concrete cases the supplementary functions appear in a natural way or can be constructed from the first pair of functions. Also, as follows from the realization result in Section 2.3 (see also Section 1.3 in [17]), co-realizations as in (4.10) and (4.11) always exist for matrix functions that are analytic in a neighborhood of the origin.

Now let the two basic assumptions be satisfied with co-realizations as in (4.10) and (4.11). Assuming additionally that

$$\operatorname{Im}(A|B) \text{ is dense in } \mathcal{X}, \text{ and } \operatorname{Ker}(C|\widetilde{A}) = \{0\},$$

it is shown in [17] that there exists a unique (possibly unbounded) operator T_{HL} with domain

$$\mathcal{D}(T_{HL}) = \operatorname{span}\{(I_{\mathcal{X}} - \lambda A)^{-k} A^j B\mathcal{U} \mid \lambda \in \Omega, j, k = 0, 1, 2, \ldots\}, \qquad (4.15)$$

such that

$$G_1(\lambda)F_1(\mu) - G_2(\lambda)F_2(\mu) = (\mu - \lambda)C(I_{\widetilde{\mathcal{X}}} - \lambda\widetilde{A})^{-1}T_{HL}(I_{\mathcal{X}} - \mu A)^{-1}B, \quad (4.16)$$

for $|\lambda|$ and $|\mu|$ sufficiently small. In [17] this operator is called the *Bezout operator* associated with the realizations (4.10), (4.11), and the equality (4.16). In [17], it is also shown that all known concrete Bezout operators can be derived from this general scheme as particular cases.

The next proposition clarifies the connection between the operator T_{HL} and the abstract Bezout operator introduced in the previous section.

Proposition 4.4. *Let F_1, F_2, G_1, G_2 be the operator-valued functions given by (4.10), (4.11), and let $G_1(\cdot)F_1(\cdot)$ and $G_2(\cdot)F_2(\cdot)$ coincide in a neighborhood of zero. Assume that $\operatorname{Im}(A|B)$ is dense in $\mathcal{X}$ and $\operatorname{Ker}(C|\widetilde{A}) = \{0\}$. Then the abstract Bezout operator T associated to $\{G_1, G_2; F_1, F_2\}$ is equal to the closure of the operator T_{HL}.*

Proof. First we apply Corollary 4.2 to $H_1(\lambda) = F_1(\lambda)F_2(\lambda)^{-1}$ and $H_2(\lambda) = G_1(\lambda)^{-1}G_2(\lambda)$. Using the identities (4.12) and (4.13) and assuming $|\lambda|$ and $|\mu|$ to be sufficiently small, this yields

$$H_2(\lambda) - H_1(\mu) = (\lambda - \mu)C(I_{\widetilde{\mathcal{X}}} - \lambda\widetilde{A}_1^{\times})^{-1}T(I_{\mathcal{X}} - \mu A_2^{\times})^{-1}B.$$

It follows that

$$G_1(\lambda)F_1(\mu) - G_2(\lambda)F_2(\mu) = G_1(\lambda)[H_1(\mu) - H_2(\lambda)]F_2(\mu)$$

$$= (\mu - \lambda)G_1(\lambda)C(I_{\widetilde{\mathcal{X}}} - \lambda\widetilde{A}_1^{\times})^{-1}T(I_{\mathcal{X}} - \mu A_2^{\times})^{-1}BF_2(\mu).$$

A straightforward calculation, using $\widetilde{A}_1^{\times} = \widetilde{A} - B_1 C$ and $A_2^{\times} = A - BC_2$, shows that

$$G_1(\lambda)C(I_{\widetilde{\mathcal{X}}} - \lambda\widetilde{A}_1^{\times})^{-1} = C(I_{\widetilde{\mathcal{X}}} - \lambda\widetilde{A})^{-1},$$

$$(I_{\mathcal{X}} - \mu A_2^{\times})^{-1}BF_2(\mu) = (I_{\mathcal{X}} - \mu A)^{-1}B.$$

We conclude that for $|\lambda|$ and $|\mu|$ sufficiently small we have

$$G_1(\lambda)F_1(\mu) - G_2(\lambda)F_2(\mu) = (\mu - \lambda)C(I_{\widetilde{\mathcal{X}}} - \lambda\widetilde{A})^{-1}T(I_{\mathcal{X}} - \mu A)^{-1}B.$$

In particular, the set $(I_{\mathcal{X}} - \mu A)^{-1} B\mathcal{U}$ is contained in $\mathcal{D}(T)$. From (4.9) it follows that $\mathcal{D}(T_{HL})$ is also a subset of $\mathcal{D}(T)$. Thus the above equality implies that $T|_{\mathcal{D}(T_{HL})} = T_{HL}$. On the other hand, in [17] it is proved that $\operatorname{Im}(A|B)$ is contained in $\mathcal{D}(T_{HL})$. Since $\operatorname{Im}(A|B)$ is a core for T, this implies that T is the closure of T_{HL}. $\qquad\square$

In spite of the remarks made in the third paragraph of this section, the implementation of the abstract scheme in a specific case may not be an easy task. For instance, getting a suitable Bezout operator for a given pair of concrete matrix functions is not always straightforward. Also, to see that a given operator is actually a Bezout operator requires to identify the corresponding matrix functions and to get appropriate co-realizations yielding the given operator as a Bezout operator in the sense of the abstract scheme. The next section illustrates the latter point for the Bezout integral operator studied in this paper.

4.4. The Bezout integral operator revisited

In this section we show that the Bezout integral operator introduced in the first chapter is a Bezout operator according the general scheme presented in Section 4.2.

Let $\mathcal{A}$, $\mathcal{B}$, $\mathcal{C}$, $\mathcal{D}$ be the entire $n \times n$ matrix functions given by (1.2) and (1.3). Assume that the quasi commutativity property (1.1) is satisfied, and let the matrices $\mathcal{A}(0)$, $\mathcal{B}(0)$, $\mathcal{C}(0)$, $\mathcal{D}(0)$ be non-singular. Put

$$E = \mathcal{A}(0)\mathcal{B}(0) = \mathcal{C}(0)\mathcal{D}(0).$$

Notice that E is invertible, and that $\mathcal{C}(0)^{-1}E\mathcal{D}(0)^{-1}$ and $\mathcal{A}(0)^{-1}E\mathcal{B}(0)^{-1}$ are both equal to the $n \times n$ identity matrix I_n. Consider the following matrix functions

$$F_{\mathcal{B}}(\lambda) = I_n + i\lambda\mathcal{B}(0)^{-1}Z_{\mathcal{B}}(I - \lambda V)^{-1}\tau, \tag{4.17}$$

$$F_{\mathcal{D}}(\lambda) = I_n + i\lambda\mathcal{D}(0)^{-1}Z_{\mathcal{D}}(I - \lambda V)^{-1}\tau, \tag{4.18}$$

$$G_{\mathcal{A}}(\lambda) = I_n + i\lambda E^{-1}\pi(I - \lambda W)^{-1}Y_{\mathcal{A}}\mathcal{A}(0)^{-1}E, \tag{4.19}$$

$$G_{\mathcal{C}}(\lambda) = I_n + i\lambda E^{-1}\pi(I - \lambda W)^{-1}Y_{\mathcal{C}}\mathcal{C}(0)^{-1}E. \tag{4.20}$$

By comparing (4.17)–(4.20) with (3.8)–(3.11), and using the properties of E, we see that

$$F_{\mathcal{B}}(\lambda) = \mathcal{B}(0)^{-1}\mathcal{B}(\lambda), \qquad\qquad F_{\mathcal{D}}(\lambda) = e^{-i\lambda\omega}\mathcal{D}(0)^{-1}\mathcal{D}(\lambda),$$

$$G_{\mathcal{A}}(\lambda) = E^{-1}\mathcal{A}(\lambda)\mathcal{A}(0)^{-1}E, \qquad G_{\mathcal{C}}(\lambda) = E^{-1}e^{i\lambda\omega}\mathcal{C}(\lambda)\mathcal{C}(0)^{-1}E.$$

Since $\mathcal{C}(0)^{-1}E\mathcal{D}(0)^{-1} = \mathcal{A}(0)^{-1}E\mathcal{B}(0)^{-1} = I_n$, the above identities can be used to show that

$$G_{\mathcal{A}}(\lambda)F_{\mathcal{B}}(\lambda) = E^{-1}\mathcal{A}(\lambda)\mathcal{B}(\lambda), \quad G_{\mathcal{C}}(\lambda)F_{\mathcal{D}}(\lambda) = E^{-1}\mathcal{C}(\lambda)\mathcal{D}(\lambda).$$

Thus quasi commutativity property (1.1) implies that

$$G_{\mathcal{A}}(\lambda)F_{\mathcal{B}}(\lambda) = G_{\mathcal{C}}(\lambda)F_{\mathcal{D}}(\lambda), \quad \lambda \in \mathbb{C}.$$

In particular, the functions $G_{\mathcal{A}}(\cdot)F_{\mathcal{B}}(\cdot)$ and $G_{\mathcal{C}}(\cdot)F_{\mathcal{D}}(\cdot)$ coincide in a neighborhood of zero.

Proposition 4.5. *Let $\mathcal{A}$, $\mathcal{B}$, $\mathcal{C}$, $\mathcal{D}$ be the entire $n \times n$ matrix functions given by (1.2) and (1.3). Assume that the quasi commutativity property (1.1) is satisfied, and let the matrices $\mathcal{A}(0)$, $\mathcal{B}(0)$, $\mathcal{C}(0)$, $\mathcal{D}(0)$ be non-singular. Then the Bezout integral operator T defined by (1.4), (1.5) is equal to the abstract Bezout operator associated to the matrix functions $\{G_\mathcal{A}, G_\mathcal{C}; F_\mathcal{B}, F_\mathcal{D}\}$ and the co-realizations (4.17)–(4.20).*

Proof. If in (4.10) and (4.11), we take

$$A = V, \qquad \widetilde{A} = W,$$

$$B = \tau, \qquad B_1 = Y_\mathcal{A}\mathcal{A}(0)^{-1}E, \qquad B_2 = Y_\mathcal{C}\mathcal{C}(0)^{-1}E,$$

$$C = iE^{-1}\pi, \quad C_1 = i\mathcal{B}(0)^{-1}Z_\mathcal{B}, \qquad C_2 = i\mathcal{D}(0)^{-1}Z_\mathcal{D},$$

then

$$F_1 = F_\mathcal{B}, \quad F_2 = F_\mathcal{D}, \quad G_1 = G_\mathcal{A}, \quad G_2 = G_\mathcal{C},$$

and $G_1(\cdot)F_1(\cdot)$ and $G_2(\cdot)F_2(\cdot)$ coincide in a neighborhood of zero.

We claim that $\operatorname{Im}(A|B)$ is dense in $\mathcal{X} = \widetilde{\mathcal{X}} = L_1^n[0, \omega]$, and $\operatorname{Ker}(C|\widetilde{A}) = \{0\}$. First notice that for $j = 0, 1, 2, \ldots$ we have

$$A^j B = V^j \tau, \quad \text{and} \quad (V^j \tau x)(t) = \frac{(-it)^j}{j!} x \quad (x \in \mathbb{C}^n). \tag{4.21}$$

Since the polynomials are dense in $L_1[0, \omega]$, this implies that $\operatorname{Im}(A|B)$ is dense in $\mathcal{X} = L_1^n[0, \omega]$. Next, observe that $f \in \operatorname{Ker}(C|\widetilde{A})$ if and only if $\pi W^n f = 0$ for $n = 0, 1, 2, \ldots$. Thus $f \in \operatorname{Ker}(C|\widetilde{A})$ implies

$$\hat{f}(\lambda) = \pi(I - \lambda W)^{-1} f = 0, \quad \lambda \in \mathbb{C},$$

which yields $f = 0$. Thus $\operatorname{Ker}(C|\widetilde{A}) = \{0\}$.

Since the Bezout integral operator T is a bounded operator on $L_1^n[0, \omega]$, the fact that $\mathcal{R} = \operatorname{Im}(A|B)$ is dense in $L_1^n[0, \omega]$ implies that $\mathcal{R}$ is a core for T. Thus in order to complete the proof it remains to show that the Bezout integral operator T satisfies the identities (i)–(iii) in Theorem 4.3. To do this we use Proposition 3.6. Indeed, (3.39) yields

$$\begin{aligned} TA - \widetilde{A}T &= TV - WT = iY_\mathcal{C}Z_\mathcal{D} - iY_\mathcal{A}Z_\mathcal{B} \\ &= B_2 E^{-1}\mathcal{C}(0)\mathcal{D}(0)C_2 - B_1 E^{-1}\mathcal{A}(0)\mathcal{B}(0)C_1 \\ &= B_2 C_2 - B_1 C_1, \end{aligned}$$

which proves (i). To get (ii) we use (3.41) to show that

$$\begin{aligned} TB &= T\tau = Y_\mathcal{C}\mathcal{D}(0) - Y_\mathcal{A}\mathcal{B}(0) \\ &= Y_\mathcal{C}\mathcal{C}(0)^{-1}E^{-1} - Y_\mathcal{A}\mathcal{A}(0)^{-1}E^{-1} = B_2 - B_1. \end{aligned}$$

Finally, (3.40) yields

$$\begin{aligned} CT &= iE^{-1}\pi T = iE^{-1}\mathcal{A}(0)Z_\mathcal{B} - iE^{-1}\mathcal{C}(0)Z_\mathcal{D} \\ &= i\mathcal{B}(0)E^{-1}Z_\mathcal{B} - i\mathcal{D}(0)E^{-1}Z_\mathcal{D} = C_1 - C_2. \end{aligned}$$

Thus (i)–(iii) hold. $\qquad\square$

4.5. The null space of the Bezout integral operator

As an addition to Theorem 1.1 we prove the following result.

Theorem 4.6. *Let $\mathcal{A}$, $\mathcal{B}$, $\mathcal{C}$, $\mathcal{D}$ be the entire $n \times n$ matrix functions given by (1.2) and (1.3), and assume that the quasi commutativity property (1.1) is satisfied. Then the null space of the Bezout integral operator T associated with $\{\mathcal{A}, \mathcal{C}; \mathcal{B}, \mathcal{D}\}$ is finite-dimensional, and a basis for this null space can be obtained in the following way. Let $\lambda_1, \ldots, \lambda_\ell$ be the set of distinct common eigenvalues of $\mathcal{B}$ and $\mathcal{D}$ in $\mathbb{C}$, and for each common eigenvalue λ_ν let*

$$x_{1,0}^\nu, \ldots, x_{1, r_1^{(\nu)}-1}^\nu, x_{2,0}^\nu, \ldots, x_{2, r_2^{(\nu)}-1}^\nu, \ldots, x_{p_\nu, 0}^\nu, \ldots, x_{p_\nu, r_{p_\nu}^{(\nu)}-1}^\nu$$

stand for a canonical set of common Jordan chains of $\mathcal{B}$ and $\mathcal{D}$ at λ_ν. Then the functions

$$\psi_{j,k}^\nu(t) = e^{-i\lambda_\nu t} \sum_{\mu=0}^{k} \frac{(-it)^{k-\mu}}{(k-\mu)!} x_{j,\mu}^\nu, \qquad k = 0, \ldots, r_j^{(\nu)} - 1, \tag{4.22}$$

$$j = 1, \ldots, p_\nu, \qquad \nu = 1, \ldots, \ell,$$

form a basis for the null space of T. In particular, $\dim \operatorname{Ker} T = \nu(\mathcal{B}, \mathcal{D})$.

Proof. We split the proof into two parts. In the first part we assume additionally that the matrices $\mathcal{A}(0)$, $\mathcal{B}(0)$, $\mathcal{C}(0)$, $\mathcal{D}(0)$ are non-singular. The general case is treated in the second part.

Part 1. Let $\mathcal{M}$ be the null space of the Bezout integral operator T associated with $\{\mathcal{A}, \mathcal{C}; \mathcal{B}, \mathcal{D}\}$. Since T is of the form $I + \Gamma$, with Γ a compact operator, the space $\mathcal{M}$ is finite-dimensional.

Assume that $\mathcal{A}(0)$, $\mathcal{B}(0)$, $\mathcal{C}(0)$, $\mathcal{D}(0)$ are non-singular, and consider the functions $F_\mathcal{B}$, $F_\mathcal{D}$, $G_\mathcal{A}$, $G_\mathcal{C}$ defined by (4.17)–(4.20). As we have seen in the previous section the operator T is equal to the abstract Bezout operator associated to the matrix functions $\{G_\mathcal{A}, G_\mathcal{C}; F_\mathcal{B}, F_\mathcal{D}\}$ and the co-realizations (4.17)–(4.20). Thus we know from Theorem 4.3 that $\mathcal{M} = \operatorname{Ker} T$ is the maximal subspace that is invariant under the operator $V - i\tau\mathcal{B}(0)^{-1}Z_\mathcal{B}$ and that is contained in the null space of the operator $i\mathcal{B}(0)^{-1}Z_\mathcal{B} - i\mathcal{D}(0)^{-1}Z_\mathcal{D}$. This allows us to apply Theorem 2.6 with $F_1 = F_\mathcal{B}$ and $F_2 = F_\mathcal{D}$. We already know that $\mathcal{M}$ is finite-dimensional. The argument used in the final paragraph of Chapter 3 shows that the operator $V - i\tau\mathcal{B}(0)^{-1}Z_\mathcal{B}$ is injective. Thus Theorem 2.6 (together with the remark made immediately after Theorem 2.6) tells us that the functions

$$u_{j,k}^\nu = \sum_{\alpha=0}^{k} (I - \lambda_\nu V)^{-(\alpha+1)} V^\alpha T x_{j, k-\alpha}^\nu, \qquad k = 0, \ldots, r_j^{(\nu)} - 1,$$

$$j = 1, \ldots, p_\nu, \qquad \nu = 1, \ldots, \ell,$$

form a basis for $\mathcal{M} = \operatorname{Ker} T$.

To complete the proof of the first part we shall show that

$$\psi_{j,k}^\nu = u_{j,k}^\nu, \qquad k = 0, \ldots, r_j^{(\nu)} - 1, \quad j = 1, \ldots, p_\nu, \quad \nu = 1, \ldots, \ell. \tag{4.23}$$

To do this, we first note that for $\alpha = 0, 1, 2, \ldots$ we have

$$
\frac{d^\alpha}{d\lambda^\alpha}(I - \lambda V)^{-1} = \frac{d^\alpha}{d\lambda^\alpha} \sum_{m=0}^\infty \lambda^m V^m
$$

$$
= \sum_{m=\alpha}^\infty m(m-1)\cdots(m-\alpha+1)\lambda^{m-\alpha} V^m
$$

$$
= \sum_{m=0}^\infty (m+\alpha)(m+\alpha-1)\cdots(m+1)\lambda^m V^{(m+\alpha)}
$$

$$
= \sum_{m=0}^\infty \frac{(m+\alpha)!}{m!}\lambda^m V^{(m+\alpha)} = \alpha! \sum_{m=0}^\infty \binom{m+\alpha}{m} \lambda^m V^{(m+\alpha)}.
$$

On the other hand,

$$
\frac{d^\alpha}{d\lambda^\alpha}(I - \lambda V)^{-1} = \alpha!(I - \lambda V)^{-(\alpha+1)} V^\alpha.
$$

It follows that

$$
(I - \lambda V)^{-(\alpha+1)} V^\alpha = \sum_{m=0}^\infty \binom{m+\alpha}{m} \lambda^m V^{(m+\alpha)}, \quad \alpha = 0, 1, 2, \ldots.
$$

Next, using the second identity in (4.21), we see that for each $x \in \mathbb{C}^n$ and $\alpha = 0, 1, 2, \ldots$ we have

$$
e^{-i\lambda t}\frac{(-it)^\alpha}{\alpha!}x = \left(\sum_{m=0}^\infty \lambda^m \frac{(-it)^m}{m!}\right)\frac{(-it)^\alpha}{\alpha!}x = \sum_{m=0}^\infty \lambda^m \frac{(-it)^{m+\alpha}}{m!\,\alpha!}x
$$

$$
= \sum_{m=0}^\infty \binom{m+\alpha}{m} \lambda^m \frac{(-it)^{m+\alpha}}{(m+\alpha)!}x
$$

$$
= \sum_{m=0}^\infty \binom{m+\alpha}{m} \lambda^m \left(V^{(m+\alpha)} \tau x\right)(t).
$$

We conclude that for each $x \in \mathbb{C}^n$ and $\alpha = 0, 1, 2, \ldots$ we have

$$
\left((I - \lambda V)^{-(\alpha+1)} V^\alpha \tau x\right)(t) = e^{-i\lambda t}\frac{(-it)^\alpha}{\alpha!}x, \quad 0 \le t \le \omega. \tag{4.24}
$$

Replacing the summation index μ in (4.22) by $\alpha = k - \mu$ and using the identity in (4.24), we see that

$$
\psi^\nu_{j,k}(t) = \sum_{\alpha=0}^k e^{-i\lambda_\nu t}\frac{(-it)^\alpha}{\alpha!}x^\nu_{j,(k-\alpha)}
$$

$$
= \sum_{\alpha=0}^k \left((I - \lambda_\nu V)^{-(\alpha+1)} V^\alpha \tau x^\nu_{j,(k-\alpha)}\right)(t) = u^\nu_{j,k}(t), \quad 0 \le t \le \omega.
$$

This proves (4.23), and we are done.

Part 2. For an arbitrary $\alpha \in \mathbb{R}$ let $\mathcal{A}_\alpha$, $\mathcal{B}_\alpha$, $\mathcal{C}_\alpha$, $\mathcal{D}_\alpha$ be the entire $n \times n$ matrix functions defined by (3.50) and (3.51). An elementary calculation, using (2.3), shows that $\lambda_1 - \alpha, \ldots, \lambda_\ell - \alpha$ are the distinct common eigenvalues of $\mathcal{B}_\alpha$ and $\mathcal{D}_\alpha$ in $\mathbb{C}$. Furthermore, the vectors

$$x^\nu_{1,0}, \ldots, x^\nu_{1,r_1^{(\nu)}-1}, x^\nu_{2,0}, \ldots, x^\nu_{2,r_2^{(\nu)}-1}, \ldots, x^\nu_{p_\nu,0}, \ldots, x^\nu_{p_\nu,r_{p_\nu}^{(\nu)}-1}$$

form a canonical set of common Jordan chains of $\mathcal{B}_\alpha$ and $\mathcal{D}_\alpha$ at $\lambda_\nu - \alpha$.

Let T_α be the Bezout integral operator associated with $\{\mathcal{A}_\alpha, \mathcal{C}_\alpha; \mathcal{B}_\alpha, \mathcal{D}_\alpha\}$, where $\alpha \in \mathbb{R}$ has been chosen in such a way that the matrices $\mathcal{A}_\alpha(0)$, $\mathcal{B}_\alpha(0)$, $\mathcal{C}_\alpha(0)$, $\mathcal{D}_\alpha(0)$ are non-singular at $\lambda = 0$. As we have seen in the Part 1 of Section 3.5 such an α always exists. The result of Part 1 of the present proof, together with the remarks in the previous paragraph, shows that the functions

$$\widetilde{\psi}^\nu_{j,k}(t) = e^{-i(\lambda_\nu - \alpha)t} \sum_{\mu=0}^{k} \frac{(-it)^{k-\mu}}{(k-\mu)!} x^\nu_{j,\mu}, \quad k = 0, \ldots, r_j^{(\nu)} - 1,$$

$$j = 1, \ldots, p_\nu, \quad \nu = 1, \ldots, \ell,$$

form a basis for the null space of T_α. Recall that $T_\alpha = M_\alpha T M_{-\alpha}$, where M_α is the operator on $L_1^n[0,\omega]$ defined by (3.52). It follows that $g \in \operatorname{Ker} T$ if and only if for some $\widetilde{g} \in \operatorname{Ker} T_\alpha$ we have

$$g(t) = e^{-i\alpha t}\widetilde{g}(t), \quad 0 \le t \le \omega.$$

Since $\psi^\nu_{j,k}(t) = e^{-i\alpha t}\widetilde{\psi}^\nu_{j,k}(t)$ for $0 \le t \le \omega$, we see that the functions in (4.22) form a basis of $\operatorname{Ker} T$. $\qquad\square$

From the proof of Theorem 4.6 it is clear that with slight modifications in the arguments this theorem could also have been proved at the end of the previous chapter.

The fact that the vectors in (4.22) are contained in the null space of the Bezout integral operator T can also be derived from Theorem 1.1 in [10], using the equivalence between T and the resultant operator proved in Section 3 of [10].

References

[1] B.D.O. Anderson and E.I. Jury, Generalized Bezoutian and Sylvester matrices in multivariable linear control, *IEEE Trans. Automatic Control*, AC-**21** (1976), 551–556.

[2] D.Z. Arov, Scattering theory with dissipation of energy, *Dokl. Akad. Nauk SSSR* **216** (4) (1974), pp. 713–716 [in Russian]; English translation with addenda: *Sov. Math. Dokl.* **15** (1974), pp. 848–854.

[3] D.Z. Arov, M.A. Kaashoek, and D.R. Pik, The Kalman-Yakubovich-Popov inequality for discrete time systems of infinite dimension, *J. Oper. Theory*, to appear.

[4] J.A. Ball, and N. Cohen, De Branges-Rovnyak operator models and systems theory: a survey, in: *Topics in Matrix and Operator Theory* (eds. H. Bart, I. Gohberg, M.A. Kaashoek), OT **50**, Birkhäuser Verlag, Basel, 1991, pp. 93–136.

[5] H. Bart, I. Gohberg, and M.A. Kaashoek, *Minimal Factorization of Matrix and Operator Functions.* OT **1**, Birkhäuser Verlag, Basel, 1979.

[6] R.R. Bitmead, S.Y. Kung, B.D.O. Anderson, and T. Kailath, Greatest common divisors via generalized Sylvester and Bezout matrices, *IEEE Trans. Autom. Control* AC-**23** (1978), 1043–1047.

[7] M.J. Corless, and A.E. Frazho, *Linear systems and control,* Marcel Dekker, Inc., New York, NY, 2003.

[8] I. Gohberg, S. Goldberg, and M.A. Kaashoek: *Classes of Linear Operators Vol. 1.* OT **49**, Birkhäuser Verlag, Basel, 1990.

[9] I. Gohberg, and G. Heinig, The resultant matrix and its generalizations, II. Continual analog of resultant matrix, *Acta Math. Acad. Sci. Hungar* **28** (1976), 198–209, [in Russian].

[10] I. Gohberg, M.A. Kaashoek, and L. Lerer, The continuous analogue of the resultant and related convolution operators, to appear.

[11] I. Gohberg, M.A. Kaashoek, and F. van Schagen, *Partially specified matrices and operators: classification, completion, applications,* OT **79** Birkhäuser Verlag, Basel, 1995.

[12] I. Gohberg, M.A. Kaashoek, and F. van Schagen, On inversion of convolution integral operators on a finite interval, in: *Operator Theoretical Methods and Applications to Mathematical Physics. The Erhard Meister Memorial Volume,* OT **147**, Birkhäuser Verlag, Basel, 2004, pp. 277–285.

[13] I. Gohberg, P. Lancaster, and L. Rodman, *Matrix Polynomials,* Academic Press, New York, 1982.

[14] I. Gohberg, and L. Lerer, Matrix generalizations of M.G. Krein theorems on matrix polynomials,in *Orthogonal Matrix-Valued Polynomials and Applications (I. Gohberg, ed.),* OT **34**, Birkhäuser Verlag, Basel, 1988, pp. 137–202.

[15] I.C. Gohberg, and E.I. Sigal, An operator generalization of the logarithmic residue theorem and the theorem of Rouché, *Mat.Sbornik* **84** (126) (1971), 607–629 [in Russian]; English transl. *Math. USSR, Sbornik* **13** (1971), 603–625.

[16] I. Haimovici, *Operator equations and Bezout operators for analytic operator functions,* Ph.D. thesis, Technion Haifa, Israel, 1991 [in Hebrew].

[17] I. Haimovici, and L. Lerer, Bezout operators for analytic operator functions, I. A general concept of Bezout operator, *Integral Equations Oper. Theory* **21** (1995), 33–70.

[18] G. Heinig, and K. Rost, *Algebraic Methods for Toeplitz-like Matrices and Operators,* OT **13**, Birkhäuser Verlag, Basel, 1984.

[19] J.W. Helton, Discrete time systems, operator models, and scattering theory, *J. Funct. Anal.* **16** (1974), 15–38.

[20] M.G. Krein, and M.A. Naimark, The method of symmetric and hermitian forms in theory of separation of the roots of algebraic equations, GNTI, Kharkov, 1936 [in Russian]; English transl. *Linear and Multilinear Algebra* **10**, (1981), 265–308.

[21] L. Lerer and L. Rodman, Bezoutians of rational matrix functions, *J. Funct. Anal.* **141** (1996), 1–38.

[22] L. Lerer and L. Rodman, Bezoutians of rational matrix functions, matrix equations and factorization, *Lin. Alg. Appl.* **302/303** (1999), 105–135.

[23] L. Lerer, L. Rodman and M. Tismenetsky, Bezoutian and Schur-Cohn problem for operator polynomials, *J. Math. Analysis & Appl.* **103** (1984), 83–102.

[24] L. Lerer and M. Tismenetsky, The eigenvalue separation problem for matrix polynomials, *Integral Equations Operator Theory* **5** (1982), 386–445.

[25] L. Rodman, *An introduction to operator polynomials*, OT **38**, Birkhäuser Verlag, Basel, 1989.

[26] L.A. Sakhnovich, Operatorial Bezoutiant in the theory of separation of roots of entire functions, *Functional Anal. Appl.* **10** (1976), 45–51 [in Russian].

[27] L.A. Sakhnovich, *Integral equations with difference kernels on finite intervals*, OT **84**, Birkhäuser Verlag, Basel, 1996.

I. Gohberg
School of Mathematical Sciences
Raymond and Beverly Faculty of Exact Sciences
Tel-Aviv University
Ramat Aviv 69978, Israel
e-mail: `gohberg@math.tau.ac.il`

I. Haimovici
Steimatzky Str. 9/9
Ramat Aviv Hahadasha
69639 Tel Aviv, Israel
e-mail: `iulianh@zahav.net.il`

M.A. Kaashoek
Afdeling Wiskunde
Faculteit der Exacte Wetenschappen
Vrije Universiteit
De Boelelaan 1081a
1081 HV Amsterdam, The Netherlands
e-mail: `ma.kaashoek@few.vu.nl`

L. Lerer
Department of Mathematics
Technion – Israel Institute of Technology
Haifa 32000, Israel
e-mail: `llerer@techunix.technion.ac.il`

Subseries

Linear Operators and Linear Systems

Subseries editors:

Daniel Alpay
Department of Mathematics
Ben Gurion University of the Negev
P.O. Box 653
Beer Sheva 84105
Israel
e-mail: dany@math.bgu.ac.il

Joseph A. Ball
Department of Mathematics
Virginia Tech
Blacksburg, VA 24061
USA
e-mail: ball@math.vt.edu

André M.C. Ran
Division of Mathematics
and Computer Science
Faculty of Sciences
Vrije Universiteit
NL-1081 HV Amsterdam
The Netherlands
e-mail: ran@cs.vu.nl

The theory of linear operators has had an important influence on the development of mathematical systems theory. On the other hand, mathematical systems theory serves as a direct source of motivation and new techniques for the theory of linear operators and its applications. The subseries *Linear Operators and Linear Systems* (LOLS) is dedicated to these connections between the theory of linear operators and the mathematical theory of linear systems. LOLS will continue in the tradition of the series *Operator Theory: Advances and Applications* and maintain the high quality of the volumes. Books published in LOLS will be either monographs or consist of essays presenting the state of the art and new results.

The volumes will be addressed to a wide range of mathematicians from beginners to experts in these fields.

Proposals for manuscripts may be sent to one of the subseries editors.

Operator Theory: Advances and Applications

Your Specialized Publisher in Mathematics
Birkhäuser 

Edited by
Gohberg, I., School of Mathematical Sciences, Tel Aviv University, Ramat Aviv, Israel

This series is devoted to the publication of current research in operator theory, with particular emphasis on applications to classical analysis and the theory of integral equations, as well as to numerical analysis, mathematical physics and mathematical methods in electrical engineering.

For orders originating from all over the world except USA/Canada/Latin America:

Birkhäuser Verlag AG
c/o Springer GmbH & Co
Haberstrasse 7
D-69126 Heidelberg
Fax: +49 / 6221 / 345 4 229
e-mail: birkhauser@springer.de
http://www.birkhauser.ch

For orders originating in the USA/Canada/Latin America:

Birkhäuser
333 Meadowland Parkway
USA-Secaucus
NJ 07094-2491
Fax: +1 201 348 4505
e-mail: orders@birkhauser.com

OT 161: Alpay, D. / Gohberg, I. (Eds.), The State Space Method. Generalizations and Applications (2005). ISBN 3-7643-7370-9

OT 160: Kaashoek, M.A. / Seatzu, S. / van der Mee, C. (Eds.), Recent Advances in Operator Theory and its Applications. The Israel Gohberg Anniversary Volume (2005). ISBN 3-7643-7290-7

OT 159: Reissig, M. / Schulze, B.-W. (Eds.), New Trends in the Theory of Hyperbolic Functions (2005). Subseries Advances in Partial Differential Equations. ISBN 3-7643-7283-4

OT 158: Eiderman, V.Ya. / Samokhin, M.V. (Eds.), Selected Topics in Complex Analysis (2005). ISBN 3-7643-7251-6

OT 157: Alpay, D. / Vinnikov, V. (Eds.), Operator Theory, Systems Theory and Scattering Theory: Multidimensional Generalizations (2005). ISBN 3-7643-7212-5

OT 156: Ebenfelt, P. / Gustafsson, B. / Khavinson, D. / Putinar, M. (Eds.), Quadrature Domains and Their Applications. The Harold S. Shapiro Anniversary Volume (2005). ISBN 3-7643-7145-5

OT 155: Ashino, R. / Boggiatto, P. / Wong, M.W. (Eds.), Advances in Pseudo-Differential Operators (2004). ISBN 3-7643-7140-4

OT 154: Janas, J. / Kurasov, P. / Naboko, S. (Eds.), Spectral Methods for Operators of Mathematical Physics (2004). ISBN 3-7643-7133-1

OT 153: Gaspar, D. / Gohberg, I. / Timotin, D. / Vasilescu, F.H. / Zsido, L. (Eds.), Recent Advances in Operator Theory, Operator Algebras, and their Applications (2004). ISBN 3-7643-7127-7

OT 152: Eidelman, S.D. / Ivasyshen, S.D. / Kochubei, A.N., Analytic Methods in the Theory of Differential and Pseudo-differential Equations of Parabolic Type (2004). ISBN 3-7643-7115-3

OT 151: Gil, J.B. / Krainer, T. / Witt, I. (Eds.), Aspects of Boundary Problems in Analysis and Geometry (2004). Subseries Advances in Partial Differential Equations. ISBN 3-7643-7069-6

OT 150: Rabinovich, V. / Roch, S. / Silbermann, B., Limit Operators and their Applications in Operator Theory (2004). ISBN 3-7643-7081-5

OT 149: Ball, J.A. / Helton, J.W. / Klaus, M. / Rodman, L. (Eds.), Current Trends in Operator Theory and its Applications (2004). ISBN 3-7643-7067-X

OT 148: Ashyralyev, A. / Sobolevskii, P.E., New Difference Schemes for Partial Differential Equations (2004). ISBN 3-7643-7054-8

OT 147: Gohberg, I. / dos Santos, A.F. / Speck, F.-O. / Teixeira, F.S. / Wendland, W. (Eds.), Operator Theoretical Methods and Applications to Mathematical Physics. The Erhard Meister Memorial Volume (2004). ISBN 3-7643-6634-6